S0-ADO-907

VW
Golf & Jetta
Automotive
Repair
Manual

by George Parise, Mark Coombs, Spencer Drayton and John H Haynes

Member of the Guild of Motoring Writers

Models covered:
VW Golf, GTI and Jetta, 1993 thru 1997
Cabrio, 1995 thru 1997
Four-cylinder gasoline and diesel engines
Does not cover information specific to models with the VR6 V6 engine

ABCDE
FGHIJ
KLMNO
PQRS

Haynes Publishing Group
Sparkford Nr Yeovil
Somerset BA22 7JJ England

Haynes North America, Inc
861 Lawrence Drive
Newbury Park
California 91320 USA

© **Haynes North America, Inc. 1997**

With permission from J.H. Haynes & Co. Ltd.

A book in the Haynes Automotive Repair Manual Series

Printed in the U.S.A.

All rights reserved. No part of this book may be reproduced or transmitted
in any form or by any means, electronic or mechanical, including photo-
copying, recording or by any information storage or retrieval system,
without permission in writing from the copyright holder.

ISBN 1 56392 244 4

Library of Congress Catalog Card Number 97-70291

While every attempt is made to ensure that the information in this manual is correct, no liability can be accepted by the authors or publishers
for loss, damage or injury caused by any errors in, or omissions from,
the information given.

97-288

Contents

1996 VW Golf GTI

1996 Jetta GL

About this manual

Its purpose

The purpose of this manual is to help you get the best value from your vehicle. It can do so in several ways. It can help you decide what work must be done, even if you choose to have it done by a dealer service department or a repair shop; it provides information and procedures for routine maintenance and servicing; and it offers diagnostic and repair procedures to follow when trouble occurs.

We hope you use the manual to tackle the work yourself. For many simpler jobs, doing it yourself may be quicker than arranging an appointment to get the vehicle into a shop and making the trips to leave it and pick it up. More importantly, a lot of money can be saved by avoiding the expense the shop must pass on to you to cover its labor and overhead costs. An added benefit is the sense of satisfaction and accomplishment that you feel after doing the job yourself.

Using the manual

The manual is divided into Chapters. Each Chapter is divided into numbered Sections, which are headed in bold type between horizontal lines. Each Section consists of consecutively numbered paragraphs.

At the beginning of each numbered Section you will be referred to any illustrations which apply to the procedures in that Section. The reference numbers used in illustration captions pinpoint the pertinent Section and the Step within that Section. That is, illustration 3.2 means the illustration refers to Section 3 and Step (or paragraph) 2 within that Section.

Procedures, once described in the text, are not normally repeated. When it's necessary to refer to another Chapter, the reference will be given as Chapter and Section number. Cross references given without use of the word "Chapter" apply to Sections and/or paragraphs in the same Chapter. For example, "see Section 8" means in the same Chapter.

References to the left or right side of the vehicle assume you are sitting in the driver's seat, facing forward.

Even though we have prepared this manual with extreme care, neither the publisher nor the author can accept responsibility for any errors in, or omissions from, the information given.

NOTE

A **Note** provides information necessary to properly complete a procedure or information which will make the procedure easier to understand.

CAUTION

A **Caution** provides a special procedure or special steps which must be taken while completing the procedure where the Caution is found. Not heeding a Caution can result in damage to the assembly being worked on.

WARNING

A **Warning** provides a special procedure or special steps which must be taken while completing the procedure where the Warning is found. Not heeding a Warning can result in personal injury.

Introduction to the VW Golf, GTI and Jetta

The new VW Golf, GTI and Jetta range was introduced in 1993. These models are available with 1.8 liter (Canada models only) and 2.0 liter gasoline engines, as well as a 1.9 liter turbocharged diesel engine. Some models are available with a 2.8 liter V6 engine, but information specific to these models is not included in this manual.

The engine is of four-cylinder overhead camshaft design, mounted transversely, with the transmission mounted on the left-hand side. All models have a five-speed manual transaxle or a four-speed automatic transaxle.

In 1995, a convertible version of the Golf, called the Cabrio, was introduced.

All models have fully independent, Macpherson strut-type front suspension. The rear suspension is semi-independent, with struts/shock absorber assemblies and trailing arms.

A wide range of standard and optional equipment is available within the range to suit most tastes, including central locking, electric windows, an electric sunroof, an anti-lock braking system, and an airbag.

Provided that regular servicing is carried out in accordance with the manufacturer's recommendations, these vehicles should prove reliable and very economical. The engine compartment is well-designed, and most of the items requiring frequent attention are easily accessible.

Vehicle identification numbers

Modifications are a continuing and unpublicized process in vehicle manufacturing. Since spare parts manuals and lists are compiled on a numerical basis, the individual vehicle numbers are essential to correctly identify the component required.

Vehicle Identification Number (VIN)

This very important identification number is located on a plate attached to the hood lock carrier (also called the radiator support) **(see illustration)**. The VIN also appears on the Vehicle Certificate of Title and Registration. It contains information such as where and when the vehicle was manufactured, the model year and the body style.

VIN model year

One particularly important piece of information found in the VIN is the model year code. Counting from the left, the model year code designation is the 10th digit.

On the models covered by this manual the model year codes are:

P... 1993
R... 1994
S... 1995
T... 1996
V... 1997

Chassis Identification Number

The chassis identification number is stamped on the firewall in the engine compartment **(see illustration)**. Like the VIN it contains valuable information about the manufacturing of the vehicle such as the destination, model variations and transaxle information.

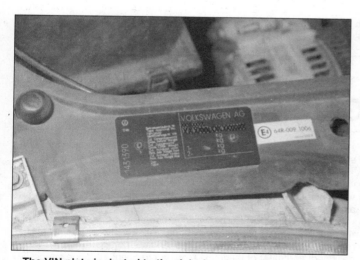

The VIN plate is riveted to the right-hand end of the hood lock carrier crossmember

The Chassis Identification Number is stamped on the engine compartment firewall

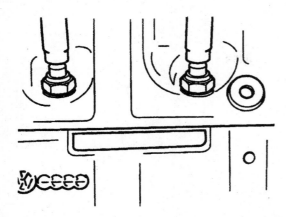

The Engine Identification Number on gasoline engines is stamped on the front side of the engine block, below the cylinder head

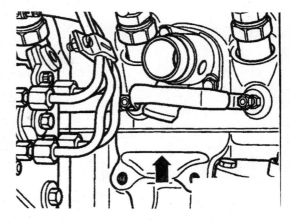

On the diesel engine it's located on a pad between the injection pump and the vacuum pump

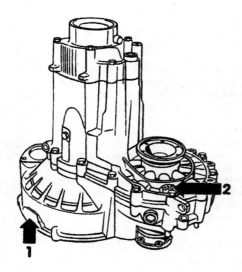

The manual transaxle ID number is located on top of the transaxle bellhousing and on the differential case

1   Code letters   2   Transaxle type

Engine identification numbers

The engine code number on gasoline engines can be found on a pad on the front (radiator) side of the cylinder block, just below the cylinder head **(see illustration)**.

On the diesel engine it's located on a pad between the injection pump and the vacuum pump **(see illustration)**.

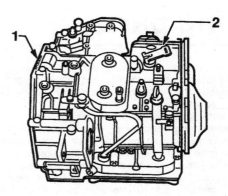

The automatic transaxle ID number is located on top of the bellhousing and on top of the case, near the shift lever

1   Code letters   2   Transaxle type

Transaxle identification numbers

The manual transaxle identification number is stamped on top of the bellhousing and on the differential case **(see illustration)**. The automatic transaxle identification number is stamped on the bellhousing and on the top of the transaxle **(see illustration)**.

Vehicle Emissions Control Information (VECI) label

The emissions control information label is found under the hood, normally on the radiator support. This label contains information on the emissions control equipment installed on the vehicle, as well as tune-up specifications.

Buying parts

Replacement parts are available from many sources, which generally fall into one of two categories - authorized dealer parts departments and independent retail auto parts stores. Our advice concerning these parts is as follows:

Retail auto parts stores: Good auto parts stores will stock frequently needed components which wear out relatively fast, such as clutch components, exhaust systems, brake parts, tune-up parts, etc. These stores often supply new or reconditioned parts on an exchange basis, which can save a considerable amount of money. Discount auto parts stores are often very good places to buy materials and parts needed for general vehicle maintenance such as oil, grease, filters, spark plugs, belts, touch-up paint, bulbs, etc. They also usually sell

tools and general accessories, have convenient hours, charge lower prices and can often be found not far from home.

Authorized dealer parts department: This is the best source for parts which are unique to the vehicle and not generally available elsewhere (such as major engine parts, transmission parts, trim pieces, etc.).

Warranty information: If the vehicle is still covered under warranty, be sure that any replacement parts purchased - regardless of the source - do not invalidate the warranty!

To be sure of obtaining the correct parts, have engine and chassis numbers available and, if possible, take the old parts along for positive identification.

Maintenance techniques, tools and working facilities

Maintenance techniques

There are a number of techniques involved in maintenance and repair that will be referred to throughout this manual. Application of these techniques will enable the home mechanic to be more efficient, better organized and capable of performing the various tasks properly, which will ensure that the repair job is thorough and complete.

Fasteners

Fasteners are nuts, bolts, studs and screws used to hold two or more parts together. There are a few things to keep in mind when working with fasteners. Almost all of them use a locking device of some type, either a lockwasher, locknut, locking tab or thread adhesive. All threaded fasteners should be clean and straight, with undamaged threads and undamaged corners on the hex head where the wrench fits. Develop the habit of replacing all damaged nuts and bolts with new ones. Special locknuts with nylon or fiber inserts can only be

used once. If they are removed, they lose their locking ability and must be replaced with new ones.

Rusted nuts and bolts should be treated with a penetrating fluid to ease removal and prevent breakage. Some mechanics use turpentine in a spout-type oil can, which works quite well. After applying the rust penetrant, let it work for a few minutes before trying to loosen the nut or bolt. Badly rusted fasteners may have to be chiseled or sawed off or removed with a special nut breaker, available at tool stores.

If a bolt or stud breaks off in an assembly, it can be drilled and removed with a special tool commonly available for this purpose. Most automotive machine shops can perform this task, as well as other repair procedures, such as the repair of threaded holes that have been stripped out.

Flat washers and lockwashers, when removed from an assembly, should always be replaced exactly as removed. Replace any damaged washers with new ones. Never use a lockwasher on any soft metal surface (such as aluminum), thin sheet metal or plastic.

Fastener sizes

For a number of reasons, automobile manufacturers are making wider and wider use of metric fasteners. Therefore, it is important to be able to tell the difference between standard (sometimes called U.S. or SAE) and metric hardware, since they cannot be interchanged.

All bolts, whether standard or metric, are sized according to diameter, thread pitch and length. For example, a standard 1/2 - 13 x 1 bolt is 1/2 inch in diameter, has 13 threads per inch and is 1 inch long. An M12 - 1.75 x 25 metric bolt is 12 mm in diameter, has a thread pitch of 1.75 mm (the distance between threads) and is 25 mm long. The two bolts are nearly identical, and easily confused, but they are not interchangeable.

In addition to the differences in diameter, thread pitch and length, metric and standard bolts can also be distinguished by examining the bolt heads. To begin with, the distance across the flats on a standard bolt head is measured in inches, while the same dimension on a metric bolt is sized in millimeters (the same is true for nuts). As a result, a standard wrench should not be used on a metric bolt and a metric

wrench should not be used on a standard bolt. Also, most standard bolts have slashes radiating out from the center of the head to denote the grade or strength of the bolt, which is an indication of the amount of torque that can be applied to it. The greater the number of slashes, the greater the strength of the bolt. Grades 0 through 5 are commonly used on automobiles. Metric bolts have a property class (grade) number, rather than a slash, molded into their heads to indicate bolt strength. In this case, the higher the number, the stronger the bolt. Property class numbers 8.8, 9.8 and 10.9 are commonly used on automobiles.

Strength markings can also be used to distinguish standard hex nuts from metric hex nuts. Many standard nuts have dots stamped into one side, while metric nuts are marked with a number. The greater the number of dots, or the higher the number, the greater the strength of the nut.

Metric studs are also marked on their ends according to property class (grade). Larger studs are numbered (the same as metric bolts), while smaller studs carry a geometric code to denote grade.

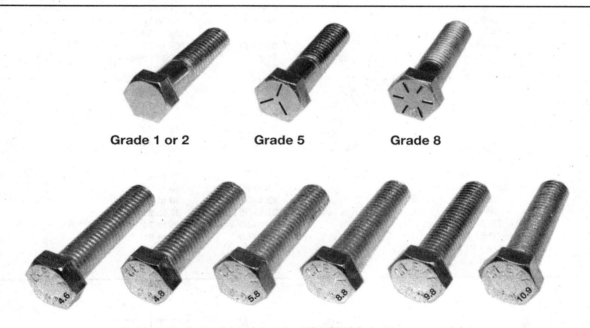

Grade 1 or 2   Grade 5   Grade 8

Bolt strength marking (standard/SAE/USS; bottom - metric)

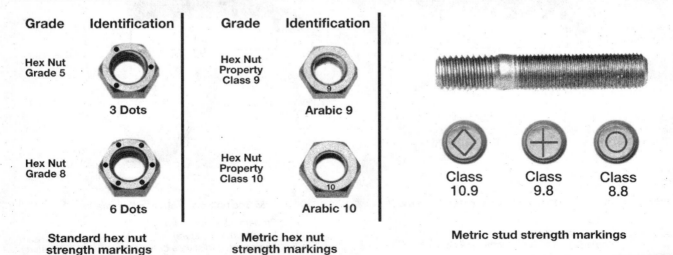

Grade	Identification
Hex Nut Grade 5	3 Dots
Hex Nut Grade 8	6 Dots

Standard hex nut strength markings

Grade	Identification
Hex Nut Property Class 9	Arabic 9
Hex Nut Property Class 10	Arabic 10

Metric hex nut strength markings

Class 10.9   Class 9.8   Class 8.8

Metric stud strength markings

It should be noted that many fasteners, especially Grades 0 through 2, have no distinguishing marks on them. When such is the case, the only way to determine whether it is standard or metric is to measure the thread pitch or compare it to a known fastener of the same size.

Standard fasteners are often referred to as SAE, as opposed to metric. However, it should be noted that SAE technically refers to a non-metric fine thread fastener only. Coarse thread non-metric fasteners are referred to as USS sizes.

Since fasteners of the same size (both standard and metric) may have different strength ratings, be sure to reinstall any bolts, studs or nuts removed from your vehicle in their original locations. Also, when replacing a fastener with a new one, make sure that the new one has a strength rating equal to or greater than the original.

Tightening sequences and procedures

Most threaded fasteners should be tightened to a specific torque value (torque is the twisting force applied to a threaded component such as a nut or bolt). Overtightening the fastener can weaken it and cause it to break, while undertightening can cause it to eventually come loose. Bolts, screws and studs, depending on the material they are made of and their thread diameters, have specific torque values, many of which are noted in the Specifications at the beginning of each Chapter. Be sure to follow the torque recommendations closely. For fasteners not assigned a specific torque, a general torque value chart is presented here as a guide. These torque values are for dry (unlubricated) fasteners threaded into steel or cast iron (not aluminum). As was previously mentioned, the size and grade of a fastener determine the amount of torque that can safely be applied to it. The figures listed

Metric thread sizes	Ft-lbs	Nm
M-6	6 to 9	9 to 12
M-8	14 to 21	19 to 28
M-10	28 to 40	38 to 54
M-12	50 to 71	68 to 96
M-14	80 to 140	109 to 154
Pipe thread sizes		
1/8	5 to 8	7 to 10
1/4	12 to 18	17 to 24
3/8	22 to 33	30 to 44
1/2	25 to 35	34 to 47
U.S. thread sizes		
1/4 - 20	6 to 9	9 to 12
5/16 - 18	12 to 18	17 to 24
5/16 - 24	14 to 20	19 to 27
3/8 - 16	22 to 32	30 to 43
3/8 - 24	27 to 38	37 to 51
7/16 - 14	40 to 55	55 to 74
7/16 - 20	40 to 60	55 to 81
1/2 - 13	55 to 80	75 to 108

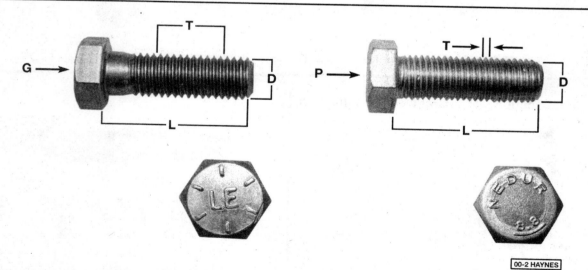

Standard (SAE and USS) bolt dimensions/grade marks

    G   Grade marks (bolt strength)
    L   Length (in inches)
    T   Thread pitch (number of threads per inch)
    D   Nominal diameter (in inches)

Metric bolt dimensions/grade marks

    P   Property class (bolt strength)
    L   Length (in millimeters)
    T   Thread pitch (distance between threads in millimeters)
    D   Diameter

here are approximate for Grade 2 and Grade 3 fasteners. Higher grades can tolerate higher torque values.

Fasteners laid out in a pattern, such as cylinder head bolts, oil pan bolts, differential cover bolts, etc., must be loosened or tightened in sequence to avoid warping the component. This sequence will normally be shown in the appropriate Chapter. If a specific pattern is not given, the following procedures can be used to prevent warping.

Initially, the bolts or nuts should be assembled finger-tight only. Next, they should be tightened one full turn each, in a criss-cross or diagonal pattern. After each one has been tightened one full turn, return to the first one and tighten them all one-half turn, following the same pattern. Finally, tighten each of them one-quarter turn at a time until each fastener has been tightened to the proper torque. To loosen and remove the fasteners, the procedure would be reversed.

Component disassembly

Component disassembly should be done with care and purpose to help ensure that the parts go back together properly. Always keep track of the sequence in which parts are removed. Make note of special characteristics or marks on parts that can be installed more than one way, such as a grooved thrust washer on a shaft. It is a good idea to lay the disassembled parts out on a clean surface in the order that they were removed. It may also be helpful to make sketches or take instant photos of components before removal.

When removing fasteners from a component, keep track of their locations. Sometimes threading a bolt back in a part, or putting the washers and nut back on a stud, can prevent mix-ups later. If nuts and bolts cannot be returned to their original locations, they should be kept in a compartmented box or a series of small boxes. A cupcake or muffin tin is ideal for this purpose, since each cavity can hold the bolts and nuts from a particular area (i.e. oil pan bolts, valve cover bolts, engine mount bolts, etc.). A pan of this type is especially helpful when working on assemblies with very small parts, such as the carburetor, alternator, valve train or interior dash and trim pieces. The cavities can be marked with paint or tape to identify the contents.

Whenever wiring looms, harnesses or connectors are separated, it is a good idea to identify the two halves with numbered pieces of masking tape so they can be easily reconnected.

Gasket sealing surfaces

Throughout any vehicle, gaskets are used to seal the mating surfaces between two parts and keep lubricants, fluids, vacuum or pressure contained in an assembly.

Many times these gaskets are coated with a liquid or paste-type gasket sealing compound before assembly. Age, heat and pressure can sometimes cause the two parts to stick together so tightly that they are very difficult to separate. Often, the assembly can be loosened by striking it with a soft-face hammer near the mating surfaces. A regular hammer can be used if a block of wood is placed between the hammer and the part. Do not hammer on cast parts or parts that could be easily damaged. With any particularly stubborn part, always recheck to make sure that every fastener has been removed.

Avoid using a screwdriver or bar to pry apart an assembly, as they can easily mar the gasket sealing surfaces of the parts, which must remain smooth. If prying is absolutely necessary, use an old broom handle, but keep in mind that extra clean up will be necessary if the wood splinters.

After the parts are separated, the old gasket must be carefully scraped off and the gasket surfaces cleaned. Stubborn gasket material can be soaked with rust penetrant or treated with a special chemical to soften it so it can be easily scraped off. A scraper can be fashioned from a piece of copper tubing by flattening and sharpening one end. Copper is recommended because it is usually softer than the surfaces to be scraped, which reduces the chance of gouging the part. Some gaskets can be removed with a wire brush, but regardless of the method used, the mating surfaces must be left clean and smooth. If for some reason the gasket surface is gouged, then a gasket sealer thick enough to fill scratches will have to be used during reassembly of the components. For most applications, a non-drying (or semi-drying) gasket sealer should be used.

Hose removal tips

Warning: *If the vehicle is equipped with air conditioning, do not disconnect any of the A/C hoses without first having the system depressurized by a dealer service department or a service station.*

Hose removal precautions closely parallel gasket removal precautions. Avoid scratching or gouging the surface that the hose mates against or the connection may leak. This is especially true for radiator hoses. Because of various chemical reactions, the rubber in hoses can bond itself to the metal spigot that the hose fits over. To remove a hose, first loosen the hose clamps that secure it to the spigot. Then, with slip-joint pliers, grab the hose at the clamp and rotate it around the spigot. Work it back and forth until it is completely free, then pull it off. Silicone or other lubricants will ease removal if they can be applied between the hose and the outside of the spigot. Apply the same lubricant to the inside of the hose and the outside of the spigot to simplify installation.

As a last resort (and if the hose is to be replaced with a new one anyway), the rubber can be slit with a knife and the hose peeled from the spigot. If this must be done, be careful that the metal connection is not damaged.

If a hose clamp is broken or damaged, do not reuse it. Wire-type clamps usually weaken with age, so it is a good idea to replace them with screw-type clamps whenever a hose is removed.

Tools

A selection of good tools is a basic requirement for anyone who plans to maintain and repair his or her own vehicle. For the owner who has few tools, the initial investment might seem high, but when compared to the spiraling costs of professional auto maintenance and repair, it is a wise one.

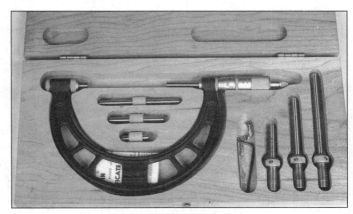

Micrometer set

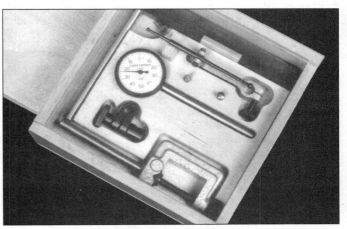

Dial indicator set

Dial caliper

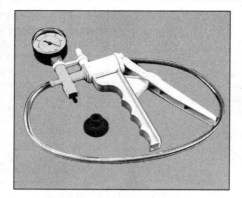

Hand-operated vacuum pump

Timing light

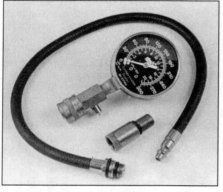

Compression gauge with spark plug hole adapter

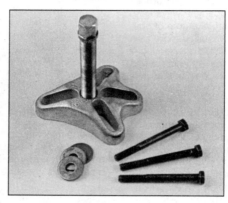

Damper/steering wheel puller

General purpose puller

Hydraulic lifter removal tool

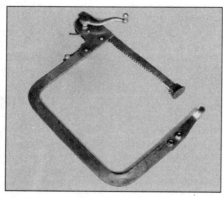

Valve spring compressor

Valve spring compressor

Ridge reamer

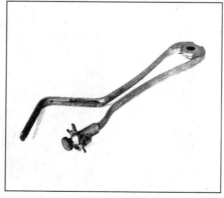

Piston ring groove cleaning tool

Ring removal/installation tool

Ring compressor

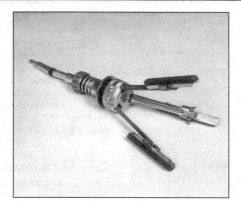

Cylinder hone

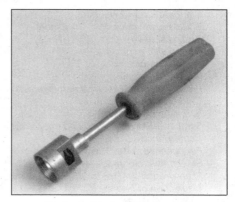

Brake hold-down spring tool

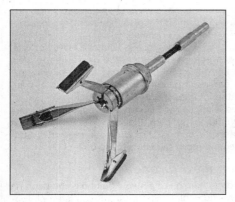

Brake cylinder hone

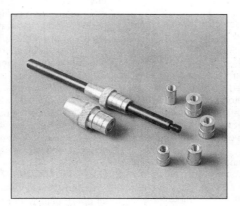

Clutch plate alignment tool

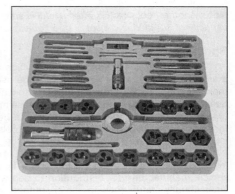

Tap and die set

To help the owner decide which tools are needed to perform the tasks detailed in this manual, the following tool lists are offered: *Maintenance and minor repair, Repair/overhaul* and *Special.*

The newcomer to practical mechanics should start off with the *maintenance and minor repair* tool kit, which is adequate for the simpler jobs performed on a vehicle. Then, as confidence and experience grow, the owner can tackle more difficult tasks, buying additional tools as they are needed. Eventually the basic kit will be expanded into the *repair and overhaul* tool set. Over a period of time, the experienced do-it-yourselfer will assemble a tool set complete enough for most repair and overhaul procedures and will add tools from the special category when it is felt that the expense is justified by the frequency of use.

Maintenance and minor repair tool kit

The tools in this list should be considered the minimum required for performance of routine maintenance, servicing and minor repair work. We recommend the purchase of combination wrenches (box-end and open-end combined in one wrench). While more expensive than open end wrenches, they offer the advantages of both types of wrench.

Combination wrench set (1/4-inch to 1 inch or 6 mm to 19 mm)
Adjustable wrench, 8 inch
Spark plug wrench with rubber insert
Spark plug gap adjusting tool
Feeler gauge set
Brake bleeder wrench
Standard screwdriver (5/16-inch x 6 inch)
Phillips screwdriver (No. 2 x 6 inch)
Combination pliers - 6 inch
Hacksaw and assortment of blades
Tire pressure gauge
Grease gun

Oil can
Fine emery cloth
Wire brush
Battery post and cable cleaning tool
Oil filter wrench
Funnel (medium size)
Safety goggles
Jackstands (2)
Drain pan

Note: *If basic tune-ups are going to be part of routine maintenance, it will be necessary to purchase a good quality stroboscopic timing light and combination tachometer/dwell meter. Although they are included in the list of special tools, it is mentioned here because they are absolutely necessary for tuning most vehicles properly.*

Repair and overhaul tool set

These tools are essential for anyone who plans to perform major repairs and are in addition to those in the maintenance and minor repair tool kit. Included is a comprehensive set of sockets which, though expensive, are invaluable because of their versatility, especially when various extensions and drives are available. We recommend the 1/2-inch drive over the 3/8-inch drive. Although the larger drive is bulky and more expensive, it has the capacity of accepting a very wide range of large sockets. Ideally, however, the mechanic should have a 3/8-inch drive set and a 1/2-inch drive set.

Socket set(s)
Reversible ratchet
Extension - 10 inch
Universal joint
Torque wrench (same size drive as sockets)
Ball peen hammer - 8 ounce
Soft-face hammer (plastic/rubber)
Standard screwdriver (1/4-inch x 6 inch)

Standard screwdriver (stubby - 5/16-inch)
Phillips screwdriver (No. 3 x 8 inch)
Phillips screwdriver (stubby - No. 2)
Pliers - vise grip
Pliers - lineman's
Pliers - needle nose
Pliers - snap-ring (internal and external)
Cold chisel - 1/2-inch
Scribe
Scraper (made from flattened copper tubing)
Centerpunch
Pin punches (1/16, 1/8, 3/16-inch)
Steel rule/straightedge - 12 inch
Allen wrench set (1/8 to 3/8-inch or 4 mm to 10 mm)
A selection of files
Wire brush (large)
Jackstands (second set)
Jack (scissor or hydraulic type)

Note: *Another tool which is often useful is an electric drill with a chuck capacity of 3/8-inch and a set of good quality drill bits.*

Special tools

The tools in this list include those which are not used regularly, are expensive to buy, or which need to be used in accordance with their manufacturer's instructions. Unless these tools will be used frequently, it is not very economical to purchase many of them. A consideration would be to split the cost and use between yourself and a friend or friends. In addition, most of these tools can be obtained from a tool rental shop on a temporary basis.

This list primarily contains only those tools and instruments widely available to the public, and not those special tools produced by the vehicle manufacturer for distribution to dealer service departments. Occasionally, references to the manufacturer's special tools are included in the text of this manual. Generally, an alternative method of doing the job without the special tool is offered. However, sometimes there is no alternative to their use. Where this is the case, and the tool cannot be purchased or borrowed, the work should be turned over to the dealer service department or an automotive repair shop.

Valve spring compressor
Piston ring groove cleaning tool
Piston ring compressor
Piston ring installation tool
Cylinder compression gauge
Cylinder ridge reamer
Cylinder surfacing hone
Cylinder bore gauge
Micrometers and/or dial calipers
Hydraulic lifter removal tool
Balljoint separator
Universal-type puller
Impact screwdriver
Dial indicator set
Stroboscopic timing light (inductive pick-up)
Hand operated vacuum/pressure pump
Tachometer/dwell meter
Universal electrical multimeter
Cable hoist
Brake spring removal and installation tools
Floor jack

Buying tools

For the do-it-yourselfer who is just starting to get involved in vehicle maintenance and repair, there are a number of options available when purchasing tools. If maintenance and minor repair is the extent of the work to be done, the purchase of individual tools is satisfactory. If, on the other hand, extensive work is planned, it would be a good idea to purchase a modest tool set from one of the large retail chain stores. A set can usually be bought at a substantial savings over the individual tool prices, and they often come with a tool box. As additional tools are

needed, add-on sets, individual tools and a larger tool box can be purchased to expand the tool selection. Building a tool set gradually allows the cost of the tools to be spread over a longer period of time and gives the mechanic the freedom to choose only those tools that will actually be used.

Tool stores will often be the only source of some of the special tools that are needed, but regardless of where tools are bought, try to avoid cheap ones, especially when buying screwdrivers and sockets, because they won't last very long. The expense involved in replacing cheap tools will eventually be greater than the initial cost of quality tools.

Care and maintenance of tools

Good tools are expensive, so it makes sense to treat them with respect. Keep them clean and in usable condition and store them properly when not in use. Always wipe off any dirt, grease or metal chips before putting them away. Never leave tools lying around in the work area. Upon completion of a job, always check closely under the hood for tools that may have been left there so they won't get lost during a test drive.

Some tools, such as screwdrivers, pliers, wrenches and sockets, can be hung on a panel mounted on the garage or workshop wall, while others should be kept in a tool box or tray. Measuring instruments, gauges, meters, etc. must be carefully stored where they cannot be damaged by weather or impact from other tools.

When tools are used with care and stored properly, they will last a very long time. Even with the best of care, though, tools will wear out if used frequently. When a tool is damaged or worn out, replace it. Subsequent jobs will be safer and more enjoyable if you do.

How to repair damaged threads

Sometimes, the internal threads of a nut or bolt hole can become stripped, usually from overtightening. Stripping threads is an all-too-common occurrence, especially when working with aluminum parts, because aluminum is so soft that it easily strips out.

Usually, external or internal threads are only partially stripped. After they've been cleaned up with a tap or die, they'll still work. Sometimes, however, threads are badly damaged. When this happens, you've got three choices:

1) *Drill and tap the hole to the next suitable oversize and install a larger diameter bolt, screw or stud.*
2) *Drill and tap the hole to accept a threaded plug, then drill and tap the plug to the original screw size. You can also buy a plug already threaded to the original size. Then you simply drill a hole to the specified size, then run the threaded plug into the hole with a bolt and jam nut. Once the plug is fully seated, remove the jam nut and bolt.*
3) *The third method uses a patented thread repair kit like Heli-Coil or Slimsert. These easy-to-use kits are designed to repair damaged threads in straight-through holes and blind holes. Both are available as kits which can handle a variety of sizes and thread patterns. Drill the hole, then tap it with the special included tap. Install the Heli-Coil and the hole is back to its original diameter and thread pitch.*

Regardless of which method you use, be sure to proceed calmly and carefully. A little impatience or carelessness during one of these relatively simple procedures can ruin your whole day's work and cost you a bundle if you wreck an expensive part.

Working facilities

Not to be overlooked when discussing tools is the workshop. If anything more than routine maintenance is to be carried out, some sort of suitable work area is essential.

It is understood, and appreciated, that many home mechanics do not have a good workshop or garage available, and end up removing an engine or doing major repairs outside. It is recommended, however, that the overhaul or repair be completed under the cover of a roof.

A clean, flat workbench or table of comfortable working height is an absolute necessity. The workbench should be equipped with a vise that has a jaw opening of at least four inches.

As mentioned previously, some clean, dry storage space is also required for tools, as well as the lubricants, fluids, cleaning solvents, etc. which soon become necessary.

Sometimes waste oil and fluids, drained from the engine or cooling system during normal maintenance or repairs, present a disposal problem. To avoid pouring them on the ground or into a sewage system, pour the used fluids into large containers, seal them with caps and take them to an authorized disposal site or recycling center. Plastic jugs, such as old antifreeze containers, are ideal for this purpose.

Always keep a supply of old newspapers and clean rags available. Old towels are excellent for mopping up spills. Many mechanics use rolls of paper towels for most work because they are readily available and disposable. To help keep the area under the vehicle clean, a large cardboard box can be cut open and flattened to protect the garage or shop floor.

Whenever working over a painted surface, such as when leaning over a fender to service something under the hood, always cover it with an old blanket or bedspread to protect the finish. Vinyl covered pads, made especially for this purpose, are available at auto parts stores.

Jacking and towing

Jacking

Warning: *The jack supplied with the vehicle should only be used for changing a tire or placing jackstands under the frame. Never work under the vehicle or start the engine while this jack is being used as the only means of support.*

The vehicle should be on level ground. Place the shift lever in Park, if you have an automatic, or Reverse if you have a manual transaxle. Block the wheel diagonally opposite the wheel being changed. Set the parking brake.

Remove the spare tire and jack from stowage. Remove the wheel cover and trim ring (if so equipped) with the tapered end of the lug nut wrench by inserting and twisting the handle and then prying against the back of the wheel cover. Loosen the wheel lug bolts about 1/4-to-1/2 turn each.

Place the scissors-type jack under the side of the vehicle and adjust the jack height until it fits in the impression in the vertical rocker panel flange nearest the wheel to be changed. There is a front and rear jacking point on each side of the vehicle **(see illustration).**

Turn the jack handle clockwise until the tire clears the ground. Remove the lug bolts and pull the wheel off. Replace it with the spare.

Install the lug bolts with the beveled edges facing in. Tighten them snugly. Don't attempt to tighten them completely until the vehicle is lowered or it could slip off the jack. Turn the jack handle counterclockwise to lower the vehicle. Remove the jack and tighten the lug bolts in a diagonal pattern.

Install the cover (and trim ring, if used) and be sure it's snapped into place all the way around.

Stow the tire, jack and wrench. Unblock the wheels.

Towing

As a general rule, the vehicle should be towed with the front (drive) wheels off the ground. If they can't be raised, place them on a dolly. The ignition key must be in the ACC position, since the steering lock mechanism isn't strong enough to hold the front wheels straight while towing.

Vehicles equipped with manual transaxle can be towed from the front only with all four wheels on the ground, provided that speeds don't exceed 50 mph and the distance is not over 50 miles.

Caution: *Never tow a vehicle with an automatic transaxle with the wheels on the ground - the transaxle could be damaged due to the lack of lubrication.*

When towing a vehicle equipped with a manual transaxle with all four wheels on the ground, be sure to place the shift lever in neutral and release the parking brake. Turn the ignition key to the first position to unlock the steering wheel.

Equipment specifically designed for towing should be used. It should be attached to the main structural members of the vehicle, not the bumpers or brackets.

Safety is a major consideration when towing and all applicable state and local laws must be obeyed. A safety chain system must be used at all times.

The jack fits over the rocker panel flange (there are two jacking points on each side of the vehicle, indicated by an impression in the rocker panel flange)

Booster battery (jump) starting

Observe these precautions when using a booster battery to start a vehicle:

a) *Before connecting the booster battery, make sure the ignition switch is in the Off position.*
b) *Turn off the lights, heater and other electrical loads.*
c) *Your eyes should be shielded. Safety goggles are a good idea.*
d) *Make sure the booster battery is the same voltage as the dead one in the vehicle.*
e) *The two vehicles MUST NOT TOUCH each other!*
f) *Make sure the transaxle is in Neutral (manual) or Park (automatic).*
g) *If the booster battery is not a maintenance-free type, remove the vent caps and lay a cloth over the vent holes.*

Connect the red jumper cable to the positive (+) terminals of each battery **(see illustration).**

Connect one end of the black jumper cable to the negative (-) terminal of the booster battery. The other end of this cable should be connected to a good ground on the vehicle to be started, such as a bolt or bracket on the body.

Start the engine using the booster battery, then, with the engine running at idle speed, disconnect the jumper cables in the reverse order of connection.

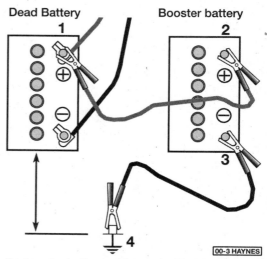

Make the booster battery cable connections in the numerical order shown (note that the negative cable of the booster battery is NOT attached to the negative terminal of the dead battery)

Automotive chemicals and lubricants

A number of automotive chemicals and lubricants are available for use during vehicle maintenance and repair. They include a wide variety of products ranging from cleaning solvents and degreasers to lubricants and protective sprays for rubber, plastic and vinyl.

Cleaners

Carburetor cleaner and choke cleaner is a strong solvent for gum, varnish and carbon. Most carburetor cleaners leave a dry-type lubricant film which will not harden or gum up. Because of this film it is not recommended for use on electrical components.

Brake system cleaner is used to remove grease and brake fluid from the brake system, where clean surfaces are absolutely necessary. It leaves no residue and often eliminates brake squeal caused by contaminants.

Electrical cleaner removes oxidation, corrosion and carbon deposits from electrical contacts, restoring full current flow. It can also be used to clean spark plugs, carburetor jets, voltage regulators and other parts where an oil-free surface is desired.

Demoisturants remove water and moisture from electrical components such as alternators, voltage regulators, electrical connectors and fuse blocks. They are non-conductive, non-corrosive and non-flammable.

Degreasers are heavy-duty solvents used to remove grease from the outside of the engine and from chassis components. They can be sprayed or brushed on and, depending on the type, are rinsed off either with water or solvent.

Lubricants

Motor oil is the lubricant formulated for use in engines. It normally contains a wide variety of additives to prevent corrosion and reduce foaming and wear. Motor oil comes in various weights (viscosity ratings) from 5 to 80. The recommended weight of the oil depends on the season, temperature and the demands on the engine. Light oil is used in cold climates and under light load conditions. Heavy oil is used in hot climates and where high loads are encountered. Multi-viscosity oils are designed to have characteristics of both light and heavy oils and are available in a number of weights from 5W-20 to 20W-50.

Gear oil is designed to be used in differentials, manual transmissions and other areas where high-temperature lubrication is required.

Chassis and wheel bearing grease is a heavy grease used where increased loads and friction are encountered, such as for wheel bearings, balljoints, tie-rod ends and universal joints.

High-temperature wheel bearing grease is designed to withstand the extreme temperatures encountered by wheel bearings in disc brake equipped vehicles. It usually contains molybdenum disulfide (moly), which is a dry-type lubricant.

White grease is a heavy grease for metal-to-metal applications where water is a problem. White grease stays soft under both low and high temperatures (usually from -100 to +190-degrees F), and will not wash off or dilute in the presence of water.

Assembly lube is a special extreme pressure lubricant, usually containing moly, used to lubricate high-load parts (such as main and rod bearings and cam lobes) for initial start-up of a new engine. The assembly lube lubricates the parts without being squeezed out or washed away until the engine oiling system begins to function.

Silicone lubricants are used to protect rubber, plastic, vinyl and nylon parts.

Graphite lubricants are used where oils cannot be used due to contamination problems, such as in locks. The dry graphite will lubricate metal parts while remaining uncontaminated by dirt, water, oil or acids. It is electrically conductive and will not foul electrical contacts in locks such as the ignition switch.

Moly penetrants loosen and lubricate frozen, rusted and corroded fasteners and prevent future rusting or freezing.

Heat-sink grease is a special electrically non-conductive grease that is used for mounting electronic ignition modules where it is essential that heat is transferred away from the module.

Sealants

RTV sealant is one of the most widely used gasket compounds. Made from silicone, RTV is air curing, it seals, bonds, waterproofs, fills surface irregularities, remains flexible, doesn't shrink, is relatively easy to remove, and is used as a supplementary sealer with almost all low and medium temperature gaskets.

Anaerobic sealant is much like RTV in that it can be used either to seal gaskets or to form gaskets by itself. It remains flexible, is solvent resistant and fills surface imperfections. The difference between an anaerobic sealant and an RTV-type sealant is in the curing. RTV cures when exposed to air, while an anaerobic sealant cures only in the absence of air. This means that an anaerobic sealant cures only after the assembly of parts, sealing them together.

Thread and pipe sealant is used for sealing hydraulic and pneumatic fittings and vacuum lines. It is usually made from a Teflon compound, and comes in a spray, a paint-on liquid and as a wrap-around tape.

Chemicals

Anti-seize compound prevents seizing, galling, cold welding, rust and corrosion in fasteners. High-temperature ant-seize, usually made with copper and graphite lubricants, is used for exhaust system and exhaust manifold bolts.

Anaerobic locking compounds are used to keep fasteners from vibrating or working loose and cure only after installation, in the absence of air. Medium strength locking compound is used for small nuts, bolts and screws that may be removed later. High-strength locking compound is for large nuts, bolts and studs which aren't removed on a regular basis.

Oil additives range from viscosity index improvers to chemical treatments that claim to reduce internal engine friction. It should be noted that most oil manufacturers caution against using additives with their oils.

Gas additives perform several functions, depending on their chemical makeup. They usually contain solvents that help dissolve gum and varnish that build up on carburetor, fuel injection and intake parts. They also serve to break down carbon deposits that form on the inside surfaces of the combustion chambers. Some additives contain upper cylinder lubricants for valves and piston rings, and others contain chemicals to remove condensation from the gas tank.

Miscellaneous

Brake fluid is specially formulated hydraulic fluid that can withstand the heat and pressure encountered in brake systems. Care must be taken so this fluid does not come in contact with painted surfaces or plastics. An opened container should always be resealed to prevent contamination by water or dirt.

Weatherstrip adhesive is used to bond weatherstripping around doors, windows and trunk lids. It is sometimes used to attach trim pieces.

Undercoating is a petroleum-based, tar-like substance that is designed to protect metal surfaces on the underside of the vehicle from corrosion. It also acts as a sound-deadening agent by insulating the bottom of the vehicle.

Waxes and polishes are used to help protect painted and plated surfaces from the weather. Different types of paint may require the use of different types of wax and polish. Some polishes utilize a chemical or abrasive cleaner to help remove the top layer of oxidized (dull) paint on older vehicles. In recent years many non-wax polishes that contain a wide variety of chemicals such as polymers and silicones have been introduced. These non-wax polishes are usually easier to apply and last longer than conventional waxes and polishes.

Conversion factors

Length (distance)

Inches (in)	X 25.4	= Millimetres (mm)	X 0.0394	= Inches (in)
Feet (ft)	X 0.305	= Metres (m)	X 3.281	= Feet (ft)
Miles	X 1.609	= Kilometres (km)	X 0.621	= Miles

Volume (capacity)

Cubic inches (cu in; in³)	X 16.387	= Cubic centimetres (cc; cm³)	X 0.061	= Cubic inches (cu in; in³)
Imperial pints (Imp pt)	X 0.568	= Litres (l)	X 1.76	= Imperial pints (Imp pt)
Imperial quarts (Imp qt)	X 1.137	= Litres (l)	X 0.88	= Imperial quarts (Imp qt)
Imperial quarts (Imp qt)	X 1.201	= US quarts (US qt)	X 0.833	= Imperial quarts (Imp qt)
US quarts (US qt)	X 0.946	= Litres (l)	X 1.057	= US quarts (US qt)
Imperial gallons (Imp gal)	X 4.546	= Litres (l)	X 0.22	= Imperial gallons (Imp gal)
Imperial gallons (Imp gal)	X 1.201	= US gallons (US gal)	X 0.833	= Imperial gallons (Imp gal)
US gallons (US gal)	X 3.785	= Litres (l)	X 0.264	= US gallons (US gal)

Mass (weight)

Ounces (oz)	X 28.35	= Grams (g)	X 0.035	Ounces (oz)
Pounds (lb)	X 0.454	= Kilograms (kg)	X 2.205	= Pounds (lb)

Force

Ounces-force (ozf; oz)	X 0.278	= Newtons (N)	X 3.6	= Ounces-force (ozf; oz)
Pounds-force (lbf; lb)	X 4.448	= Newtons (N)	X 0.225	= Pounds-force (lbf; lb)
Newtons (N)	X 0.1	= Kilograms-force (kgf; kg)	X 9.81	= Newtons (N)

Pressure

Pounds-force per square inch (psi; lbf/in²; lb/in²)	X 0.070	= Kilograms-force per square centimetre (kgf/cm²; kg/cm²)	X 14.223	= Pounds-force per square inch (psi; lbf/in²; lb/in²)
Pounds-force per square inch (psi; lbf/in²; lb/in²)	X 0.068	= Atmospheres (atm)	X 14.696	= Pounds-force per square inch (psi; lbf/in²; lb/in²)
Pounds-force per square inch (psi; lbf/in²; lb/in²)	X 0.069	= Bars	X 14.5	= Pounds-force per square inch (psi; lbf/in²; lb/in²)
Pounds-force per square inch (psi; lbf/in²; lb/in²)	X 6.895	= Kilopascals (kPa)	X 0.145	= Pounds-force per square inch (psi; lbf/in²; lb/in²)
Kilopascals (kPa)	X 0.01	= Kilograms-force per square centimetre (kgf/cm²; kg/cm²)	X 98.1	= Kilopascals (kPa)

Torque (moment of force)

Pounds-force inches (lbf in; lb in)	X 1.152	= Kilograms-force centimetre (kgf cm; kg cm)	X 0.868	= Pounds-force inches (lbf in; lb in)
Pounds-force inches (lbf in; lb in)	X 0.113	= Newton metres (Nm)	X 8.85	= Pounds-force inches (lbf in; lb in)
Pounds-force inches (lbf in; lb in)	X 0.083	= Pounds-force feet (lbf ft; lb ft)	X 12	= Pounds-force inches (lbf in; lb in)
Pounds-force feet (lbf ft; lb ft)	X 0.138	= Kilograms-force metres (kgf m; kg m)	X 7.233	= Pounds-force feet (lbf ft; lb ft)
Pounds-force feet (lbf ft; lb ft)	X 1.356	= Newton metres (Nm)	X 0.738	= Pounds-force feet (lbf ft; lb ft)
Newton metres (Nm)	X 0.102	= Kilograms-force metres (kgf m; kg m)	X 9.804	= Newton metres (Nm)

Power

Horsepower (hp)	X 745.7	= Watts (W)	X 0.0013	= Horsepower (hp)

Velocity (speed)

Miles per hour (miles/hr; mph)	X 1.609	= Kilometres per hour (km/hr; kph)	X 0.621	= Miles per hour (miles/hr; mph)

Fuel consumption*

Miles per gallon, Imperial (mpg)	X 0.354	= Kilometres per litre (km/l)	X 2.825	= Miles per gallon, Imperial (mpg)
Miles per gallon, US (mpg)	X 0.425	= Kilometres per litre (km/l)	X 2.352	= Miles per gallon, US (mpg)

Temperature

Degrees Fahrenheit = (°C x 1.8) + 32

Degrees Celsius (Degrees Centigrade; °C) = (°F - 32) x 0.56

*It is common practice to convert from miles per gallon (mpg) to litres/100 kilometres (l/100km), where mpg (Imperial) x l/100 km = 282 and mpg (US) x l/100 km = 235

Safety first!

Regardless of how enthusiastic you may be about getting on with the job at hand, take the time to ensure that your safety is not jeopardized. A moment's lack of attention can result in an accident, as can failure to observe certain simple safety precautions. The possibility of an accident will always exist, and the following points should not be considered a comprehensive list of all dangers. Rather, they are intended to make you aware of the risks and to encourage a safety conscious approach to all work you carry out on your vehicle.

Essential DOs and DON'Ts

DON'T rely on a jack when working under the vehicle. Always use approved jackstands to support the weight of the vehicle and place them under the recommended lift or support points.

DON'T attempt to loosen extremely tight fasteners (i.e. wheel lug nuts) while the vehicle is on a jack - it may fall.

DON'T start the engine without first making sure that the transmission is in Neutral (or Park where applicable) and the parking brake is set.

DON'T remove the radiator cap from a hot cooling system - let it cool or cover it with a cloth and release the pressure gradually.

DON'T attempt to drain the engine oil until you are sure it has cooled to the point that it will not burn you.

DON'T touch any part of the engine or exhaust system until it has cooled sufficiently to avoid burns.

DON'T siphon toxic liquids such as gasoline, antifreeze and brake fluid by mouth, or allow them to remain on your skin.

DON'T inhale brake lining dust - it is potentially hazardous (see *Asbestos* below).

DON'T allow spilled oil or grease to remain on the floor - wipe it up before someone slips on it.

DON'T use loose fitting wrenches or other tools which may slip and cause injury.

DON'T push on wrenches when loosening or tightening nuts or bolts. Always try to pull the wrench toward you. If the situation calls for pushing the wrench away, push with an open hand to avoid scraped knuckles if the wrench should slip.

DON'T attempt to lift a heavy component alone - get someone to help you.

DON'T rush or take unsafe shortcuts to finish a job.

DON'T allow children or animals in or around the vehicle while you are working on it.

DO wear eye protection when using power tools such as a drill, sander, bench grinder, etc. and when working under a vehicle.

DO keep loose clothing and long hair well out of the way of moving parts.

DO make sure that any hoist used has a safe working load rating adequate for the job.

DO get someone to check on you periodically when working alone on a vehicle.

DO carry out work in a logical sequence and make sure that everything is correctly assembled and tightened.

DO keep chemicals and fluids tightly capped and out of the reach of children and pets.

DO remember that your vehicle's safety affects that of yourself and others. If in doubt on any point, get professional advice.

Asbestos

Certain friction, insulating, sealing, and other products - such as brake linings, brake bands, clutch linings, torque converters, gaskets, etc. - may contain asbestos. Extreme care must be taken to avoid inhalation of dust from such products, since it is hazardous to health. If in doubt, assume that they do contain asbestos.

Fire

Remember at all times that gasoline is highly flammable. Never smoke or have any kind of open flame around when working on a vehicle. But the risk does not end there. A spark caused by an electrical short circuit, by two metal surfaces contacting each other, or even by static electricity built up in your body under certain conditions, can ignite gasoline vapors, which in a confined space are highly explosive. Do not, under any circumstances, use gasoline for cleaning parts. Use an approved safety solvent.

Always disconnect the battery ground (-) cable at the battery before working on any part of the fuel system or electrical system. Never risk spilling fuel on a hot engine or exhaust component. It is strongly recommended that a fire extinguisher suitable for use on fuel and electrical fires be kept handy in the garage or workshop at all times. Never try to extinguish a fuel or electrical fire with water.

Fumes

Certain fumes are highly toxic and can quickly cause unconsciousness and even death if inhaled to any extent. Gasoline vapor falls into this category, as do the vapors from some cleaning solvents. Any draining or pouring of such volatile fluids should be done in a well ventilated area.

When using cleaning fluids and solvents, read the instructions on the container carefully. Never use materials from unmarked containers.

Never run the engine in an enclosed space, such as a garage. Exhaust fumes contain carbon monoxide, which is extremely poisonous. If you need to run the engine, always do so in the open air, or at least have the rear of the vehicle outside the work area.

If you are fortunate enough to have the use of an inspection pit, never drain or pour gasoline and never run the engine while the vehicle is over the pit. The fumes, being heavier than air, will concentrate in the pit with possibly lethal results.

The battery

Never create a spark or allow a bare light bulb near a battery. They normally give off a certain amount of hydrogen gas, which is highly explosive.

Always disconnect the battery ground (-) cable at the battery before working on the fuel or electrical systems.

If possible, loosen the filler caps or cover when charging the battery from an external source (this does not apply to sealed or maintenance-free batteries). Do not charge at an excessive rate or the battery may burst.

Take care when adding water to a non maintenance-free battery and when carrying a battery. The electrolyte, even when diluted, is very corrosive and should not be allowed to contact clothing or skin.

Always wear eye protection when cleaning the battery to prevent the caustic deposits from entering your eyes.

Household current

When using an electric power tool, inspection light, etc., which operates on household current, always make sure that the tool is correctly connected to its plug and that, where necessary, it is properly grounded. Do not use such items in damp conditions and, again, do not create a spark or apply excessive heat in the vicinity of fuel or fuel vapor.

Secondary ignition system voltage

A severe electric shock can result from touching certain parts of the ignition system (such as the spark plug wires) when the engine is running or being cranked, particularly if components are damp or the insulation is defective. In the case of an electronic ignition system, the secondary system voltage is much higher and could prove fatal.

Troubleshooting

Contents

This section provides an easy reference guide to the more common problems which may occur during the operation of your vehicle. These problems and their possible causes are grouped under headings denoting various components or systems, such as Engine, Cooling system, etc. They also refer you to the chapter and/or section which deals with the problem.

Remember that successful troubleshooting is not a mysterious black art practiced only by professional mechanics. It is simply the result of the right knowledge combined with an intelligent, systematic approach to the problem. Always work by a process of elimination, starting with the simplest solution and working through to the most complex - and never overlook the obvious. Anyone can run the gas tank dry or leave the lights on overnight, so don't assume that you are exempt from such oversights.

Finally, always establish a clear idea of why a problem has occurred and take steps to ensure that it doesn't happen again. If the electrical system fails because of a poor connection, check the other connections in the system to make sure that they don't fail as well. If a particular fuse continues to blow, find out why - don't just replace one fuse after another. Remember, failure of a small component can often be indicative of potential failure or incorrect functioning of a more important component or system.

Engine

1   Engine will not rotate when attempting to start

1   Battery terminal connections loose or corroded (Chapter 1).
2   Battery discharged or faulty (Chapter 1).
3   Automatic transaxle not completely engaged in Park (Chapter 7B) or clutch pedal not completely depressed.
4   Broken, loose or disconnected wiring in the starting circuit (Chapters 5A and 12).
5   Starter motor pinion jammed in flywheel ring gear (Chapter 5A).
6   Starter solenoid faulty (Chapter 5A).
7   Starter motor faulty (Chapter 5A).
8   Ignition switch faulty (Chapter 12).
9   Starter pinion or flywheel teeth worn or broken (Chapter 5A).

2   Engine rotates but will not start

1   Fuel tank empty.
2   Battery discharged (engine rotates slowly) (Chapter 5A).
3   Battery terminal connections loose or corroded (Chapter 1).
4   Leaking fuel injector(s), faulty fuel pump, pressure regulator, etc. (Chapter 4).
5   Broken or stripped timing belt (Chapter 2).
6   Ignition components damp or damaged (Chapter 5B).
7   Worn, faulty or incorrectly gapped spark plugs (Chapter 1).
8   Broken, loose or disconnected wiring in the starting circuit (Chapter 5A).
9   Loose distributor is changing ignition timing (Chapter 5B).
10   Broken, loose or disconnected wires at the ignition coil or faulty coil (Chapter 5B).

3   Engine hard to start when cold

1   Battery discharged or low (Chapter 1).
2   Malfunctioning fuel system (Chapter 4).
3   Faulty coolant temperature sensor or intake air temperature sensor (Chapter 4).
4   Injector(s) leaking (Chapter 4).
5   Faulty ignition system (Chapter 5B).

4   Engine hard to start when hot

1   Air filter clogged (Chapter 1).
2   Fuel not reaching the fuel injection system (Chapter 4).
3   Corroded battery connections, especially ground (Chapter 1).
4   Faulty coolant temperature sensor or intake air temperature sensor (Chapter 4).
5   Low cylinder compression (Chapter 2).

5   Starter motor noisy or excessively rough in engagement

1   Pinion or flywheel gear teeth worn or broken (Chapter 5A).

2   Starter motor mounting bolts loose or missing (Chapter 5A).
3   Starter motor internal components worn or damaged (Chapter 5A).

6   Engine starts but stops immediately

1   Loose or faulty electrical connections at distributor or coil (Chapter 5B).
2   Insufficient fuel reaching the fuel injector(s) (Chapters 1 and 4).
3   Vacuum leak at the gasket between the intake manifold/plenum and throttle body (Chapters 1 and 4).
4   Idle speed incorrect (Chapter 1).
5   Intake air leaks, broken vacuum lines.

7   Oil puddle under engine

1   Oil pan gasket and/or oil pan drain bolt washer leaking (Chapter 2).
2   Oil pressure sending unit leaking (Chapter 2).
3   Camshaft cover leaking (Chapter 2).
4   Engine oil seals leaking (Chapter 2).
5   Oil pump housing leaking (Chapter 2).

8   Engine lopes while idling or idles erratically

1   Vacuum leakage.
2   Leaking EGR valve (Chapter 6).
3   Air filter clogged (Chapter 1).
4   Fuel pump not delivering sufficient fuel to the fuel injection system (Chapter 4).
5   Leaking head gasket (Chapter 2).
6   Timing belt and/or sprockets worn (Chapter 2).
7   Camshaft lobes worn (Chapter 2).

9   Engine misses at idle speed

1   Spark plugs worn or not gapped properly (Chapter 1).
2   Faulty spark plug wires (Chapter 1).
3   Vacuum leaks (Chapters 2 and 4).
4   Incorrect ignition timing (Chapter 1).
5   Uneven or low compression (Chapter 2).
6   Problem with the fuel injection system (Chapter 4).

10   Engine misses throughout driving speed range

1   Fuel filter clogged and/or impurities in the fuel system (Chapter 1).
2   Low fuel output at the fuel injector(s) (Chapter 4).
3   Faulty or incorrectly gapped spark plugs (Chapter 1).
4   Incorrect ignition timing (Chapter 5B).
5   Cracked distributor cap, disconnected distributor wires or damaged distributor components (Chapters 1 and 5B).
6   Leaking spark plug wires (Chapters 1 or 5B).
7   Faulty emission system components (Chapter 6).
8   Low or uneven cylinder compression pressures (Chapter 2).
9   Weak or faulty ignition system (Chapter 5B).
10   Vacuum leak in fuel injection system (Chapter 4).

11   Engine stumbles on acceleration

1   Spark plugs fouled, worn or incorrectly gapped (Chapter 1).
2   Problem with fuel injection system (Chapter 4).
3   Fuel filter clogged (Chapters 1 and 4).
4   Intake manifold air leak (Chapter 4).

5   EGR system malfunction (Chapter 6).
6   Faulty injector(s) - diesel models (Chapter 4C).

12   Engine surges while holding accelerator steady

1   Intake air leak (Chapter 4).
2   Fuel pump or fuel pressure regulator faulty (Chapter 4).
3   Problem with fuel injection system (Chapter 4).
4   Problem with the emissions control system (Chapter 6).

13   Engine stalls

1   Idle speed incorrect (Chapter 1).
2   Fuel filter clogged and/or water and impurities in the fuel system (Chapters 1 and 4).
3   Distributor components damp or damaged (Chapter 5).
4   Faulty emissions system components (Chapter 6).
5   Faulty or incorrectly gapped spark plugs (Chapter 1).
6   Faulty spark plug wires (Chapter 1).
7   Vacuum leak in the fuel injection system, intake manifold or vacuum hoses (Chapters 2 and 4).

14   Engine lacks power

1   Vacuum leak at intake manifold (Chapter 4).
2   Excessive play in distributor shaft (Chapter 5).
3   Worn rotor, distributor cap, spark plug wires or faulty coil (Chapters 1 and 5).
4   Worn, faulty or incorrectly gapped spark plugs (Chapter 1).
5   Problem with the fuel injection system (Chapter 4).
6   Plugged air filter (Chapter 1).
7   Brakes binding (Chapter 9).
8   Automatic transaxle fluid level incorrect (Chapter 1).
9   Clutch slipping (Chapter 8).
10   Fuel filter clogged and/or impurities in the fuel system (Chapters 1 and 4).
11   Emission control system not functioning properly (Chapter 6).
12   Low or uneven cylinder compression pressures (Chapter 2).
13   Obstructed exhaust system (Chapters 2 and 4).
14   Timing belt incorrectly installed or tensioned (Chapter 2).

15   Engine backfires

1   Emission control system not functioning properly (Chapter 6).
2   Timing belt incorrectly installed or tensioned (Chapter 2) .
3   Faulty secondary ignition system (cracked spark plug insulator, faulty plug wires, distributor cap and/or rotor) (Chapters 1 and 5B).
4   Problem with the fuel injection system (Chapter 4).
5   Vacuum leak at fuel injector(s), intake manifold, air control valve or vacuum hoses (Chapters 2 and 4).
6   Valve sticking.

16   Pinging or knocking engine sounds during acceleration or uphill

1   Incorrect grade of fuel.
2   Knock sensor or circuit faulty (Chapter 4B).
3   Fuel injection system faulty (Chapter 4).
4   Improper or damaged spark plugs or wires (Chapter 1).
5   Worn or damaged distributor components (Chapter 5).
6   EGR valve not functioning (Chapter 6).
7   Vacuum leak (Chapters 2 and 4).

17   Engine runs with oil pressure light on

1   Low oil level (Chapter 1).
2   Idle rpm below specification (Chapter 1).
3   Short in wiring circuit (Chapter 12).
4   Faulty oil pressure sender (Chapter 2).
5   Worn engine bearings and/or oil pump (Chapter 2).
6   Oil pick-up strainer clogged (Chapter 2).
7   Oil pressure relief valve defective (Chapter 2).

18   Engine diesels (continues to run) after switching off

1   Idle speed too high (Chapter 1).
2   Excessive engine operating temperature (Chapter 3).
3   Excessive carbon deposits on valves and pistons (see Chapter 2).

Engine electrical system

19   Battery will not hold a charge

1   Alternator drivebelt defective or not adjusted properly (Chapter 1).
2   Battery electrolyte level low (Chapter 1).
3   Battery terminals loose or corroded (Chapter 1).
4   Alternator not charging properly (Chapter 5A).
5   Loose, broken or faulty wiring in the charging circuit (Chapter 5A).
6   Short in vehicle wiring (Chapter 12).
7   Internally defective battery (Chapters 1 and 5A).

20   Alternator light fails to go out

1   Faulty alternator or charging circuit (Chapter 5A).
2   Alternator drivebelt defective or out of adjustment (Chapter 1).
3   Alternator voltage regulator inoperative (Chapter 5A).

21   Alternator light fails to come on when key is turned on

1   Warning light bulb defective (Chapter 12).
2   Fault in the printed circuit, dash wiring or bulb holder (Chapter 12).

Fuel system

22   Excessive fuel consumption

1   Dirty or clogged air filter element (Chapter 1).
2   Emissions system not functioning properly (Chapter 6).
3   Fuel injection system not functioning properly (Chapter 4).
4   Low tire pressure or incorrect tire size (Chapter 1).
5   Brakes dragging (Chapter 9)

23   Fuel leakage and/or fuel odor

1   Leaking fuel feed or return line (Chapters 1 and 4).
2   Tank overfilled.
3   Charcoal canister filter clogged (Chapters 1 and 6).
4   Problem with fuel injection system (Chapter 4).

Cooling system

24 Overheating

1 Insufficient coolant in system (Chapter 1).
2 Water pump drivebelt defective or out of adjustment (Chapter 1).
3 Radiator core blocked or grille restricted (Chapter 3).
4 Thermostat faulty (Chapter 3).
5 Electric coolant fan inoperative or blades broken (Chapter 3).
6 Radiator cap not maintaining proper pressure (Chapter 3).
7 Ignition timing incorrect (Chapter 5).

25 Overcooling

1 Faulty thermostat (Chapter 3).
2 Inaccurate temperature gauge sending unit (Chapter 3).

26 External coolant leakage

1 Deteriorated/damaged hoses; loose clamps (Chapters 1 and 3).
2 Water pump defective (Chapter 3).
3 Leakage from radiator core or coolant expansion tank (Chapter 3).
4 Engine drain or water jacket core plugs leaking (Chapter 2).

27 Internal coolant leakage

1 Leaking cylinder head gasket (Chapter 2).
2 Cracked cylinder bore or cylinder head (Chapter 2).

28 Coolant loss

1 Too much coolant in system (Chapter 1).
2 Coolant boiling away because of overheating (Chapter 3).
3 Internal or external leakage (Chapter 3).
4 Faulty radiator cap (Chapter 3).

29 Poor coolant circulation

1 Inoperative water pump (Chapter 3).
2 Restriction in cooling system (Chapters 1 and 3).
3 Water pump drivebelt defective/out of adjustment (Chapter 1).
4 Thermostat sticking (Chapter 3).

Clutch

30 Pedal travels to floor - no pressure or very little resistance

1 Clutch cable broken (Chapter 8).
2 Broken release bearing or fork (Chapter 8).
3 Pressure plate defective (Chapter 8).

31 Unable to select gears

1 Faulty transaxle (Chapter 7).
2 Faulty clutch disc or pressure plate (Chapter 8).
3 Faulty release lever or release bearing (Chapter 8).
4 Faulty shift lever assembly or rods (Chapter 8).

32 Clutch slips (engine speed increases with no increase in vehicle speed)

1 Clutch plate worn (Chapter 8).
2 Clutch plate is oil soaked by leaking rear main seal (Chapter 8).
3 Clutch plate not seated (Chapter 8).
4 Warped pressure plate or flywheel (Chapter 8).
5 Weak diaphragm springs in pressure plate (Chapter 8).
6 Clutch plate overheated. Allow to cool.
7 Clutch cable sticking (Chapter 8).

33 Grabbing (chattering) as clutch is engaged

1 Oil on clutch plate lining, burned or glazed facings (Chapter 8).
2 Worn or loose engine or transaxle mounts (Chapters 2 and 7).
3 Worn splines on clutch plate hub (Chapter 8).
4 Warped pressure plate or flywheel (Chapter 8).
5 Burned or smeared resin on flywheel or pressure plate (Chapter 8).

34 Transaxle rattling (clicking)

1 Release fork loose (Chapter 8).
2 Low engine idle speed (Chapter 1).

35 Noise in clutch area

Faulty release bearing (Chapter 8).

36 Clutch pedal stays on floor

1 Broken release bearing or fork (Chapter 8).
2 Clutch cable sticking (Chapter 8).

37 High pedal effort

1 Clutch cable sticking (Chapter 8).
2 Pressure plate faulty (Chapter 8).

Manual transaxle

38 Knocking noise at low speeds

1 Worn driveaxle constant velocity (CV) joints (Chapter 8).
2 Worn side gear shaft counterbore in differential case (Chapter 7A).*

39 Noise most pronounced when turning

Differential gear noise (Chapter 7A).*

40 Clunk on acceleration or deceleration

1 Loose engine or transaxle mounts (Chapters 2 and 7A).
2 Worn differential pinion shaft in case.*
3 Worn side gear shaft counterbore in differential case (Chapter 7A).*
4 Worn or damaged driveaxle inboard CV joints (Chapter 8).

41   Clicking noise in turns

Worn or damaged outboard CV joint (Chapter 8).

42   Vibration

1   Rough wheel bearing (Chapters 1 and 10).
2   Damaged driveaxle (Chapter 8).
3   Out of round tires (Chapter 1).
4   Tire out of balance (Chapters 1 and 10).
5   Worn CV joint (Chapter 8).

43   Noisy in neutral with engine running

1   Damaged input gear bearing (Chapter 7A).*
2   Damaged clutch release bearing (Chapter 8).

44   Noisy in one particular gear

1   Damaged or worn constant mesh gears (Chapter 7A).*
2   Damaged or worn synchronizers (Chapter 7A).*
3   Bent reverse fork (Chapter 7A).*
4   Damaged fourth speed gear or output gear (Chapter 7A).*
5   Worn or damaged reverse idler gear or idler bushing (Chapter 7A).*

45   Noisy in all gears

1   Insufficient lubricant (Chapter 7A).
2   Damaged or worn bearings (Chapter 7A).*
3   Worn or damaged input gear shaft and/or output gear shaft (Chapter 7A).*

46   Slips out of gear

1   Worn or improperly adjusted linkage (Chapter 7A).
2   Transaxle loose on engine (Chapter 7A).
3   Shift linkage does not work freely, binds (Chapter 7A).
4   Input gear bearing retainer broken or loose (Chapter 7A).*
5   Worn shift fork (Chapter 7A).*

47   Leaks lubricant

1   Side gear shaft seals worn (Chapter 7A).
2   Excessive amount of lubricant in transaxle (Chapters 1 and 7A).
3   Loose or broken input gear shaft bearing retainer (Chapter 7A).*
4   Input gear bearing retainer O-ring and/or lip seal damaged (Chapter 7A).*
5   Vehicle speed sensor O-ring leaking (Chapter 7A).

48   Hard to shift

Shift linkage loose or worn (Chapter 7A).

Although the corrective action necessary to remedy the symptoms described is beyond the scope of this manual, the above information should be helpful in isolating the cause of the condition so that the owner can communicate clearly with a professional mechanic.

Automatic transaxle

Note: *Due to the complexity of the automatic transaxle, it is difficult for the home mechanic to properly diagnose and service this component. For problems other than the following, the vehicle should be taken to a dealer or transaxle shop.*

49   Fluid leakage

1   Automatic transaxle fluid is a deep red color on 1993 and 1994 models; on 1995 and later models, it's yellow. Fluid leaks should not be confused with engine oil, which can easily be blown onto the transaxle by air flow.
2   To pinpoint a leak, first remove all built-up dirt and grime from the transaxle housing with degreasing agents and/or steam cleaning. Then drive the vehicle at low speeds so air flow will not blow the leak far from its source. Raise the vehicle and determine where the leak is coming from. Common areas of leakage are:

a)   *Pan (Chapters 1 and 7)*
b)   *Dipstick tube (Chapters 1 and 7)*
c)   *Transaxle oil lines (Chapter 7)*
d)   *Speed sensor (Chapter 7)*
e)   *Driveaxle oil seals (Chapter 7).*

50   Transaxle fluid brown or has a burned smell

Transaxle fluid overheated (Chapter 1).

51   General shift mechanism problems

1   Chapter 7, Part B, deals with checking and adjusting the shift linkage on automatic transaxles. Common problems which may be attributed to poorly adjusted linkage are:

a)   *Engine starting in gears other than Park or Neutral.*
b)   *Indicator on shifter pointing to a gear other than the one actually being used.*
c)   *Vehicle moves when in Park.*
2   Refer to Chapter 7B for the shift linkage adjustment procedure.

52   Transaxle will not downshift with accelerator pedal pressed to the floor

The transaxle is electronically controlled. This type of problem - which is caused by a malfunction in the control unit, a sensor or solenoid, or the circuit itself - is beyond the scope of this book. Take the vehicle to a dealer service department or a competent automatic transmission shop.

53   Engine will start in gears other than Park or Neutral

Neutral start switch out of adjustment or malfunctioning (Chapter 7B).

54   Transaxle slips, shifts roughly, is noisy or has no drive in forward or reverse gears

There are many probable causes for the above problems, but the home mechanic should be concerned with only one possibility - fluid level. Before taking the vehicle to a repair shop, check the level and condition of the fluid as described in Chapter 1. Correct the fluid level as necessary or change the fluid and filter if needed. If the problem persists, have a professional diagnose the cause.

Driveaxles

55 Clicking noise in turns

Worn or damaged outboard CV joint (Chapter 8).

56 Shudder or vibration during acceleration

1 Excessive toe-in (Chapter 10).
2 Worn or damaged inboard or outboard CV joints (Chapter 8).
3 Sticking inboard CV joint assembly (Chapter 8).

57 Vibration at highway speeds

1 Out of balance front wheels and/or tires (Chapters 1 and 10).
2 Out of round front tires (Chapters 1 and 10).
3 Worn CV joint(s) (Chapter 8).

Brakes

Note: *Before assuming that a brake problem exists, make sure that:*
 a) *The tires are in good condition and properly inflated (Chapter 1).*
 b) *The front end alignment is correct (Chapter 10).*
 c) *The vehicle is not loaded with weight in an unequal manner.*

58 Vehicle pulls to one side during braking

1 Incorrect tire pressures (Chapter 1).
2 Front end out of alignment (have the front end aligned).
3 Front, or rear, tire sizes not matched to one another.
4 Restricted brake lines or hoses (Chapter 9).
5 Malfunctioning drum brake or caliper assembly (Chapter 9).
6 Loose suspension parts (Chapter 10).
7 Loose calipers (Chapter 9).
8 Excessive wear of brake shoe or pad material or disc/drum on one side.

59 Noise (high-pitched squeal when the brakes are applied)

Front and/or rear disc brake pads worn out. The noise comes from the wear sensor rubbing against the disc (does not apply to all vehicles). Replace pads with new ones immediately (Chapter 9).

60 Brake roughness or chatter (pedal pulsates)

1 Excessive lateral runout (Chapter 9).
2 Uneven pad wear (Chapter 9).
3 Defective disc (Chapter 9).
4 Out-of-round brake drum (Chapter 9).

61 Excessive brake pedal effort required to stop vehicle

1 Malfunctioning power brake booster (Chapter 9).
2 Partial system failure (Chapter 9).
3 Excessively worn pads or shoes (Chapter 9).
4 Piston in caliper or wheel cylinder stuck or sluggish (Chapter 9).
5 Brake pads or shoes contaminated with oil or grease (Chapter 9).
6 Brake disc grooved and/or glazed (Chapter 1).
7 New pads or shoes installed and not yet seated. It will take a while for the new material to seat against the disc or drum.

62 Excessive brake pedal travel

1 Partial brake system failure (Chapter 9).
2 Insufficient fluid in master cylinder (Chapters 1 and 9).
3 Air trapped in system (Chapters 1 and 9).

63 Dragging brakes

1 Incorrect adjustment of brake light switch (Chapter 9).
2 Master cylinder pistons not returning correctly (Chapter 9).
3 Restricted brakes lines or hoses (Chapters 1 and 9).
4 Incorrect parking brake adjustment (Chapter 9).
5 Sticking caliper or wheel cylinder piston (Chapter 9).

64 Grabbing or uneven braking action

1 Malfunction of proportioning valve (Chapter 9).
2 Malfunction of power brake booster unit (Chapter 9).
3 Binding brake pedal mechanism (Chapter 9).
4 Brake lining contaminated with grease or brake fluid (Chapter 9).

65 Brake pedal feels spongy when depressed

1 Air in hydraulic lines (Chapter 9).
2 Master cylinder mounting bolts loose (Chapter 9).
3 Master cylinder defective (Chapter 9).

66 Brake pedal travels to the floor with little resistance

1 Little or no fluid in the master cylinder reservoir caused by leaking caliper piston(s) (Chapter 9).
2 Loose, damaged or disconnected brake lines (Chapter 9).
3 Defective master cylinder (Chapter 9).

67 Parking brake does not hold

Parking brake improperly adjusted (Chapters 1 and 9).

Suspension and steering systems

Note: *Before attempting to diagnose the suspension and steering systems, perform the following preliminary checks:*
 a) *Tires for wrong pressure and uneven wear.*
 b) *Steering universal joints from the column to the rack and pinion for loose connectors or wear.*
 c) *Front and rear suspension and the rack-and-pinion assembly for loose or damaged parts.*
 d) *Out-of-round or out-of-balance tires, bent rims and loose and/or rough wheel bearings.*

68 Vehicle pulls to one side

1 Mismatched or uneven tires (Chapter 10).
2 Sagging springs (Chapter 10).
3 Wheel alignment out-of-specifications (Chapter 10).
4 Front brake dragging (Chapter 9).

69 Abnormal or excessive tire wear

1 Wheel alignment out-of-specifications (Chapter 10).

2   Sagging springs (Chapter 10).
3   Tire out-of-balance (Chapter 10).
4   Worn strut damper (Chapter 10).
5   Overloaded vehicle.
6   Tires not rotated regularly.

70   Wheel makes a thumping noise

1   Blister or bump on tire (Chapter 10).
2   Improper strut damper action (Chapter 10).

71   Shimmy, shake or vibration

1   Tire or wheel out-of-balance or out-of-round (Chapter 10).
2   Loose or worn wheel bearings (Chapters 1, 8 and 10).
3   Worn tie-rod ends (Chapter 10).
4   Worn lower balljoints (Chapters 1 and 10).
5   Excessive wheel runout (Chapter 10).
6   Blister or bump on tire (Chapter 10).

72   Hard steering

1   Lack of lubrication at balljoints, tie-rod ends and rack and pinion assembly (Chapter 10).
2   Front wheel alignment out-of-specifications (Chapter 10).
3   Low tire pressure(s) (Chapters 1 and 10).

73   Poor returnability of steering to center

1   Lack of lubrication at balljoints and tie-rod ends (Chapter 10).
2   Binding in balljoints (Chapter 10).
3   Binding in steering column (Chapter 10).
4   Lack of lubricant in steering gear assembly (Chapter 10).
5   Front wheel alignment out-of-specifications (Chapter 10).

74   Abnormal noise at the front end

1   Lack of lubrication at balljoints and tie-rod ends (Chapters 1 and 10).
2   Damaged strut mounting (Chapter 10).
3   Worn control arm bushings or tie-rod ends (Chapter 10).
4   Loose stabilizer bar (Chapter 10).
5   Loose wheel nuts (Chapters 1 and 10).
6   Loose suspension bolts (Chapter 10)

75   Wander or poor steering stability

1   Mismatched or uneven tires (Chapter 10).
2   Lack of lubrication at balljoints and tie-rod ends (Chapters 1 and 10).
3   Worn strut assemblies (Chapter 10).
4   Loose stabilizer bar (Chapter 10).
5   Broken or sagging springs (Chapter 10).
6   Wheels out of alignment (Chapter 10).

76   Erratic steering when braking

1   Wheel bearings worn (Chapter 10).
2   Sagging springs (Chapter 10).

3   Leaking wheel cylinder or caliper (Chapter 10).
4   Warped rotors or drums (Chapter 10).

77   Excessive pitching and/or rolling around corners or during braking

1   Loose stabilizer bar (Chapter 10).
2   Worn strut dampers or mountings (Chapter 10).
3   Sagging springs (Chapter 10).
4   Overloaded vehicle.

78   Suspension bottoms

1   Overloaded vehicle.
2   Worn strut dampers (Chapter 10).
3   Incorrect or sagging springs (Chapter 10).

79   Cupped tires

1   Front wheel or rear wheel alignment out-of-specifications (Chapter 10).
2   Worn strut dampers (Chapter 10).
3   Wheel bearings worn (Chapter 10).
4   Excessive tire or wheel runout (Chapter 10).
5   Worn balljoints (Chapter 10).

80   Excessive tire wear on outside edge

1   Inflation pressures incorrect (Chapter 1).
2   Excessive speed in turns.
3   Front end alignment incorrect (excessive toe-in or camber). Have professionally aligned.
4   Suspension arm bent or twisted (Chapter 10).

81   Excessive tire wear on inside edge

1   Inflation pressures incorrect (Chapter 1).
2   Front end alignment incorrect (toe-out). Have professionally aligned.
3   Loose or damaged steering components (Chapter 10).

82   Tire tread worn in one place

1   Tires out-of-balance.
2   Damaged or buckled wheel. Inspect and replace if necessary.
3   Defective tire (Chapter 1).

83   Excessive play or looseness in steering system

1   Wheel bearing(s) worn (Chapter 10).
2   Tie-rod end loose (Chapter 10).
3   Steering gear loose (Chapter 10).
4   Worn or loose steering intermediate shaft (Chapter 10).

84   Rattling or clicking noise in steering gear

1   Steering gear loose (Chapter 10).
2   Steering gear defective.

Chapter 1
Tune-up and routine maintenance

Contents

1

Specifications

Recommended lubricants and fluids

Engine oil
  Type
    Gasoline engine .... API grade SH or SH/CD multi-grade and fuel efficient oil
    Diesel engine.. API grade CG-4 multi-grade and low sulfated ash limit engine oil
  Viscosity .... See accompanying chart

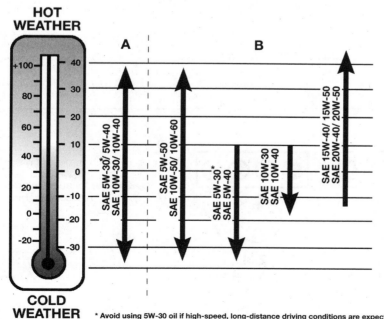

Engine oil viscosity chart - for best fuel economy and cold starting, select the lowest SAE viscosity grade for the expected temperature range

A   Energy conserving oil
B   Multi-grade oil

LOOK FOR ONE OF THESE LABELS

96017-1A-A HAYNES

* Avoid using 5W-30 oil if high-speed, long-distance driving conditions are expected.

Recommended lubricants and fluids

Engine coolant	50/50 mixture of water and ethylene glycol-based antifreeze
Automatic transaxle fluid type	
1993 and 1994 models	DEXRON II automatic transmission fluid
1995 and later models	VW automatic transmission fluid only (part no. G 052 162 A1 or A2)
Manual transaxle lubricant	SAE 75W/90 gear oil, meeting VW specification G50
Differential lubricant (models with automatic transaxle)	
1993 and 1994 models	SAE 75W/90 gear oil, meeting VW specification G50
1995 and later models	VW automatic transmission fluid only (part no. G 052 162 A1 or A2)
Brake and clutch fluid	DOT 4 brake fluid
Power steering fluid	VW power steering fluid, part no. G002000 or G002012 or equivalent

Capacities*

Engine oil (including filter)	
1.8L gasoline engine (code ACC)	4.2 quarts
2.0L gasoline engine (code ABA)	5.3 quarts
1.9L diesel engine (code AAZ)	4.5 quarts
Cooling system (all models)	6.6 quarts
Transaxle	
Manual transaxle	4.0 pints
Automatic transaxle	3.2 quarts
Power-assisted steering	1.6 quarts

All capacities approximate. Add as necessary to bring to the appropriate level.

Ignition system

Firing order	1-3-4-2
Ignition timing	See Chapter 5B
Spark plugs	
Type	
1.8L engine (code ACC)	Bosch W7DTC
2.0L engine (code ABA)	Bosch FR8DS
Gap	
1.8L engine (code ACC)	0.028 to 0.035 inch
2.0L engine (code ABA)	0.024 inch

Brakes

Brake pad minimum thickness (including backing plate)	9/32 inch
Brake shoe friction material minimum thickness	3/32 inch

Torque specifications

	Ft-lbs
Wheel bolts	81
Spark plugs:	
1.8L engine (code ACC)	18
2.0L engine (code ABA)	22
Oil pan drain plug	22

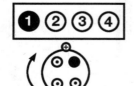

1.8L

2.0L

Cylinder location and distributor rotation diagram

96017-1A-SPECS HAYNES

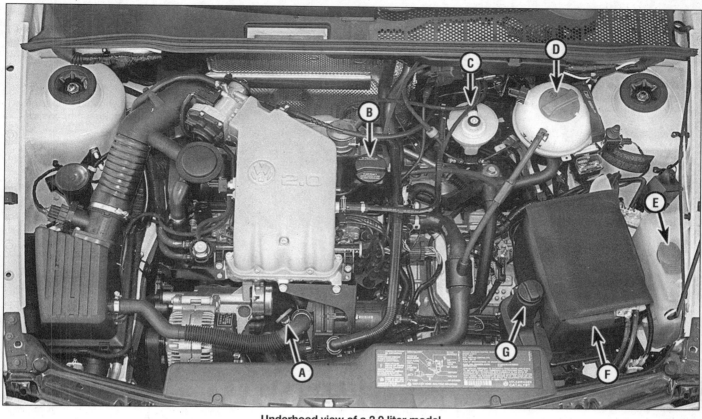

Underhood view of a 2.0 liter model

A   Engine oil level dipstick
B   Engine oil filler cap
C   Brake fluid reservoir

D   Coolant expansion tank
E   Windshield washer fluid reservoir

F   Battery
G   Power steering fluid reservoir

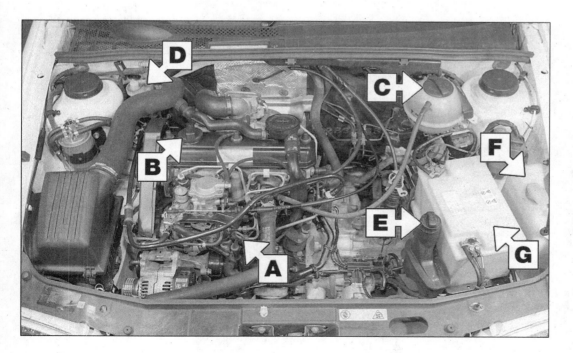

Underhood of a 1.9L diesel model (right-hand drive model shown; left-hand drive models similar)

A   Engine oil level dipstick
B   Engine oil filler cap
C   Coolant expansion tank
D   Brake fluid reservoir (mounted on left side of engine compartment on left-hand drive models)
E   Power steering fluid reservoir
F   Windshield washer fluid reservoir
G   Battery

Typical engine compartment underside components

1   Engine oil filter
2   Engine oil drain plug
3   Alternator
4   Front suspension subframe
5   Exhaust front pipe
6   Catalytic converter
7   Front brake caliper
8   Front suspension control arm
9   Tie-rod end
10 Oxygen sensor
11 Brake lines

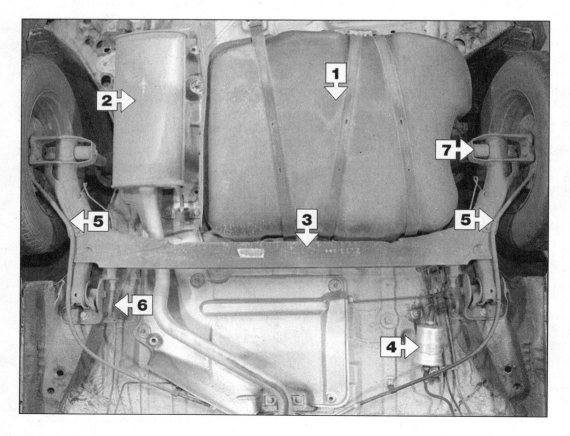

Typical rear underside components

1   Fuel tank
2   Muffler
3   Rear axle assembly
4   Fuel filter (gasoline-engine models)
5   Parking brake cable
6   Brake hose
7   Shock absorber/coil spring assembly lower mount

1 Maintenance schedule

1   The maintenance intervals in this manual are provided with the assumption that you, not the dealer, will be carrying out the work. These are the minimum maintenance intervals recommended by us for vehicles driven daily. If you wish to keep your vehicle in peak condition at all times, you may wish to perform some of these procedures more often. We encourage frequent maintenance, because it enhances the efficiency, performance and resale value of your vehicle.

2   When the vehicle is new, it should be serviced by a factory-authorized dealer service department, in order to preserve the factory warranty.

3   All models covered by this manual are equipped with a Service Reminder Indicator in the instrument panel. Every time the engine is started the panel will illuminate for a few seconds, displaying either of the following. This provides a handy reminder of when the next service is required:

Display shows "IN 00" - no service required
Display shows "OEL" - 7,500 mile service required
Display shows "IN 01" - 15,000 mile service required
Display shows "IN 02" - 30,000 mile service required

Every 250 miles or weekly

Check the engine oil level (Section 4)
Check the engine coolant level (Section 4)
Check the windshield washer fluid level (Section 4)
Check the tires and tire pressures (Section 5)

Every 7,500 miles or 6 months, whichever comes first - "OEL" on Service Reminder Indicator

All items listed above, plus . . .

Reset the Service Reminder Indicator (Section 3)
Replace the engine oil and filter (Section 6)
Check the front brake pad thickness (Section 7)
Check and, if necessary, replace the windshield/rear glass wiper blades (Section 8)
Rotate the tires (Section 9)
Check and service the battery (Section 10)
Drain the water from the fuel filter (diesel-engine models) (Section 11)

Every 15,000 miles or 12 months, whichever comes first - "OEL," "IN 01" on Service Reminder Indicator

All items listed above, plus . . .

Check the automatic transaxle fluid and differential lubricant level (Section 12)
Check all underhood components and hoses for fluid leaks (Section 13)
Check the rear brake pad thickness - rear disc brake models (Section 14)

Check the rear brake shoe lining thickness - rear drum brake models (Section 15)
Check the operation of the parking brake (See Chapter 9)
Check the steering and suspension components for condition and security (Section 16)
Check and, if necessary, adjust the engine idle speed (diesel-engine models) (Section 17)
Check the condition of the driveaxle boots (Section 18)
Check the condition of the exhaust system and its mountings (Section 19)
Check the headlight beam adjustment (Section 20)
Check the operation of the windshield/tailgate glass washer system(s) (as applicable) (Section 21)
Check the condition of the airbag unit(s) (Section 22)
Lubricate all hinges and locks (Section 23)
Perform a road test (Section 24)
Reset the Service Reminder Indicator (Section 3)

Every 30,000 miles or 24 months, whichever comes first - "OEL," "IN 01" and "IN 02" on Service Reminder Indicator

All items listed above, plus . . .

Replace the spark plugs (Section 25)
Replace the air filter element (Section 26)
Replace the automatic transaxle fluid (Section 27)*
Replace the pollen filter element (Section 28)
Check the ignition system (Section 29)
Check the manual transaxle oil level (Section 30)
Check the condition of the drivebelt(s), and replace if necessary (Section 31)
Replace the fuel filter (diesel-engine models) (Section 32)
Check the condition of the timing belt and adjust, if necessary (diesel-engine models) (Section 33)
Reset the Service Reminder Indicator (Section 3)

Every 60,000 miles

All items listed above, plus . . .

Replace the timing belt (See Chapter 2A or 2B)
Replace the fuel filter (gasoline-engine models) (Section 34)

Every 2 years (regardless of mileage)

Replace the coolant (Section 35)
Replace the brake fluid (Section 36)
Check the engine management system (gasoline-engine models) (Section 37)
Check the exhaust emissions (diesel-engine models) (Section 38)

*   *If the vehicle is operated in hilly or mountainous terrain, extremely high temperatures, frequent stop-and-go driving conditions or pulls a trailer often, the automatic transaxle fluid should be changed every 30,000 miles*

2 Introduction

General information

This Chapter is designed to help the home mechanic maintain his/her vehicle for safety, economy, long life and peak performance.

The Chapter contains a master maintenance schedule, followed by Sections dealing specifically with each task in the schedule. Visual checks, adjustments, component replacement and other helpful items are included. Refer to the accompanying illustrations of the engine compartment and the underside of the vehicle for the locations of the various components.

Servicing your vehicle in accordance with the mileage/time maintenance schedule and the following Sections will provide a planned maintenance program, which should result in a long and reliable service life. This is a comprehensive plan, so maintaining some items but not others at the specified service intervals, will not produce the same results.

As you service your vehicle, you will discover that many of the procedures can - and should - be grouped together, because of the particular procedure being performed, or because of the proximity of two otherwise unrelated components to one another. For example, if the vehicle is raised for any reason, the exhaust can be inspected at the same time as the suspension and steering components.

The first step in this maintenance program is to prepare yourself before the actual work begins. Read through all the Sections relevant to the work to be carried out, then make a list and gather all the parts and tools required. If a problem is encountered, seek advice from an auto parts store or a dealer parts department.

3 Tune-up general information and Service Reminder Indicator resetting

General information

1   If, from the time the vehicle is new, the routine maintenance schedule is followed closely, and frequent checks are made of fluid levels and high-wear items, as suggested throughout this manual, the engine will be kept in relatively good running condition, and the need for additional work will be minimized.

2   It is possible that there will be times when the engine is running poorly due to the lack of regular maintenance. This is even more likely if a used vehicle, which has not received regular and frequent maintenance checks, is purchased. In such cases, additional work may need to be performed, outside of the regular maintenance intervals.

3   If engine wear is suspected, a compression test (see Chapter 2) will provide valuable information regarding the overall performance of the main internal components. Such a test can be used as a basis to decide on the extent of the work to be carried out. If, for example, a compression test indicates serious internal engine wear, conventional maintenance as described in this Chapter will not greatly improve the performance of the engine, and may prove a waste of time and money, unless extensive overhaul work is carried out first.

4   The following series of operations are those most often required to improve the performance of a generally poor-running engine:

Primary operations

a) Clean, inspect and test the battery (Section 10).
b) Check all the engine-related fluids (Section 4).
c) Check the condition and tension of the drivebelts (Section 31).
d) Replace the spark plugs (Section 25).
e) Inspect the distributor cap and rotor (Section 29).
f) Check the condition of the air filter, and replace it if necessary (Section 26).
g) Replace the fuel filter (gasoline-engine models) (Section 34).
h) Check the condition of all hoses, and check for fluid leaks (Section 13).
i) Check the engine management system (gasoline-engine models) (Section 37)

j) Check the exhaust gas emissions (diesel-engine models) (Section 38).
k) Drain the water from the fuel filter (diesel-engine models) (Section 11)
l) Check and, if necessary, adjust the engine idle speed (diesel-engine models) (Section 17)

5   If the above operations do not prove fully effective, perform the following secondary operations:

Secondary operations

All items listed under "Primary operations," plus the following:

a) Check the charging system (see Chapter 5A).
b) Check the ignition system (see Chapter 5B).
c) Check the fuel system (see Chapter 4A or 4B).
d) Replace the distributor cap and rotor (Section 29).
e) Replace the spark plug wires (Section 29)
f) Replace the fuel filter (diesel-engine models) (Section 32)

Resetting the Service Reminder Indicator

6   After all necessary maintenance work has been completed, the relevant Service Reminder Indicator code must be reset. If more than one service schedule is performed, note that the relevant display interval codes must be reset individually.

7   The display is reset using the reset button on the left-hand side of the instrument panel (below the speedometer) and the clock setting button on the right-hand side of the panel (below the clock/tachometer); on models with a digital clock the lower (minute) button is used. Resetting is performed as follows.

8   Turn the ignition switch to the On position and check that the speedometer mileage indicator is set to the mileage setting and not the trip meter setting. Press and hold in the button on the left of the instrument panel. Keeping the button depressed, switch off the ignition and release the button. The word "OEL" should be shown on the display. By depressing the left-hand button again the display will change to "IN 01" followed by "IN 02". Set the display to the relevant service which has just been performed, then depress the clock adjustment button briefly until "——-" is displayed; this indicates that the Service Reminder Indicator has been reset. Repeat the reset procedure for all the relevant service display intervals.

9   On completion, switch On; when "IN 01" is shown in the display, turn the ignition Off.

4 Fluid level checks (every 250 miles or weekly)

Note: *The following are fluid level checks to be done on a 250 mile or weekly basis. Additional fluid level checks can be found in specific maintenance procedures which follow. Regardless of intervals, be alert to fluid leaks under the vehicle which would indicate a fault to be corrected immediately.*

1   Fluids are an essential part of the lubrication, cooling, brake and windshield washer systems. Because the fluids gradually become depleted and/or contaminated during normal operation of the vehicle, they must be periodically replenished. See *Recommended lubricants and fluids* at the beginning of this Chapter before adding fluid to any of the following components. **Note:** *The vehicle must be on level ground when fluid levels are checked.*

Engine oil

Refer to illustrations 4.2, 4.4 and 4.6

2   The engine oil level is checked with a dipstick that extends through a tube and into the oil pan at the bottom of the engine **(see illustration)**.

3   The oil level should be checked before the vehicle has been driven, or about 15 minutes after the engine has been shut off. If the oil is checked immediately after driving the vehicle, some of the oil will remain in the upper engine components, resulting in an inaccurate reading on the dipstick.

4.2 The dipstick is located at the right front of the engine

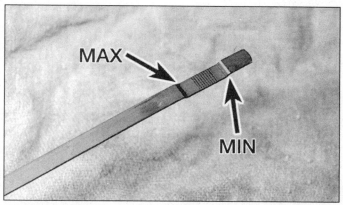

4.4 Note the oil level on the end of the dipstick; it should be between the upper (MAX) and lower (MIN) marks (it takes about one quart to raise the oil level from the lower mark to the upper mark)

1

4   Pull the dipstick out of the tube and wipe all the oil from the end with a clean rag or paper towel. Insert the clean dipstick all the way back into the tube, then pull it out again. Note the oil at the end of the dipstick. Add oil as necessary to keep the level between the ADD and FULL marks on the dipstick **(see illustration)**.

5   Do not overfill the engine by adding too much oil since this may result in oil fouled spark plugs, oil leaks or oil seal failures.

6   Oil is added to the engine after removing a cap from the valve cover **(see illustration)**. A funnel may help to reduce spills.

7   Checking the oil level is an important preventive maintenance step. A consistently low oil level indicates oil leakage through damaged seals, defective gaskets or past worn rings or valve guides. If the oil looks milky or has water droplets in it, the cylinder head gasket(s) may be blown or the head(s) or block may be cracked. The engine should be checked immediately. The condition of the oil should also be checked. Whenever you check the oil level, slide your thumb and index finger up the dipstick before wiping off the oil. If you see small dirt or metal particles clinging to the dipstick, the oil should be changed (see Section 6).

Engine coolant

Refer to illustration 4.10

Warning: *Do not allow antifreeze to come in contact with your skin or painted surfaces of the vehicle. Flush contaminated areas immediately with plenty of water. Don't store new coolant or leave old coolant lying*

around where it's accessible to children or pets - they're attracted by its sweet smell and may drink it. Ingestion of even a small amount of coolant can be fatal! Wipe up garage floor and drip pan coolant spills immediately. Keep antifreeze containers covered and repair leaks in the cooling system as soon as they are noted.

8   All vehicles covered by this manual are equipped with a pressurized coolant recovery system. A white plastic expansion tank located at the left rear corner of the engine compartment is connected by hoses to the engine. As the engine heats up, the coolant level in the expansion tank will rise. As the engine cools, the level will recede. **Warning:** *Never remove the expansion tank cap when the engine is warm.*

9   The coolant level should be checked regularly. When the engine is cold, the coolant level should be at or slightly above the MIN mark on the tank. Once the engine has warmed up, the level should be at or near the MAX mark **(see illustration)**. If it isn't, allow the engine to cool, then slowly unscrew the cap (to release any residual pressure in the system) from the tank and add a 50/50 mixture of ethylene glycol-based antifreeze and water until the level is half-way between the marks. **Warning:** *If you hear pressure (or see steam) escaping as you unscrew the cap, wait until it stops before continuing to remove the cap.*

10   Drive the vehicle and recheck the coolant level. If only a small amount of coolant is required to bring the system up to the proper level, water can be used. However, repeated additions of water will dilute the antifreeze and water solution. In order to maintain the proper

4.6 To add oil, unscrew the cap on the valve cover (make sure the area around the cap is clean before removing it)

4.9 The coolant level varies with the temperature of the engine. When the engine is cold, the level should be between the MIN and MAX marks; when the engine is hot, the level should rise to the MAX mark

4.13 The windshield washer fluid reservoir is located in the front left corner of the engine compartment, next to the battery

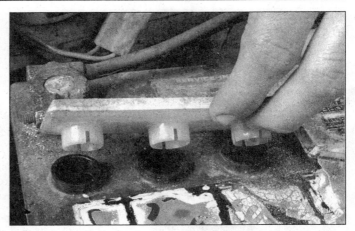

4.16 Remove the cell caps to check the water level in the battery - if the level is low, add distilled water only

ratio of antifreeze and water, always top up the coolant level with the correct mixture. An empty plastic milk jug or bleach bottle makes an excellent container for mixing coolant. Do not use rust inhibitors or additives.

11   If the coolant level drops consistently, there may be a leak in the system. Inspect the radiator, hoses, filler cap, expansion tank, drain plugs and water pump (see Section 13). If no leaks are noted, have the system pressure tested by a service station.

12   Check the condition of the coolant as well. It should be relatively clear. If it's brown or rust colored, the system should be drained, flushed and refilled. Even if the coolant appears to be normal, the corrosion inhibitors wear out, so it must be replaced at the specified intervals.

Windshield washer fluid

Refer to illustration 4.13

13   Fluid for the windshield washer system is located in a plastic reservoir at the left front corner of the engine compartment **(see illustration)**.

14   In milder climates, plain water can be used in the reservoir, but it should be kept no more than 2/3 full to allow for expansion if the water freezes. In colder climates, use windshield washer system antifreeze, available at any auto parts store, to lower the freezing point of the fluid. Mix the antifreeze with water in accordance with the manufacturer's directions on the container. **Caution:** *Don't use cooling system antifreeze, it will damage the vehicle's paint.*

15   To help prevent icing in cold weather, warm the windshield with the defroster before using the washer.

Battery electrolyte

Refer to illustration 4.16

16   Most vehicles with which this manual is concerned are equipped with a maintenance-free type battery that has no filler caps. Water doesn't have to be added to these batteries at any time. If, however, a battery with removable cell caps has been installed, the caps on the top of the battery should be removed periodically to check for a low electrolyte level **(see illustration)**. This check is most critical during the warm summer months. **Warning:** *Hydrogen gas is always present in the battery cells. See the* Warning *in Section 10.*

Brake fluid

Refer to illustration 4.18

17   The brake master cylinder is mounted in the left rear corner of the engine compartment.

18   The fluid level can be checked visually from the outside by looking at the translucent plastic reservoir **(see illustration)**. Be sure to wipe the top of the reservoir cover with a clean rag to prevent contamination of the brake system before removing the cover.

19   When adding fluid, pour it carefully into the reservoir to avoid spilling it on surrounding painted surfaces. Be sure the specified fluid is used, since mixing different types of brake fluid can cause damage to the system. See *Recommended lubricants and fluids* at the front of

4.18 The brake fluid level can be viewed through the translucent reservoir - the level must be kept between the MAX and MIN marks at all times

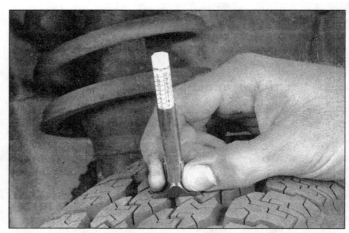

5.2 Use a tread depth gauge to monitor tire wear - they are available at auto parts stores and service stations and cost very little

UNDERINFLATION

CUPPING

Cupping may be caused by:
- Underinflation and/or mechanical irregularities such as out-of-balance condition of wheel and/or tire, and bent or damaged wheel.
- Loose or worn steering tie-rod or steering idler arm.
- Loose, damaged or worn front suspension parts.

OVERINFLATION

INCORRECT TOE-IN OR EXTREME CAMBER

FEATHERING DUE TO MISALIGNMENT

5.3 This chart will help you determine the condition of the tires, the probable cause(s) of abnormal wear and the corrective action necessary

this Chapter or your owner's manual. **Warning:** *Brake fluid can harm your eyes and damage painted surfaces, so use extreme caution when handling or pouring it. Do not use brake fluid that has been standing open or is more than one year old. Brake fluid absorbs moisture from the air. Excess moisture can cause a dangerous loss of brake performance.*

20   At this time, the fluid and master cylinder can be inspected for contamination. The system should be drained and refilled if deposits, dirt particles or water droplets are seen in the fluid.

21   After filling the reservoir to the proper level, make sure the cover or cap is on tight to prevent fluid leakage.

22   The brake fluid level in the master cylinder will drop slightly as the pads at the front wheels wear down during normal operation. If the master cylinder requires repeated additions to keep it at the proper level, it's an indication of leakage in the brake system, which should be corrected immediately. Check all brake lines and connections.

23   If, upon checking the master cylinder fluid level, you discover one or both reservoirs empty or nearly empty, the brake system should be bled (see Chapter 9).

5   Tire and tire pressure checks (every 250 miles or weekly)

Refer to illustrations 5.2, 5.3, 5.4a, 5.4b and 5.8

1   Periodic inspection of the tires may spare you the inconvenience of being stranded with a flat tire. It can also provide you with vital information regarding possible problems in the steering and suspension systems before major damage occurs.

2   The original tires on this vehicle are equipped with 1/2-inch wide wear bands that will appear when tread depth reaches 1/16-inch. Tread wear can be monitored with a simple, inexpensive device known as a tread depth indicator **(see illustration)**.

3   Note any abnormal tread wear **(see illustration)**. Tread pattern

irregularities such as cupping, flat spots and more wear on one side than the other are indications of front end alignment and/or balance problems. If any of these conditions are noted, take the vehicle to a tire shop or service station to correct the problem.

4   Look closely for cuts, punctures and embedded nails or tacks. Sometimes a tire will hold air pressure for a short time or leak down very slowly after a nail has embedded itself in the tread. If a slow leak persists, check the valve stem core to make sure it's tight **(see illustration)**. Examine the tread for an object that may have embedded itself in the tire or for a "plug" that may have begun to leak (radial tire punctures are repaired with a plug that's installed in a puncture). If a puncture is suspected, it can be easily verified by spraying a solution

5.4a If a tire loses air on a steady basis, check the valve core first to make sure it's snug (special inexpensive wrenches are commonly available at auto parts stores)

5.4b If the valve core is tight, raise the corner of the vehicle with the low tire and spray a soapy water solution onto the tread as the tire is turned slowly - leaks will cause small bubbles to appear

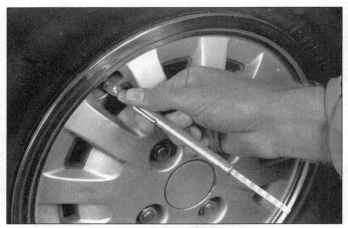

5.8 To extend the life of the tires, check the air pressure at least once a week with an accurate gauge (don't forget the spare!)

of soapy water onto the puncture area **(see illustration)**. The soapy solution will bubble if there's a leak. Unless the puncture is unusually large, a tire shop or service station can usually repair the tire.

5   Carefully inspect the inner sidewall of each tire for evidence of brake fluid leakage. If you see any, inspect the brakes immediately.

6   Correct air pressure adds miles to the lifespan of the tires, improves mileage and enhances overall ride quality. Tire pressure cannot be accurately estimated by looking at a tire, especially if it's a radial. A tire pressure gauge is essential. Keep an accurate gauge in the vehicle. The pressure gauges attached to the nozzles of air hoses at gas stations are often inaccurate.

7   Always check tire pressure when the tires are cold. Cold, in this case, means the vehicle has not been driven over a mile in the three hours preceding a tire pressure check. A pressure rise of four to eight pounds is not uncommon once the tires are warm.

8   Unscrew the valve cap protruding from the wheel or hubcap and push the gauge firmly onto the valve stem **(see illustration)**. Note the reading on the gauge and compare the figure to the recommended tire pressure shown on the tire sidewall or on the placard on the driver's side door jamb. Be sure to reinstall the valve cap to keep dirt and moisture out of the valve stem mechanism. Check all four tires and, if necessary, add enough air to bring them up to the recommended pressure.

9   Don't forget to keep the spare tire inflated to the specified pressure (refer to your owner's manual or the tire sidewall).

---

6   Engine oil and filter change (every 7,500 miles or 6 months

---

Refer to illustrations 6.2, 6.3, 6.7 and 6.9

1   Frequent oil and filter changes are the most important preventative maintenance procedures which can be undertaken. As engine oil ages, it becomes diluted and contaminated, which leads to premature engine wear.

2   Before starting this procedure, gather all the necessary tools and materials **(see illustration)**. Also make sure that you have plenty of clean rags and newspapers handy, to mop up any spills. Ideally, the engine oil should be warm, as it will drain better, and more built-up sludge will be removed with it. Take care, however, not to touch the exhaust or any other hot parts of the engine when working under the vehicle. To avoid any possibility of scalding, and to protect yourself from possible skin irritants and other harmful contaminants in used engine oils, it is advisable to wear gloves when carrying out this work. Access to the underside of the vehicle will be greatly improved if it can be raised on a lift, driven onto ramps, or raised and supported on jackstands (see *Jacking and Towing* at the front of this manual). Whichever

method is chosen, make sure that the vehicle remains level, or if it is at an angle, that the drain plug is at the lowest point.

3   Using a socket a or a box-end wrench, loosen the drain plug about half a turn **(see illustration)**. Position the drain pan under the drain plug, then remove the plug completely. Recover the sealing ring from the drain plug.

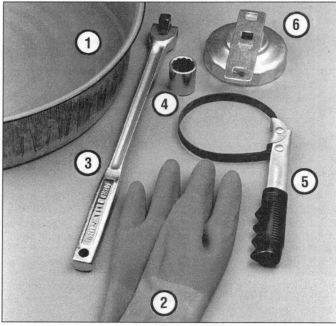

6.2 These tools are required when changing the engine oil and filter

1   *Drain pan* - It should be fairly shallow in depth, but wide to prevent spills
2   *Rubber gloves* - When removing the drain plug and filter, you will get oil on your hands (the gloves will prevent burns)
3   *Breaker bar* - Sometimes the oil drain plug is tight, and a long breaker bar is needed to loosen it
4   *Socket* - To be used with the breaker bar or a ratchet (must be the correct size to fit the drain plug - six-point preferred)
5   *Filter wrench* - This is a metal band-type wrench, which requires clearance around the filter to be effective
6   *Filter wrench* - This type fits on the bottom of the filter and can be turned with a ratchet or breaker bar (different-size wrenches are available for different types of filters)

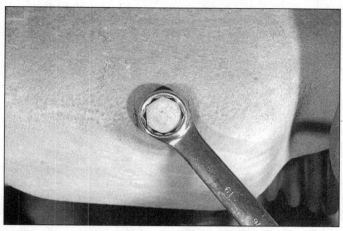

6.3 Loosening the oil pan drain plug

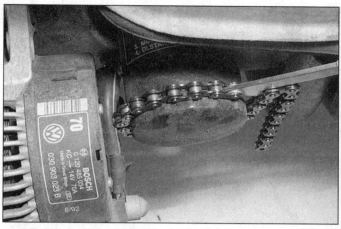

6.7 A chain wrench can also be used to loosen the oil filter

4   Allow some time for the old oil to drain, noting that it may be necessary to reposition the container as the oil flow slows to a trickle.

5   After all the oil has drained, wipe off the drain plug with a clean rag, and install a new sealing washer. Clean the area around the drain plug opening and install the plug. Tighten the plug to the torque listed in this Chapter's Specifications.

6   Move the container into position under the oil filter, which is located on the front side of the cylinder block.

7   Using an oil filter removal tool if necessary, loosen the filter initially, then unscrew it by hand the rest of the way **(see illustration)**. Empty the oil in the old filter into the container.

8   Use a clean rag to remove all oil, dirt and sludge from the filter sealing area on the engine. Check the old filter to make sure that the rubber sealing ring has not stuck to the engine. If it has, remove it.

9   Apply a film of clean engine oil to the sealing ring on the new filter **(see illustration)**, then screw it into position on the engine. Tighten the filter firmly by hand only - do not use any tools.

10   Remove the old oil and all tools from under the car then lower the car to the ground (if applicable).

11   Remove the dipstick, then unscrew the oil filler cap from the valve cover or oil filler/breather neck (as applicable). Fill the engine, using the correct grade and type of oil (see *"Lubricants, fluids and capacities"*). A funnel may help to reduce spillage. Pour in half the specified quantity of oil first, then wait a few minutes for the oil to flow into the oil pan. Continue adding oil a small quantity at a time until the level is up to the lower mark on the dipstick. Add oil a little at a time until the oil is up to the MAX mark.

12   Start the engine and run it for a few minutes; check for leaks around the oil filter seal and the oil pan drain plug. Note that there may be a delay of a few seconds before the oil pressure warning light goes out when the engine is first started, as the oil circulates through the engine oil galleries and the new oil filter before the pressure builds up.

13   Switch off the engine and wait a few minutes for the oil to settle in the oil pan once more. With the new oil circulated and the filter completely full, recheck the level on the dipstick, and add more oil as necessary.

14   The old oil drained from the engine cannot be reused in its present state and should be disposed of. Check with your local refuse disposal company, disposal facility or environmental agency to see if they will accept the oil for recycling. Don't pour used oil into drains or onto the ground. After the oil has cooled, it can be drained into a suitable container (capped plastic jugs, topped bottles, milk cartons, etc.) for transport to one of these disposal sites.

---

7   Front brake pad check (every 7,500 miles or 6 months)

Refer to illustration 7.3

1   Apply the parking brake. Loosen the wheel lug bolts, raise the front of the vehicle and support it securely on jackstands. Remove the front wheels.

2   For a comprehensive check, the brake pads should be removed and cleaned. The caliper can then also be checked more thoroughly. Refer to Chapter 9 for further information.

3   Measure the thickness of the brake pads (including the steel backing plate) **(see illustration)**. If any pad is worn to the specified

6.9 Lubricate the oil filter gasket with clean engine oil before installing the filter on the engine

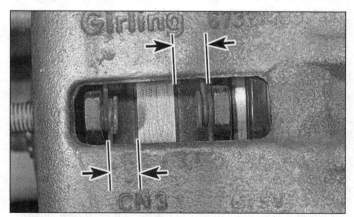

7.3 The thickness of the friction material on each brake pad can be measured through the aperture in the caliper body

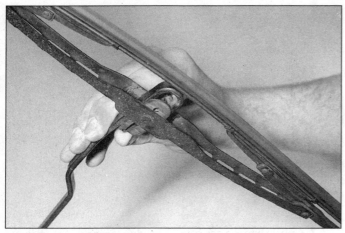

8.5 To remove a wiper blade, pull the arm away from the glass until it locks, swivel the blade 90-degrees, depress the locking tab with your fingers and slide the blade out of the arm's hooked end

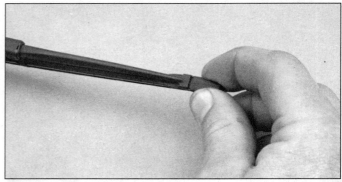

8.6 To remove the element from the blade, squeeze the tabs and pull the element out of the metal frame

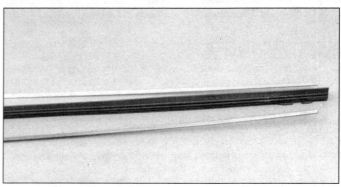

8.7 The metal retainers must be inserted into the slots in the rubber before installation

thickness or less, *all four pads must be replaced as a set.*

4   Install the wheels and lug bolts. Lower the vehicle and tighten the lug bolts to the torque listed in this Chapter's Specifications.

8   Wiper blade inspection and replacement (every 7,500 miles or 6 months)

Refer to illustrations 8.5, 8.6 and 8.7

1   The windshield and rear wiper and blade assemblies should be inspected periodically for damage, loose components and cracked or worn blade elements.

2   Road film can build up on the wiper blades and affect their efficiency, so they should be washed regularly with a mild detergent solution.

3   The action of the wiping mechanism can loosen the bolts, nuts and fasteners, so they should be checked and tightened, as necessary, at the same time the wiper blades are checked.

4   If the wiper blade elements (sometimes called inserts) are cracked, worn or warped, they should be replaced with new ones.

5   Remove the wiper blade assembly from the wiper arm by swiveling the blade 90-degrees to the arm, then depress the locking tab and slide the blade out of the arm's hooked end **(see illustration)**.

6   With the blade removed from the vehicle, you can remove the rubber element from the blade. Grasp the end of the wiper bridge securely with one hand and the element with the other. Detach the end of the element from the bridge claw and slide to free it, then slide the element out **(see illustration)**.

7   Compare the new element with the old for length, design, etc. Remove the metal retainers from the element and install them in the new element **(see illustration)**.

8   Slide the new element into place, notched end last, and secure the claw into the notches.

9   Reinstall the blade assembly on the arm, wet the glass and test for proper operation.

9   Tire rotation (every 7,500 miles or 6 months)

Refer to illustration 9.2

1   The tires should be rotated at the specified intervals and whenever uneven wear is noticed. Since the vehicle will be raised and the tires removed anyway, check the brakes (see Section 7) at this time.

2   Radial tires must be rotated in a specific pattern **(see illustration)**.

3   Refer to the information in *Jacking and towing* at the front of this

manual for the proper procedures to follow when raising the vehicle and changing a tire. If the brakes are to be checked, do not apply the parking brake as stated. Make sure the tires are blocked to prevent the vehicle from rolling.

4   Preferably, the entire vehicle should be raised at the same time. This can be done on a hoist or by jacking up each corner and then lowering the vehicle onto jackstands placed under the frame rails. Always use four jackstands and make sure the vehicle is firmly supported.

5   After rotation, check and adjust the tire pressures as necessary and be sure to check the lug bolt tightness.

6   For further information on the wheels and tires, refer to Chapter 10.

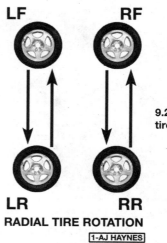

9.2 The recommended tire rotation pattern for these models

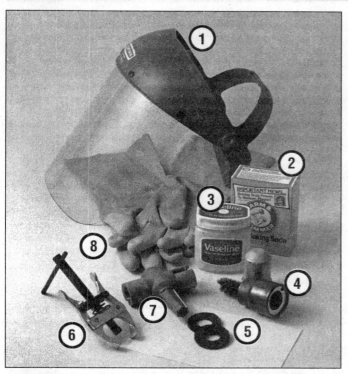

10.1 Tools and materials required for battery maintenance

1 **Face shield/safety goggles** - When removing corrosion with a brush, the acidic particles can easily fly up into your eyes
2 **Baking soda** - A solution of baking soda and water can be used to neutralize corrosion
3 **Petroleum jelly** - A layer of this on the battery posts will help prevent corrosion
4 **Battery post/cable cleaner** - This wire brush cleaning tool will remove all traces of corrosion from the battery posts and cable clamps
5 **Treated felt washers** - Placing one of these on each post, directly under the cable clamps, will help prevent corrosion
6 **Puller** - Sometimes the cable clamps are very difficult to pull off the posts, even after the nut/bolt has been completely loosened. This tool pulls the clamp straight up and off the post without damage
7 **Battery post/cable cleaner** - Here is another cleaning tool which is a slightly different version of Number 4 above, but it does the same thing
8 **Rubber gloves** - Another safety item to consider when servicing the battery; remember that's acid inside the battery!

10 Battery check, maintenance and charging (every 7,500 miles or 6 months)

Refer to illustrations 10.1, 10.6a, 10.6b, 10.7a and 10.7b
Warning: *Certain precautions must be followed when checking and servicing the battery. Hydrogen gas, which is highly flammable, is always present in the battery cells, so keep lighted tobacco and all other open flames and sparks away from the battery. The electrolyte inside the battery is actually dilute sulfuric acid, which will cause injury if splashed on your skin or in your eyes. It will also ruin clothes and painted surfaces. When removing the battery cables, always detach the negative cable first and hook it up last!*

1   A routine preventive maintenance program for the battery in your vehicle is the only way to ensure quick and reliable starts. But before performing any battery maintenance, make sure that you have the proper equipment necessary to work safely around the battery **(see illustration)**.

10.6a Battery terminal corrosion usually appears as light, fluffy powder

10.6b Removing the cable from a battery post with a wrench - sometimes special battery pliers are required for this procedure if corrosion has caused deterioration of the nut hex (always remove the ground cable first and hook it up last!)

2   There are also several precautions that should be taken whenever battery maintenance is performed. Before servicing the battery, always turn the engine and all accessories off and disconnect the cable from the negative terminal of the battery. **Caution:** *If the stereo in your vehicle is equipped with an anti-theft system, make sure you have the correct activation code before disconnecting the battery.*
3   The battery produces hydrogen gas, which is both flammable and explosive. Never create a spark, smoke or light a match around the battery. Always charge the battery in a ventilated area.
4   Electrolyte contains poisonous and corrosive sulfuric acid. Do not allow it to get in your eyes, on your skin or on your clothes. Never ingest it. Wear protective safety glasses when working near the battery. Keep children away from the battery.
5   Note the external condition of the battery. If the positive terminal and cable clamp on your vehicle's battery is equipped with a rubber protector, make sure it isn't torn or damaged. It should completely cover the terminal. Look for any corroded or loose connections, cracks in the case or cover or loose hold-down clamps. Also check the entire length of each cable for cracks and frayed conductors.

Cleaning

6   If corrosion, which looks like white, fluffy deposits **(see illustration)** is evident, particularly around the terminals, the battery should be removed for cleaning. Loosen the cable clamp bolts, being careful to remove the ground cable first, and slide them off the terminals **(see illustration)**. Then disconnect the hold-down clamp bolt and nut, remove the clamp and lift the battery from the engine compartment.

10.7a Regardless of the type of tool used on the battery posts, a clean, shiny surface should be the result

10.7b When cleaning the cable clamps, all corrosion must be removed (the inside of the clamp is tapered to match the taper on the post, so don't remove too much material)

7   Clean the cable clamps thoroughly with a battery brush or a terminal cleaner and a solution of warm water and baking soda **(see illustration)**. Wash the terminals and the top of the battery case with the same solution but make sure that the solution doesn't get into the battery. When cleaning the cables, terminals and battery top, wear safety goggles and rubber gloves to prevent any solution from coming in contact with your eyes or hands. Wear old clothes too - even diluted, sulfuric acid splashed onto clothes will burn holes in them. If the terminals have been extensively corroded, clean them up with a terminal cleaner **(see illustration)**. Thoroughly wash all cleaned areas with plain water.
8   Make sure the battery tray is in good condition and the hold-down clamp bolt or nut is tight. If the battery is removed from the tray, make sure no parts remain in the bottom of the tray when the battery is reinstalled. When reinstalling the hold-down clamp bolt or nut, do not over-tighten it.
9   Corrosion on the hold-down components, battery case and surrounding areas can be removed with a solution of water and baking soda. Thoroughly rinse all cleaned areas with plain water.
10   Any metal parts of the vehicle damaged by corrosion should be covered with a zinc-based primer, then painted.
11   Information on jump starting can be found at the front of this manual. For more detailed battery checking procedures, refer to the *Haynes Automotive Electrical Manual*.

Charging

Warning: *When batteries are being charged, hydrogen gas, which is very explosive and flammable, is produced. Do not smoke or allow open flames near a charging or a recently charged battery. Wear eye protection when near the battery during charging. Also, make sure the charger is unplugged before connecting or disconnecting the battery from the charger.*

12   Slow-rate charging is the best way to restore a battery that's discharged to the point where it will not start the engine. It's also a good way to maintain the battery charge in a vehicle that's only driven a few miles between starts. Maintaining the battery charge is particularly important in the winter when the battery must work harder to start the engine and electrical accessories that drain the battery are in greater use.
13   It's best to use a one or two-amp battery charger (sometimes called a "trickle" charger). They are the safest and put the least strain on the battery. They are also the least expensive. For a faster charge, you can use a higher amperage charger, but don't use one rated more than 1/10th the amp/hour rating of the battery. Rapid boost charges that claim to restore the power of the battery in one to two hours are hardest on the battery and can damage batteries not in good condition. This type of charging should only be used in emergency situations.
14   The average time necessary to charge a battery should be listed in the instructions that come with the charger. As a general rule, a trickle charger will charge a battery in 12 to 16 hours.

11   Fuel filter water draining (diesel-engine models) (every 15,000 miles or 12 months)

Refer to illustration 11.5
1   From time to time, the water collected from the fuel by the filter unit must be drained out.
2   The fuel filter is mounted on the inner fender, above the right hand wheel housing. At the top of the filter unit, pull out the clip and lift off the control valve, leaving the fuel hoses attached **(see illustrations 32.2a and 32.2b)**.
3   Loosen the clamp screw and raise the filter in its retaining bracket **(see illustrations 32.4a and 32.4b)**.
4   Position a container below the filter unit and pad the surrounding area with rags to absorb any fuel that may spill.
5   Unscrew the drain valve at the base of the filter unit, until fuel starts to run out into the container **(see illustration)**. Keep the valve open until about 100 cc (1.7 ounces) of fuel has been drained.
6   Using a new O-ring, install the control valve to the top of the filter and insert the retaining clip. Close the drain valve and wipe off any surplus fuel from the nozzle.
7   Remove the container and rags, then push the filter unit back into the retaining bracket and tighten the bracket securing screw.
8   Run the engine at idle and check around the fuel filter for fuel leaks.
9   Raise the engine speed to about 2000 rpm several times, then allow the engine to idle again. Observe the fuel flow through the transparent hose leading to the fuel injection pump and check that it is free of air bubbles.

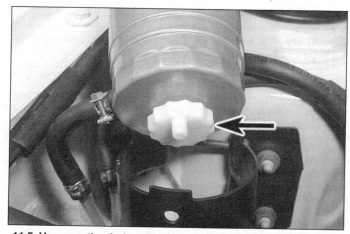

11.5 Unscrew the drain valve (arrow) at the base of the fuel filter

12 Automatic transaxle fluid and differential lubricant level check (every 15,000 miles or 12 months)

Automatic transaxle fluid

1993 and 1994 models

1   Take the vehicle on a short drive to warm the transaxle up to normal operating temperature, then park the vehicle on level ground. The fluid level is checked using the dipstick located at the front of the engine compartment, on the front of the transaxle unit.
2   With the engine idling and the selector lever in the "P" (Park) position, withdraw the dipstick from the tube and wipe all the fluid from its end with a clean rag or paper towel. Insert the clean dipstick back into the tube as far as it will go, then withdraw it once more. Note the fluid level on the end of the dipstick; it should be between the MAX and MIN marks.
3   If adding fluid is necessary, add the required quantity of the specified fluid to the transaxle through the dipstick tube. Use a funnel with a fine mesh gauze, to avoid spillage, and to ensure that no foreign matter enters the transaxle. **Note:** *Never overfill the transaxle so that the fluid level is above the upper mark.*
4   After adding fluid, take the vehicle on a short run to distribute the fresh fluid, then recheck the level again, adding more fluid if necessary.
5   Always maintain the level between the two dipstick marks. If the level is allowed to fall below the lower mark, fluid starvation may result, which could lead to severe transaxle damage. If the level is too high, the excess fluid may be ejected. In either case, an incorrect level will adversely affect the operation of the transaxle.
6   Frequent need for adding fluid indicates that there is a leak, which should be found and corrected before it becomes serious.

1995 and later models

7   On these models the transmission fluid can only be checked within a certain fluid temperature range. The only way to accurately monitor this temperature is with a VW 1551 scan tool. A special fixture is also required to add fluid to the transaxle. Since the transaxle is a sealed unit, there should be no reason for the level to become inadequate unless it is leaking externally. Considering all this, we recommend taking the vehicle to a dealer service department or other qualified repair shop (equipped with the necessary special tools) for transaxle fluid level checks or changes.

Differential lubricant

8   The manufacturer does not recommend periodic checking of the differential lubricant, but if you have reason to believe the level may be inadequate (noise or lubricant leakage), you can check the level as follows.
9   Unscrew the speedometer driveshaft from the transaxle case. Check the level of the fluid on the end of the driveshaft; it should be between the end of the driveshaft and the first step. If not, add the specified lubricant (see this Chapter's Specifications), a little at a time, until the lubricant is up to the bottom of the step on the driveshaft. **Note:** *It only takes approximately 0.1 quart to raise the level from the bottom of the shaft to the first step.*
10   Install the speedometer drive gear and tighten it securely.

13 Hose and fluid leak check (every 15,000 miles or 12 months)

Refer to illustration 13.1

1   Visually inspect the engine mating surfaces, gaskets and seals for any signs of water or oil leaks. Pay particular attention to the areas around the valve cover, cylinder head, oil filter and oil pan mating surfaces. Bear in mind that, over a period of time, some very slight seepage from these areas is to be expected - what you are really looking for is any indication of a serious leak **(see illustration)**. Should a leak be found, replace the offending gasket or oil seal by referring to the

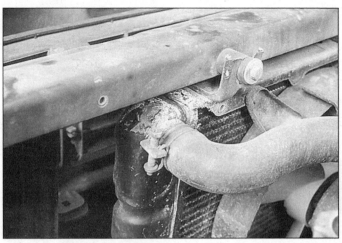

13.1 A leak in the cooling system will usually show up as white or rust-colored deposits on the area adjoining the leak

appropriate Chapters in this manual.
2   Also check the security and condition of all the engine-related pipes and hoses. Ensure that all cable-ties or securing clips are in place and in good condition. Clips which are broken or missing can lead to chafing of the hoses, pipes or wiring, which could cause more serious problems in the future.
3   Carefully check the radiator hoses and heater hoses along their entire length. Replace any hose which is cracked, swollen or deteriorated. Cracks will show up better if the hose is squeezed. Pay close attention to the hose clamps that secure the hoses to the cooling system components. Hose clamps can pinch and puncture hoses, resulting in cooling system leaks.
4   Inspect all the cooling system components (hoses, joint faces etc.) for leaks. A leak in the cooling system will usually show up as white- or rust-colored deposits on the area adjoining the leak. Where any problems of this nature are found on system components, replace the component or gasket, referring to Chapter 3 if necessary.
5   Where applicable, inspect the automatic transaxle fluid cooler hoses for leaks or deterioration.
6   With the vehicle raised, inspect the fuel tank and filler neck for punctures, cracks and other damage. The connection between the filler neck and tank is especially critical. Sometimes a rubber filler neck or connecting hose will leak due to loose retaining clamps or deteriorated rubber.
7   Carefully check all rubber hoses and metal fuel lines leading away from the fuel tank. Check for loose connections, deteriorated hoses, crimped lines, and other damage. Pay particular attention to the vent pipes and hoses, which often loop up around the filler neck and can become blocked or crimped. Follow the lines to the front of the vehicle, carefully inspecting them all the way. Replace damaged sections as necessary.
8   From within the engine compartment, check the security of all fuel hose attachments and pipe unions, and inspect the fuel hoses and vacuum hoses for kinks, chafing and deterioration.
9   Where applicable, check the condition of the power steering fluid hoses and pipes.

14 Rear brake pad check - models with rear disc brakes (every 15,000 miles or 12 months)

1   Block the front wheels to prevent the vehicle from rolling. Loosen the rear wheel lug bolts, raise the rear of the vehicle and support it securely on jackstands. Remove the rear wheels.
2   For a quick check, the thickness of friction material remaining on each brake pad can be measured through the top of the caliper body **(see illustration 7.3)**. If any pad's friction material is worn to the speci-

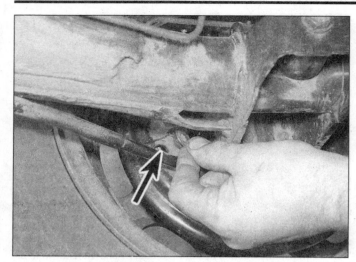

15.2 Remove the rubber plug and check the brake lining thickness through the backing plate aperture (arrow)

16.4 Check for wear in the hub bearings by grasping the wheel and trying to rock it

fied thickness or less, all four pads must be replaced as a set.

3   For a comprehensive check, the brake pads should be removed and cleaned. This will permit the caliper to be checked more thoroughly. Refer to Chapter 9 for further information.

4   Install the wheels and lug bolts. Lower the vehicle and tighten the lug bolts to the torque listed in this Chapter's Specifications.

15   Rear brake shoe check - models with rear drum brakes (every 15,000 miles or 12 months)

Refer to illustration 15.2

1   Block the front wheels, then jack up the rear of the vehicle and support it securely on jackstands.

2   For a quick check, the thickness of friction material remaining on one of the brake shoes can be observed through the hole in the brake backing plate which is exposed by prying out the sealing grommet **(see illustration)**. If a rod of the same diameter as the specified minimum friction material thickness is placed against the shoe friction material, the amount of wear can be assessed. A flashlight or inspection light will probably be required. If the friction material on any shoe is worn down to the specified minimum thickness or less, all four shoes must be replaced as a set.

3   For a comprehensive check, the brake drum should be removed and cleaned. This will allow the wheel cylinders to be checked, and the condition of the brake drum itself to be fully examined (see Chapter 9).

16   Steering and suspension check (every 15,000 miles or 12 months)

Refer to illustration 16.4

Front suspension and steering check

1   Raise the front of the vehicle and support it securely on jackstands.

2   Visually inspect the balljoint dust covers and the steering rack-and-pinion boots for splits, chafing or deterioration. Any wear of these components will cause loss of lubricant, together with dirt and water entry, resulting in rapid deterioration of the balljoints or steering gear.

3   On models with power steering, check the fluid hoses for chafing or deterioration and the pipe and hose unions for fluid leaks. Also check for signs of fluid leakage under pressure from the steering gear rubber boots, which would indicate failed fluid seals within the steering gear.

4   Grasp the wheel at the 12 o'clock and 6 o'clock positions, and try

to rock it **(see illustration)**. Very slight freeplay may be felt, but if the movement is appreciable, further investigation is necessary to determine the source. Continue rocking the wheel while an assistant depresses the brake pedal. If the movement is now eliminated or significantly reduced, it is likely that the hub bearings are at fault. If the free play is still evident with the brake pedal depressed, then there is wear in the suspension joints or mountings.

5   Now grasp the wheel at the 9 o'clock and 3 o'clock positions, and try to rock it as before. Any movement felt now may again be caused by wear in the hub bearings or the steering tie-rod ends.

6   Using a large screwdriver or prybar, check for wear in the suspension mounting bushings by prying between the relevant suspension component and its attachment point. Some movement is to be expected as the mounts are made of rubber, but excessive wear should be obvious. Also check the condition of any visible rubber bushings, looking for splits, cracks or contamination of the rubber.

7   With the car standing on its wheels, have an assistant turn the steering wheel back and forth about an eighth of a turn each way. There should be very little, if any, lost movement between the steering wheel and wheels. If this is not the case, closely observe the joints and mounts previously described, but in addition, check the steering column universal joints for wear, and the rack-and-pinion steering gear itself.

Suspension strut/shock absorber check

8   Check for any signs of fluid leakage around the suspension strut/shock absorber body, or from the rubber boot around the piston rod. Should any fluid be noticed, the suspension strut/shock absorber is defective internally and should be replaced. **Note:** *Suspension struts/shock absorbers should always be replaced in pairs on the same axle.*

9   The efficiency of the suspension strut/shock absorber may be checked by bouncing the vehicle at each corner. Generally speaking, the body will return to its normal position and stop after being depressed. If it rises and returns on a rebound, the suspension strut/shock absorber is probably suspect. Examine also the suspension strut/shock absorber upper and lower mountings for any signs of wear.

17   Idle speed check and adjustment (diesel-engine models) (every 15,000 miles or 12 months)

Refer to illustration 17.3

1   Start the engine and run it until it reaches its normal operating temperature. With the parking brake applied and the transaxle in neutral, allow the engine to idle. Check that the cold start knob is pushed in to the fully 'off' position.

17.3 Adjust the engine idle speed by turning the adjustment screw at the fuel injection pump (arrow)

18.1 Check the condition of the driveaxle boots (arrow)

2   Using a diesel tachometer, check the idle speed against the Chapter 4C Specifications.
3   If necessary, adjust the engine idle speed by rotating the adjustment knob at the fuel injection pump **(see illustration).**

18   Driveaxle boot check (every 15,000 miles or 12 months)

Refer to illustration 18.1
    With the vehicle raised and securely supported on jackstands, turn the steering onto full lock, then slowly rotate the wheel. Inspect the condition of the outer constant velocity (CV) joint rubber boots, squeezing the boots to open out the folds. Check for signs of cracking, splits or deterioration of the rubber, which may allow the grease to escape, and lead to water and grit entry into the joint. Also check the security and condition of the retaining clips. Repeat these checks on the inner CV joints **(see illustration).** If any damage or deterioration is found, the boots should be replaced (see Chapter 8).
    At the same time, check the general condition of the CV joints themselves by first holding the driveaxle and attempting to rotate the wheel. Repeat this check by holding the inner joint and attempting to rotate the driveaxle. Any appreciable movement indicates wear in the joints, wear in the driveaxle splines, or a loose driveaxle retaining nut.

19   Exhaust system check (every 15,000 miles or 12 months)

1   With the engine cold (at least an hour after the vehicle has been driven), check the complete exhaust system from the engine to the end of the tailpipe. The exhaust system is most easily checked with the vehicle raised on a hoist, or suitably supported on jackstands, so that the exhaust components are readily visible and accessible.
2   Check the exhaust pipes and connections for evidence of leaks, severe corrosion and damage. Make sure that all brackets and mountings are in good condition, and that all relevant nuts and bolts are tight. Leakage at any of the joints or in other parts of the system will usually show up as a black sooty stain in the vicinity of the leak.
3   Rattles and other noises can often be traced to the exhaust system, especially the brackets and mounts. Try to move the pipes and mufflers. If the components are able to come into contact with the body or suspension parts, secure the system with new mountings. Otherwise separate the joints (if possible) and twist the pipes as necessary to provide additional clearance.

20   Headlight beam alignment check (every 15,000 miles or 12 months)

    Accurate adjustment of the headlight beam is only possible using optical beam-setting equipment, and this work should therefore be carried out by a VW dealer service department or service station with the necessary facilities.
    Basic adjustments can be carried out in an emergency, and further details are given in Chapter 12.

21   Windshield/rear glass washer system(s) check (every 15,000 miles or 12 months)

    Check that each of the washer jet nozzles are clear and that each nozzle provides a strong jet of washer fluid. The rear glass and headlight jets should be aimed to spray at a point slightly above the center of the windshield or rear glass. On the windshield washer nozzles where there are two jets, aim one of the jets slightly above and center of the windshield and aim the other just below to ensure complete coverage of the screen. If necessary, adjust the jets using a pin.

22   Airbag unit check (every 15,000 miles or 12 months)

    On models so equipped, inspect the airbag(s) exterior condition, checking for signs of damage or deterioration. If an airbag shows signs of damage, it must be replaced (see Chapter 12).

23   Hinge and lock lubrication (every 15,000 miles or 12 months)

    Lubricate the hinges of the hood, doors and liftgate with a light general-purpose oil. Similarly, lubricate all latches, locks and lock strikers. At the same time, check the security and operation of all the locks, adjusting them if necessary (see Chapter 11).
    Lightly lubricate the hood release mechanism and cable with multi-purpose grease.

24   Road test (every 15,000 miles or 12 months)

Instruments and electrical equipment

1   Check the operation of all instruments and electrical equipment.

2   Make sure that all instruments read correctly, and switch on all electrical equipment in turn, to check that it functions properly.

Steering and suspension

3   Check for any abnormalities in the steering, suspension, handling or road "feel."

4   Drive the vehicle, and check that there are no unusual vibrations or noises.

5   Check that the steering feels positive, with no excessive "sloppiness" or roughness, and check for any suspension noises when cornering and driving over bumps.

Drivetrain

6   Check the performance of the engine, clutch (where applicable), transaxle and driveaxles.

7   Listen for any unusual noises from the engine, clutch and transaxle.

8   Make sure that the engine runs smoothly when idling, and that there is no hesitation when accelerating.

9   Check that, where applicable, the clutch action is smooth and progressive, that the vehicle takes off smoothly, and that the pedal travel is not excessive. Also listen for any noises when the clutch pedal is depressed.

10   On manual transaxle models, check that all gears can be engaged smoothly without noise, and that the gear lever action is not abnormally vague or "notchy".

11   On automatic transaxle models, make sure that all gearshifts occur smoothly, and without an increase in engine speed between changes. Check that all the gear positions can be selected with the vehicle at rest. If any problems are found, they should be referred to a VW dealer service department or other qualified shop.

12   Listen for a metallic clicking sound from the front of the vehicle, as the vehicle is driven slowly in a circle with the steering on full-lock. Carry out this check in both directions. If a clicking noise is heard, this indicates wear in a driveaxle CV joint (see Chapter 8).

Check the operation and performance of the braking system

13   Make sure that the vehicle does not pull to one side when braking, and that, on models without an Anti-lock Braking System (ABS) the wheels do not lock prematurely when braking hard. On models with ABS, the wheels should not lock up, even under severe braking.

14   Check that there is no vibration through the steering when braking.

15   Check that the parking brake operates correctly without excessive movement of the lever, and that it holds the vehicle stationary on a slope.

16   Test the operation of the brake booster unit as follows. With the engine off, depress the brake pedal four or five times to exhaust the vacuum. Hold the brake pedal depressed, then start the engine. As the engine starts, there should be a noticeable "give" in the brake pedal as vacuum builds up. Allow the engine to run for at least two minutes, and then switch it off. If the brake pedal is depressed now, it should be possible to detect a hiss from the booster as the pedal is depressed. After about four or five applications, no further hissing should be heard, and the pedal should feel considerably harder.

25   Spark plug replacement (every 30,000 miles or 24 months)

Refer to illustrations 25.2, 25.5a, 25.5b, 25.6, 25.7, 25.9 and 25.10

1   The spark plugs are located in the front side of the cylinder head.

2   In most cases, the tools necessary for spark plug replacement include a spark plug socket which fits onto a ratchet (spark plug sockets are padded inside to prevent damage to the porcelain insulators on the new plugs), various extensions and a gap gauge to check and adjust the gaps on the new plugs **(see illustration)**. A special plug wire removal tool is available for separating the wire boots from the spark

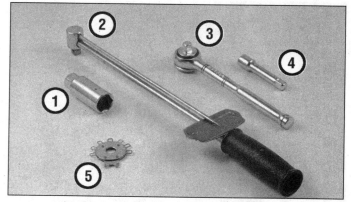

25.2 Tools required for changing spark plugs

1   *Spark plug socket - This will have special padding inside to protect the spark plug's porcelain insulator*

2   *Torque wrench - Although not mandatory, using this tool is the best way to ensure the plugs are tightened properly*

3   *Ratchet - Standard hand tool to fit the spark plug socket*

4   *Extension - Depending on model and accessories, you may need special extensions and universal joints to reach one or more of the plugs*

5   *Spark plug gap gauge - This gauge for checking the gap comes in a variety of styles. Make sure the gap for your engine is included*

plugs. A torque wrench should be used to tighten the new plugs. It is a good idea to allow the engine to cool before removing or installing the spark plugs.

3   The best approach when replacing the spark plugs is to purchase the new ones in advance, adjust them to the proper gap and replace the plugs one at a time. When buying the new spark plugs, be sure to obtain the correct plug type for your particular engine. The plug type can be found in the Specifications at the front of this Chapter.

4   Allow the engine to cool completely before attempting to remove any of the plugs. While you are waiting for the engine to cool, check the new plugs for defects and adjust the gaps.

5   Check the gap by inserting the proper thickness gauge between the electrodes at the tip of the plug **(see illustration)**. The gap between the electrodes should be the same as the one specified on the Emissions Control Information label or listed in this Chapter's Specifications. The wire should slide between the electrodes with a slight amount of drag. If the gap is incorrect, use the adjuster on the gauge

25.5a Spark plug manufacturers recommend using a wire-type gauge when checking the gap - if the wire does not slide between the electrodes with a slight drag, adjustment is required

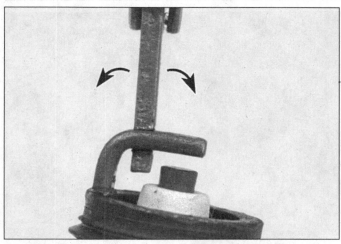

25.5b To change the gap, bend the side electrode only, as
indicated by the arrows, and be very careful not to crack or chip
the porcelain insulator surrounding the center electrode

25.6 Pull the spark plug wires from the plugs by gripping the end
fitting or boot, not the wire, otherwise the wire connection
may be fractured

body to bend the curved side electrode slightly until the proper gap is
obtained (see illustration). If the side electrode is not exactly over the
center electrode, bend it with the adjuster until it is. Check for cracks in
the porcelain insulator (if any are found, the plug should not be used).

6   With the engine cool, remove the spark plug wire from one spark
plug. Pull only on the boot at the end of the wire - do not pull on the
wire (see illustration).

7   If compressed air is available, use it to blow any dirt or foreign
material away from the spark plug hole. A brush will also work (see
illustration). The idea here is to eliminate the possibility of debris
falling into the cylinder as the spark plug is removed.

8   Place the spark plug socket over the plug and remove it from the
engine by turning it in a counterclockwise direction.

9   Compare the spark plug with the chart shown on the inside back
cover of this manual to get an indication of the general running condition of the engine. Before installing the new plugs, it is a good idea to
apply a thin coat of anti-seize compound to the threads (see illustration).

10   Thread one of the new plugs into the hole until you can no longer
turn it with your fingers, then tighten it with a torque wrench (if available) or the ratchet. It's a good idea to slip a short length of rubber
hose over the end of the plug to use as a tool to thread it into place
(see illustration). The hose will grip the plug well enough to turn it, but
will start to slip if the plug begins to cross-thread in the hole - this will
prevent damaged threads and the accompanying repair costs.

11   Before pushing the spark plug wire onto the end of the plug,

25.7 Using a brush to remove the dirt from the
spark plug recesses

inspect it following the procedures outlined in Section 29.

12   Attach the plug wire to the new spark plug, again using a twisting
motion on the boot until it's seated on the spark plug.

13   Repeat the procedure for the remaining spark plugs, replacing
them one at a time to prevent mixing up the spark plug wires.

25.9 A light coat of anti-seize compound applied to the threads of
the spark plugs will keep the threads in the cylinder head from
being damaged the next time the plugs are removed

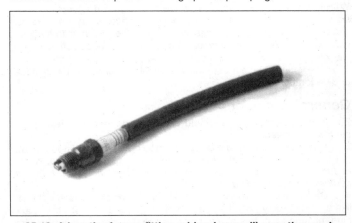

25.10 A length of snug-fitting rubber hose will save time and
prevent damaged threads when installing the spark plugs

26 Air filter replacement (every 30,000 miles or 24 months)

Refer to illustration 26.2

1   Pry open the spring clips and lift off the air cleaner top cover. **Caution:** *On certain models, the airflow meter is integral with the air cleaner top cover. Handle the airflow meter very carefully, as it is easily damaged.*
2   Lift out the air filter element **(see illustration)**.
3   Remove any debris that may have collected inside the air cleaner.
4   Install the new air filter element, ensuring that the edges are securely seated in the air cleaner housing.
5   Install the air cleaner top cover and snap the retaining clips into position.

27 Automatic transaxle fluid replacement (every 30,000 miles or 24 months)

As the automatic transaxle is not equipped with a drain plug; the transaxle fluid must be pumped out using a special adapter. For this reason, it is recommended that automatic transaxle fluid replacement is carried out by a VW dealer service department or other qualified repair shop equipped with the required tools.

28 Pollen filter replacement (every 30,000 miles or 24 months)

1   The pollen filter (on models so equipped) is located beneath the windshield wiper motor cover panels on the right-hand side.
2   Unclip the rubber seal from right side of the top of the engine compartment firewall.
3   Unscrew the retaining fastener screws and pull out the fasteners securing the right half of the windshield wiper motor cover panel. Release the half of the cover panel and remove it from the vehicle.
4   Pivot the pollen filter cover upwards and away then release the retaining clips and withdraw the cover from its housing.
5   Wipe clean the filter housing then install the new filter. Clip the filter securely in position and install the cover.
6   Install the trim cover, securing it in position with the fasteners, and seat the rubber seal on the firewall.

29 Ignition system check (every 30,000 miles or 24 months)

Warning: *Voltages produced by an electronic ignition system are considerably higher than those produced by conventional ignition systems. Extreme care must be taken when working on the system with the ignition switched on. Persons with cardiac pacemaker devices should keep well clear of the ignition circuits, components and test equipment.*
1   The ignition system components should be checked for damage or deterioration as follows.

General component check

2   The spark plug wires should be checked whenever new spark plugs are installed.
3   Pull the wires from the plugs by gripping the end fitting or boot, not the wire, otherwise the lead connection may be fractured. **Note:** *Ensure that the wires are numbered before removing them, to avoid confusion when installing.*
4   Check inside the end fitting for signs of corrosion, which will look like a white crusty powder. Push the end fitting back onto the spark plug, ensuring that it is a tight fit on the plug. If not, remove the wire again and use pliers to carefully crimp the metal connector inside the end fitting until it fits securely on the end of the spark plug.
5   Using a clean rag, wipe the entire length of the wire to remove any

26.2 Pry open the clips, lift off the air cleaner cover and remove the filter element. When installing the new air filter element, ensure that the edges are securely seated in the housing

built-up dirt and grease. Once the wire is clean, check for burns, cracks and other damage. Do not bend the wire excessively, or pull the wire lengthwise - the conductor inside might break.
6   Disconnect the other end of the wire from the distributor cap. Again, pull only on the end fitting. Check for corrosion and a tight fit in the same manner as the spark plug end. If an ohmmeter is available, check the resistance of the wire by connecting the meter between the spark plug end of the wire and the segment inside the distributor cap. Install the wire securely on completion.
7   Check the remaining wires one at a time, in the same way.
8   If new spark plug wires are required, purchase a set for your specific car and engine.
9   Unscrew its retaining screws and remove the distributor cap. Wipe it clean, and carefully inspect it inside and out for signs of cracks, black carbon tracks (tracking) and worn, burned or loose contacts; check that the cap's carbon brush is unworn, free to move against spring pressure, and making good contact with the rotor. Also inspect the cap seal for signs of wear or damage, and replace it if necessary. Remove the rotor from the distributor shaft and inspect the rotor. It is common practice to replace the cap and rotor whenever new spark plug wires are installed. **Note:** *When installing a new cap, remove the wires from the old cap one at a time, and plug them into the new cap in the same location.*
10   Do not simultaneously remove all the leads from the old cap, or firing order confusion may occur. When installing, ensure that the rotor is securely pressed onto the shaft, and tighten the cap retaining screws securely.
11   Even with the ignition system in first-class condition, some engines may still occasionally experience poor starting attributable to damp ignition components. To disperse moisture, a water-dispersant aerosol should be liberally applied.

Ignition timing - check and adjustment

12   This procedure requires special tools that are generally not available to the home mechanic. Take the vehicle to a dealer service department or other qualified repair shop to have the ignition timing checked.

30 Manual transaxle oil level check (every 30,000 miles or 24 months)

Refer to illustration 30.3

1   Park the car on a level surface. The oil level must be checked before the car is driven, or at least 5 minutes after the engine has been

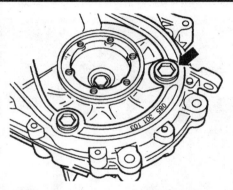

30.3 Location of the manual transaxle check/fill plug

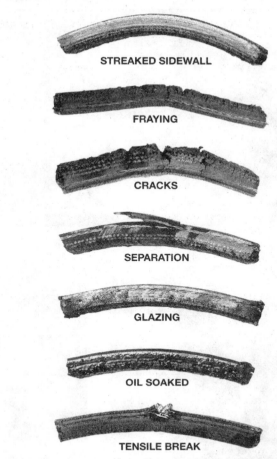

STREAKED SIDEWALL

FRAYING

CRACKS

SEPARATION

GLAZING

OIL SOAKED

TENSILE BREAK

31.5a Here are some of the more common problems associated with V-belts (check the belts very carefully to prevent an untimely breakdown)

switched off. If the oil is checked immediately after driving the car, some of the oil will remain distributed around the transaxle components, resulting in an inaccurate level reading.

2   Remove the retaining clips and screws, then lower the engine undercover.

3   Wipe the area around the check/fill plug clean, which is situated on the differential casing **(see illustration)**. Unscrew the plug.

4   The oil level should reach the lower edge of the check/fill hole. A certain amount of oil will trickle out when it is removed; this does not necessarily indicate that the level is correct. To ensure that a true level is established, wait until the initial trickle has stopped, then add oil as necessary until a trickle of new oil can be seen emerging. The level will be correct when the flow ceases; use only good-quality oil of the specified type.

5   Filling the transaxle with oil can be an awkward operation; above all, allow plenty of time for the oil level to settle properly before checking it. If a large amount is added to the transaxle, and a large amount flows out on checking the level, install the check/fill plug and take the vehicle on a short journey so that the new oil is distributed fully around the transaxle components, then recheck the level when it has settled again.

6   If the transaxle has been overfilled so that oil flows out when the check/fill plug is removed, check that the car is completely level (front-to-rear and side-to-side), and allow the surplus to drain off into a suitable container.

7   When the level is correct, install the plug, tightening it securely. Wash off any spilled oil then install the engine undercover.

31 Drivebelt check and replacement (every 30,000 miles or 24 months)

Refer to illustrations 31.5a and 31.5b

Check

1   Disconnect the battery negative cable and position it away from the terminal. **Caution:** *If the stereo in your vehicle is equipped with an anti-theft system, make sure you have the correct activation code before disconnecting the battery.*

2   Park the vehicle on a level surface, apply the parking brake and block the rear wheels.

3   Loosen the wheel lug bolts, raise the front of the vehicle and support it securely on jackstands. Remove the wheels.

4   Turn the steering to full right hand lock, then remove the screws and clips and lower the engine undercover.

5   Using a socket and ratchet or breaker bar on the crankshaft sprocket bolt, rotate the crankshaft so the full length of the drivebelts can be examined. Look for cracks, splitting and fraying on the surface of the belt; check also for signs of glazing (shiny patches) and separation of the belt plies **(see illustrations)**. If damage or wear is visible, the belt should be replaced.

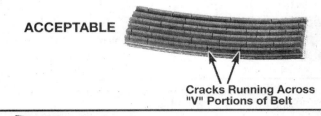

ACCEPTABLE

Cracks Running Across "V" Portions of Belt

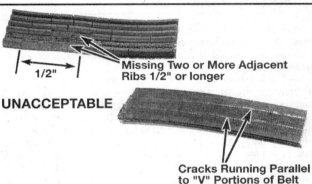

1/2" Missing Two or More Adjacent Ribs 1/2" or longer

UNACCEPTABLE

Cracks Running Parallel to "V" Portions of Belt

31.5b Small cracks in the underside of a V-ribbed belt are acceptable - lengthwise cracks, or missing pieces that cause the belt to make noise, are cause for replacement

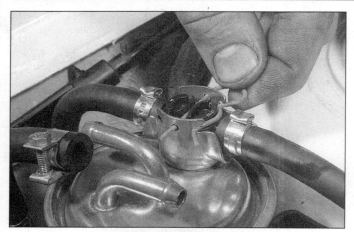

32.2a Remove the clip . . .

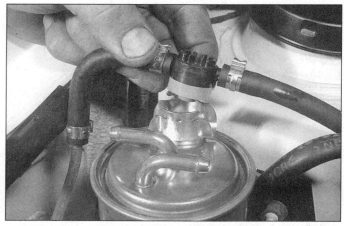

32.2b . . . and lift off the control valve, leaving the fuel hoses attached to it

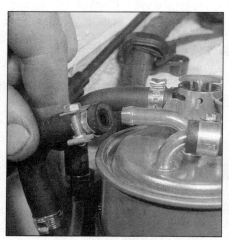

32.3 Loosen the clamps and detach the hoses from the filter

32.4a Loosen the securing screw . . .

32.4b . . . and raise the filter out of its bracket

Replacement

6   For details of auxiliary drivebelt replacement, refer to the relevant part of Chapter 2 Part A or B.

32 Fuel filter replacement (diesel-engine models) (every 30,000 miles or 24 months)

Refer to illustrations 32.2a, 32.2b, 32.3, 32.4a and 32.4b
Warning: *Diesel fuel is flammable, so take extra precautions when you work on any part of the fuel system. See the* **Warning** *at the beginning of Section 34.*
1   The fuel filter is mounted on the inner fender, above the right hand wheel housing. Position a container underneath the filter unit and pad the surrounding area with rags to absorb any fuel that may spill.
2   At the top of the filter unit, pull out the clip and lift off the control valve, leaving the fuel hoses attached to it **(see illustrations)**.
3   Loosen the hose clamps and pull the fuel hoses from the ports on the top of the filter unit **(see illustration)**. If crimp-type clamps are used, cut them off using diagonal cutting pliers, and use equivalent size worm-drive clamps during installation. Note the installed position of each hose. **Warning:** *Be prepared for fuel spillage.*
4   Loosen the securing screw and raise the filter out of its retaining bracket **(see illustrations)**.
5   Install the new fuel filter into the retaining bracket and tighten the securing screw.

6   Install the control valve to the top of the filter and insert the retaining clip.
7   Reconnect the fuel hoses to their proper ports and tighten the clamps securely. Note the fuel flow arrow markings next to each port. Remove the container and rags from the engine compartment.
8   Start and run the engine at idle, then check around the fuel filter for leaks. **Note:** *It may take a few seconds of cranking before the engine starts.*
9   Raise the engine speed to about 2000 rpm several times, then allow the engine to idle again. Observe the fuel flow through the transparent hose leading to the fuel injection pump and check that it is free of air bubbles.

33 Timing belt check and adjustment (diesel-engine models) (every 30,000 miles or 24 months)

1   Referring to Chapter 2 Part B, remove the timing belt cover and inspect the timing belt for signs of damage or deterioration. Check the timing belt carefully for any signs of uneven wear, splitting, or oil contamination. Pay particular attention to the roots of the teeth. Replace the belt if there is the slightest doubt about its condition. The cost of a new belt is nothing when compared to the cost of repairs, should the belt break in service.
2   If signs of oil contamination are found, trace the source of the leak and repair it. Clean the timing belt area and all related components to remove all traces of oil.

3   Check and, if necessary, adjust the belt tension as described in Chapter 2 Part B. On completion, install the belt cover.

34   Fuel filter replacement (gasoline-engine models) (every 60,000 miles)

Warning: *Gasoline is extremely flammable, so take extra precautions when you work on any part of the fuel system. Don't smoke or allow open flames or bare light bulbs near the work area, and don't work in a garage where a natural gas-type appliance (such as a water heater or a clothes dryer) with a pilot light is present. Since gasoline is carcinogenic, wear latex gloves when there's a possibility of being exposed to fuel, and, if you spill any fuel on your skin, rinse it off immediately with soap and water. Mop up any spills immediately and do not store fuel-soaked rags where they could ignite. The fuel system is under constant pressure, so, if any fuel lines are to be disconnected, the fuel pressure in the system must be relieved first. When you perform any kind of work on the fuel system, wear safety glasses and have a Class B type fire extinguisher on hand.*

1   The fuel filter is situated underneath the rear of the vehicle, on the right-hand side in front of the fuel tank. To gain access to the filter, block the front wheels, then raise the rear of the vehicle and support it securely on jackstands.

2   Refer to of Chapter 4 Part A or B as applicable and depressurize the fuel system.

3   Loosen the hose clamps and disconnect the fuel lines from either side of the filter. If the clamps are of the crimp type, cut them off with diagonal cutting pliers and replace them with equivalent size worm-drive clamps upon reconnection.

4   Release the filter retaining clip, remove the cover bracket and lower the filter unit away from its mounting bracket.

5   Place the new filter in position, ensuring that the arrow on the filter body is pointing towards the front of the vehicle, in the direction of the fuel flow.

6   Connect the fuel hoses to the filter, securing them in position with their retaining clips.

7   Install the cover bracket then tighten the filter strap screws securely.

8   Start the engine and let it run for a few seconds, then turn the engine off. Check the filter hose connections for leaks, then lower the vehicle to the ground.

35   Coolant replacement (every 2 years, regardless of mileage)

Cooling system draining

Warning: *Do not allow antifreeze to come in contact with your skin or painted surfaces of the vehicle. Rinse off spills immediately with plenty of water. Antifreeze is highly toxic if ingested. Never leave antifreeze lying around in an open container or in puddles on the floor; children and pets are attracted by it's sweet smell and may drink it. Check with local authorities about disposing of used antifreeze. Many communities have collection centers which will see that antifreeze is disposed of safely. Never dump used antifreeze on the ground or into drains.*

Note: *Non-toxic antifreeze is now manufactured and available at local auto parts stores, but even these types should be disposed of properly.*

1   With the engine completely cold, cover the expansion tank cap with a rag, then slowly turn the cap counterclockwise to relieve the pressure in the cooling system (a hissing sound will normally be heard). Wait until any pressure remaining in the system is released, then continue to turn the cap until it can be removed.

2   Position a large container beneath the radiator lower hose connection, then loosen the retaining clamp and ease the hose from the radiator fitting. If the hose joint has not been disturbed for some time, it will be necessary to twist the hose to break the joint. Do not use excessive force, or the radiator fitting could be damaged. Allow the coolant to drain into the container.

3   If the coolant has been drained for a reason other than replacement, then provided it is clean and less than two years old it can be re-used, though this is not recommended.

4   Once all the coolant has drained, reconnect the hose to the radiator and secure it in position with the retaining clamp.

Cooling system flushing

5   If coolant replacement has been neglected, or if the antifreeze mixture has become diluted, then in time, the cooling system may gradually lose efficiency, as the coolant passages become restricted due to rust, scale deposits, and other sediment. The cooling system efficiency can be restored by flushing the system clean.

6   The radiator should be flushed independently of the engine, to avoid unnecessary contamination.

Radiator flushing

7   To flush the radiator, disconnect the top and bottom hoses and any other relevant hoses from the radiator, with reference to Chapter 3.

8   Insert a garden hose into the radiator top inlet. Direct a flow of clean water through the radiator and continue flushing until clean water emerges from the radiator bottom outlet.

9   If after a reasonable period the water still does not run clear, the radiator can be flushed with a cooling system cleaning agent. It is important that the manufacturer's instructions are followed carefully. If the contamination is particularly bad, insert the hose in the radiator bottom outlet, and reverse-flush the radiator.

Engine flushing

10   To flush the engine, remove the thermostat as described in Chapter 3, then temporarily install the thermostat cover.

11   With the top and bottom hoses disconnected from the radiator, insert a garden hose into the radiator top hose. Direct a flow of clean water through the engine and continue flushing until clean water emerges from the radiator bottom hose.

12   On completion of flushing, install the thermostat and reconnect the hoses (see Chapter 3).

Cooling system filling

13   Before attempting to fill the cooling system, make sure that all hoses and clamps are in good condition, and that the clamps are tight. Note that an antifreeze mixture must be used all year round, to prevent corrosion of the engine components.

14   Remove the expansion tank filler cap and fill the system by *slowly* pouring the coolant into the expansion tank to prevent airlocks from forming.

15   If the coolant is being replaced, begin by pouring in a couple of liters of water, followed by the correct quantity of antifreeze, then top-up with more water.

16   Once the level in the expansion tank starts to rise, squeeze the radiator upper and lower hoses to help expel any trapped air in the system. Once all the air is expelled, top-up the coolant level to the "MAX" mark and install the expansion tank cap.

17   Start the engine and run it until it reaches normal operating temperature, then stop the engine and allow it to cool.

18   Check for leaks, particularly around disturbed components. Check the coolant level in the expansion tank and add some, if necessary. Note that the system must be cold before an accurate level is indicated in the expansion tank. If the expansion tank cap is removed while the engine is still warm, cover the cap with a thick cloth, and unscrew the cap slowly to gradually relieve the system pressure (a hissing sound will normally be heard). Wait until any pressure remaining in the system is released, then continue to turn the cap until it can be removed.

Antifreeze mixture

19   The antifreeze should always be replaced at the specified intervals. This is necessary not only to maintain the antifreeze properties, but also to prevent corrosion which would otherwise occur as the cor-

rosion inhibitors become progressively less effective.

20 Always use an ethylene-glycol based antifreeze which is suitable for use in mixed-metal cooling systems. The quantity of antifreeze and levels of protection are indicated in the Specifications.

21 Before adding antifreeze, the cooling system should be completely drained, preferably flushed, and all hoses checked for condition and security.

22 After filling with antifreeze, a label should be attached to the expansion tank, stating the type and concentration of antifreeze used, and the date installed. Any subsequent topping-up should be made with the same type and concentration of antifreeze.

23 Do not use engine antifreeze in the windshield/rear glass washer systems, as it will cause damage to the vehicle paint. A washer additive should be added to the washer system in the quantities stated on the bottle.

36 Brake fluid replacement (every 2 years, regardless of mileage)

Warning: *Brake fluid can harm your eyes and damage painted surfaces, so use extreme caution when handling and pouring it. Do not use fluid that has been standing open for some time, as it absorbs moisture from the air. Excess moisture can cause a dangerous loss of braking effectiveness.*

1 The procedure is similar to that for the bleeding of the hydraulic system as described in Chapter 9, except that the brake fluid reservoir should be emptied by siphoning, using a clean poultry baster or similar device before starting, and allowance should be made for the old fluid to be expelled when bleeding a section of the circuit. **Warning:** *If a poultry baster is used, never again use it for preparing food!*

2 Working as described in Chapter 9, open the first bleed screw in the sequence, and pump the brake pedal gently until nearly all the old fluid has been emptied from the master cylinder reservoir. **Note:** *Old brake fluid is invariably much darker in color than the new, making it easy to distinguish the two.*

3 Fill the reservoir up to the "MAX" level with new fluid, then continue pumping until only the new fluid remains in the reservoir, and new fluid can be seen emerging from the bleed screw. Tighten the screw,

and fill the reservoir up to the "MAX" level line.

4 Work through all the remaining bleed screws in the proper sequence until new fluid can be seen at all of them. Be careful to keep the master cylinder reservoir topped-up to above the "MIN" level at all times, or air may enter the system and greatly increase the length of the task.

5 When the operation is complete, check that all bleed screws are securely tightened, and that their dust caps are installed. Wash off all traces of spilled fluid with clean water, and recheck the master cylinder reservoir fluid level.

6 Check the operation of the brakes before taking the car on the road.

37 Engine management system check (gasoline-engine models) (every 2 years, regardless of mileage)

1 This check is part of the manufacturer's maintenance schedule, and involves testing the engine management system using special test equipment. Because of this, the check will have to be performed by a dealer service department or other qualified repair shop equipped with the necessary tools. Such testing will allow the test equipment to read any fault codes stored in the electronic control unit memory.

2 Unless a fault is suspected, this test is not essential, although it should be noted that it is recommended by the manufacturer.

3 If access to the test equipment is not possible, make a thorough check of all ignition, fuel and emission control system components, hoses, and wiring, for security and obvious signs of damage. Further details of the fuel system, emission control system and ignition system can be found in Chapter 4A and 4B, and in Chapter 5B.

38 Exhaust gas emissions check (diesel-engine models) (every 2 years, regardless of mileage)

This task must be left to a dealer service department or other qualified repair shop equipped with the necessary exhaust gas analyzer needed to check diesel exhaust gas emissions.

Chapter 2 Part A
Gasoline engines

Contents

2A

Specifications

General

Engine code
  1781cc (1.8L)
    Bosch Mono-Motronic fuel injection, 90 hp .... ACC
  1984 cc (2.0L)
    Bosch Motronic fuel injection, 115 hp .... ABA
Bore
  1.8L engine .... 3.191 inches
  2.0L engine .... 3.250 inches
Stroke
  1.8L engine .... 3.404 inches
  2.0L engine .... 3.656 inches
Compression ratio:
  1.8L engine 9.0:1
  2.0L engine 9.6:1
Compression pressures (wear limit):
  1.8L engine .... 102 psi
  2.0L engine .... 109 psi
Firing order .... 1 - 3 - 4 - 2
No. 1 cylinder location .... Timing belt end

Cylinder location and distributor rotation diagram

Lubrication system

Normal operating oil pressure .... 29 psi minimum (at 2000 rpm, oil temperature 176-degrees F)
Oil pump backlash .... 0.008 inch (wear limit)
Oil pump axial clearance .... 0.006 inch (wear limit)

Torque specifications

        Ft-lbs (unless otherwise indicated)

Alternator mounting bolts 18
Drivebelt pulley bolts 15
Camshaft cover retaining screws/nuts 84-in-lbs
Camshaft sprocket bolt 59
Crankshaft oil seal housing bolts 84-in-lbs
Crankshaft sprocket bolt
  Stage 1 66
  Stage 2 .... Angle-tighten a further 90-degrees
Cylinder head bolts
  Stage 1 30
  Stage 2 44
  Stage 3 .... Angle-tighten a further 90-degrees
  Stage 4 .... Angle-tighten a further 90-degrees

Torque specifications (continued)

	Ft-lbs (unless otherwise indicated)
Engine mounts	
Front block bolt	37
Front bracket bolts	37
Left rear mounting bracket bolts	18
Rear block-to-body bolts	18
Right rear mounting bracket bolts	18
Through-bolts	37
Exhaust manifold nuts, M8 nuts	18
Exhaust manifold nuts, M10 nuts	30
Flywheel mounting bolts	
Stage 1	44
Stage 2	Angle-tighten a further 90-degrees
Oil pickup-to-oil pump bolts	84 in-lbs
Oil pump cover bolts	84 in-lbs
Oil pump drive chain guide rail-to-crankcase bolts	84 in-lbs
Oil pump-to-crankcase bolts	15
Power steering pump mounting bolts	18
Oil pan retaining bolts	15
Timing belt tensioner center nut/bolt	33
Torque converter-to-driveplate bolts	
Stage 1	44
Stage 2	Angle-tighten a further 90-degrees

1   General information

Using this Chapter

Chapter 2 is divided into three Parts; A, B and C. Repair operations that can be carried out with the engine in the vehicle are described in Parts A (gasoline engines) and B (diesel engines). Part C covers the removal of the engine/transaxle as a unit, and describes the engine disassembly and overhaul procedures.

In Parts A and B, the assumption is made that the engine is installed in the vehicle, with all components connected. If the engine has been removed for overhaul, the preliminary disassembly information which precedes each operation may be ignored.

Engine description

The engines are water-cooled, single overhead camshaft, in-line four-cylinder units, with cast-iron cylinder blocks and aluminum-alloy cylinder heads. All are mounted transversely at the front of the vehicle, with the transaxle bolted to the left-hand side of the engine.

The cylinder head carries the camshaft, which is driven by a toothed timing belt. It also houses the intake and exhaust valves, which are closed by single or double coil springs, and which run in guides pressed into the cylinder head. The camshaft actuates the valves directly via hydraulic lifters, mounted in the cylinder head. The cylinder head contains integral oilways which supply and lubricate the lifters.

The crankshaft is supported by five main bearings, and endplay is controlled by a thrust bearing installed between cylinder Nos. 2 and 3.

Engine coolant is circulated by a pump, driven by the drivebelt. For details of the cooling system, refer to Chapter 3.

The engine is equipped with a timing-belt-driven intermediate shaft, which provides drive for the distributor and the oil pump.

Oil is circulated under pressure by a pump, driven either by the crankshaft or by the intermediate shaft, depending on engine type. Oil is drawn from the oil pan through a strainer, and then forced through an externally-mounted, replaceable screw-on filter. From there, it is distributed to the cylinder head, where it lubricates the camshaft journals and hydraulic lifters, and also to the crankcase, where it lubricates the main bearings, connecting rods, wrist pins and cylinder bores. Oil jets are mounted at the base of each cylinder - these spray oil onto the underside of the pistons to improve cooling. An oil cooler, supplied with engine coolant, reduces the temperature of the oil before it re-enters the engine.

Repairs possible with the engine installed in the vehicle

The following operations can be performed without removing the engine:-

a) Drivebelts - removal and installation.
b) Camshaft(s) - removal and installation. *
c) Camshaft oil seal - replacement.
d) Camshaft sprocket - removal and installation.
e) Water pump - removal and installation (refer to Chapter 3)
f) Crankshaft oil seals - replacement.
g) Crankshaft sprocket - removal and installation.
h) Cylinder head - removal and installation. *
i) Engine mountings - inspection and replacement.
j) Intermediate shaft oil seal - replacement.
k) Oil pump and pickup assembly - removal and installation.
l) Oil pan - removal and installation.
m) Timing belt, sprockets and cover - removal, inspection and installation.

*Cylinder head dismantling procedures are detailed in Chapter 2C, with details of camshaft and hydraulic lifter removal.
Note: It is possible to remove the pistons and connecting rods (after removing the cylinder head and oil pan) without removing the engine. However, this is not recommended. Work of this nature is more easily and thoroughly completed with the engine removed, as described in Chapter 2C.

2   Valve timing marks and Top Dead Center (TDC) - general information

Refer to illustrations 2.5, 2.6 and 2.7

General information

Note: This sub-section has been written with the assumption that the distributor, spark plug wires and timing belt are correctly installed.
1   The crankshaft, camshaft and intermediate shaft sprockets are driven by the timing belt and rotate in phase with each other. When the timing belt is removed during servicing or repair, it is possible for the shafts to rotate independently of each other, and the correct phasing is then lost.
2   The design of the engines covered in this Chapter is such that

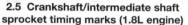

2.5 Crankshaft/intermediate shaft sprocket timing marks (1.8L engine)

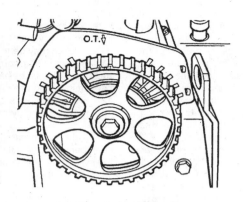

2.6 Camshaft timing marks

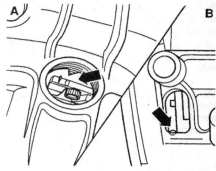

2.7 Flywheel/driveplate timing marks (2.0L engine)

A *Flywheel* B *Driveplate*

2A

potentially damaging piston-to-valve contact may occur if the camshaft is rotated when any of the pistons are stationary at, or near, the top of its stroke.

3   For this reason, it is important that the correct phasing between the camshaft, crankshaft and intermediate shaft is preserved while the timing belt is off the engine. This is achieved by setting the engine in a reference condition (known as Top Dead Center or TDC) before the timing belt is removed, and then preventing the shafts from rotating until the belt is installed. Similarly, if the engine has been disassembled for overhaul, the engine can be set to TDC during reassembly to ensure that the correct shaft phasing is restored.

4   TDC is the highest position a piston reaches within its respective cylinder - in a four-stroke engine, each piston reaches TDC twice per cycle; once on the compression stroke, and once on the exhaust stroke. In general, TDC normally refers to No. 1 cylinder on the compression stroke. (Note that the cylinders are numbered one to four, starting from the timing belt end of the engine.)

5   The crankshaft sprocket on the 1.8L engine (ACC) is equipped with a mark which, when aligned with a reference marking on the intermediate shaft sprocket indicates that No. 1 cylinder (and also No. 4 cylinder) is at TDC **(see illustration)**. The pulley for the ribbed drivebelt must be temporarily installed to position the crankshaft.

6   The camshaft sprocket is also equipped with a timing mark - when this is similarly aligned, the engine is correctly synchronized, and the timing belt can then be installed and tensioned **(see illustration)**.

7   On the 2.0L engine (code ABA) the flywheel/driveplate has markings which can be observed by removing a protective cap from the transaxle bellhousing. When this mark is aligned with a corresponding reference mark on the bellhousing casting, it indicates that No. 1 cylinder is at TDC **(see illustration)**. Note, however, that these markings cannot be used if the transaxle has been removed from the engine for repair or overhaul.

8   The following sub-Sections describe setting the engine to TDC on No. 1 cylinder.

Setting TDC on No. 1 cylinder - timing belt installed

9   Before starting work, disconnect the battery negative cable. **Caution:** *If the stereo in your vehicle is equipped with an anti-theft system, make sure you have the correct activation code before disconnecting the battery.* Prevent any vehicle movement by putting the transaxle in neutral, applying the parking brake and chocking the rear wheels.

10   On the distributor cap, note the position of the No. 1 cylinder spark plug wire terminal with respect to the distributor body. On some models, the manufacturer provides a marking in the form of a small cut-out . If the terminal is not marked, follow the spark plug wire from the No. 1 cylinder spark plug back to the distributor cap - No. 1 cylinder is at the timing-belt end of the engine - and using chalk or a pen (**not** a pencil), place a mark on the distributor body directly under the terminal.

11   Remove the distributor cap, as described in Chapter 5B.

12   Disconnect the spark plug wires from the spark plugs, noting their order of connection. Remove the spark plugs, since this will make the crankshaft easier to rotate.

13   To bring any piston up to TDC, it will be necessary to rotate the crankshaft manually. This can be done by using a wrench and socket on the bolt that retains the crankshaft pulley.

14   Rotate the crankshaft in its normal direction of rotation until the distributor rotor electrode begins to approach the mark that was made on the distributor body.

15   With reference to Section 4, remove the upper timing belt outer covers to expose the camshaft timing belt sprocket beneath.

16   Identify the timing marks on both the camshaft sprocket and the inner section of the timing belt cover (or valve cover, as applicable) - refer to the accompanying illustrations. Continue turning the crankshaft clockwise until these marks are exactly aligned with each other.

17   At this point, identify the timing marks on the crankshaft sprocket (or pulley, as applicable) and the timing belt cover (or intermediate shaft, as applicable) and check that they are correctly aligned; refer to the illustrations in *General Information*.

18   Check that the center of the distributor rotor electrode is now aligned with the No. 1 terminal mark on the distributor body. If it proves impossible to align the rotor with the No. 1 terminal while maintaining the alignment of the camshaft timing marks, refer to Chapter 5B and check that the distributor has been installed correctly.

19   When all the above steps have been completed successfully, the engine will be set to TDC on No. 1 cylinder. **Caution:** *If the timing belt is to be removed, ensure that the crankshaft, camshaft and intermediate shaft alignment is preserved by preventing the sprockets from rotating with respect to each other.*

Setting TDC on No. 1 cylinder - timing belt removed

20   This procedure has been written with the assumption that the timing belt has been removed and that the alignment between the camshaft, crankshaft and where applicable, intermediate shaft has been lost, for example following engine removal and overhaul.

21   On all the engines covered in this manual, it is possible for damage to be caused by the piston crowns striking the valve heads, if the camshaft is rotated with the timing belt removed and the crankshaft set to TDC. For this reason, the TDC setting procedure must be carried out in a particular order, as described in the following Steps.

22   Before the cylinder head is installed, use a wrench and socket on the crankshaft pulley center bolt to turn the crankshaft in its normal direction of rotation, until all four pistons are positioned **halfway down** their bores, with No. 1 piston on its upstroke (around 90-degrees before TDC).

23   With the cylinder head and camshaft sprocket installed, identify the timing marks on both the camshaft sprocket and the inner section of the timing belt cover or valve cover, as applicable; refer to the illus-

trations in *General Information.*

24   Turn the camshaft sprocket in its normal direction of rotation until the timing marks on the sprocket and timing belt inner cover are exactly aligned.

25   Check that the center of the rotor electrode is lined up with the No. 1 cylinder terminal marking on the distributor; if this is not the case, rotate the intermediate shaft sprocket to bring them into alignment.

26   On 1.8L engines (code ACC) identify the timing marks on the crankshaft pulley and the intermediate shaft **(see illustration 2.5)**. Using a socket and wrench on the crankshaft sprocket retaining bolt, turn the crankshaft through 90-degrees (quarter of a turn) in its normal direction of rotation, to bring the timing marks into alignment.

27   On 2.0L engines (code ABA) the crankshaft alignment can be verified by observing the timing marks on the flywheel and transaxle bell-housing (if the transaxle is connected to the engine). Remove the protective cap from the timing inspection hole on the bellhousing and check that the marks are aligned as described in Step 7. **Note:** *Observe from directly above the inspection hole to ensure correct alignment.*

28   Check that the center of the distributor rotor electrode is now aligned with No. 1 cylinder terminal marking on the distributor body. If it proves impossible to align the rotor with the No. 1 terminal while maintaining the alignment of the camshaft timing marks, refer to Chapter 5B and check that the distributor has been installed correctly.

29   When all the above steps have been completed successfully, the engine will be set at TDC on No. 1 cylinder. The timing belt can now be installed as described in Section 4.

Caution: *Until the timing belt is installed, ensure that the crankshaft, camshaft and intermediate shaft alignment is preserved by preventing the sprockets from rotating with respect to each other.*

3   Cylinder compression check

1   When engine performance is down, or if misfiring occurs which cannot be attributed to the ignition or fuel systems, a compression test can provide diagnostic clues as to the engine's condition. If the test is performed regularly, it can give warning of trouble before any other symptoms become apparent.

2   The engine must be fully warmed-up to normal operating temperature, the battery must be fully charged, and all the spark plugs must be removed (refer to Chapter 1). The aid of an assistant will also be required.

3   Disable the ignition system by disconnecting the ignition coil wire from the distributor cap and grounding it on the cylinder block. Use a jumper lead or similar wire to make a good connection. Disable the fuel system by removing the fuel pump relay.

4   Install a compression tester to the No. 1 cylinder spark plug hole - the type of tester which screws into the plug thread is preferable.

5   Have an assistant hold the throttle wide open, then crank the engine on the starter motor; after one or two revolutions, the compression pressure should build up to a maximum figure, and then stabilize. Record the highest reading obtained.

6   Repeat the test on the remaining cylinders, recording the pressure in each. Keep the throttle wide open.

7   All cylinders should produce very similar pressures; a difference of more than 20 psi between any two cylinders indicates a fault. Note that the compression should build up quickly in a healthy engine; low compression on the first stroke, followed by gradually-increasing pressure on successive strokes, indicates worn piston rings. A low compression reading on the first stroke, which does not build up during successive strokes, indicates leaking valves or a blown head gasket (a cracked head could also be the cause). Deposits on the undersides of the valve heads can also cause low compression.

8   Refer to the Specifications section of this Chapter, and compare the recorded compression figures with those stated by the manufacturer.

9   If the pressure in any cylinder is low, carry out the following test to isolate the cause. Introduce a teaspoonful of clean oil into that cylinder

through its spark plug hole, and repeat the test.

10   If the addition of oil temporarily improves the compression pressure, this indicates that piston ring wear is responsible for the pressure loss. No improvement suggests that leaking or burnt valves, or a blown head gasket, may be to blame.

11   A low reading from two adjacent cylinders is almost certainly due to the head gasket having blown between them; the presence of coolant in the engine oil will confirm this.

12   If one cylinder is about 20 percent lower than the others and the engine has a slightly rough idle, a worn camshaft lobe could be the cause.

13   If the compression reading is unusually high, the combustion chambers are probably coated with carbon deposits. If this is the case, the cylinder head should be removed and decarbonized.

14   On completion of the test, install the spark plugs and restore the ignition system.

4   Timing belt - removal and installation

Refer to illustrations 4.12 and 4.23

General information

1   The primary function of the toothed timing belt is to drive the camshaft(s), but it is also used to drive the intermediate shaft. Should the belt slip or break in service, the valve timing will be disturbed and piston-to-valve contact may occur, resulting in serious engine damage.

2   For this reason, it is important that the timing belt is tensioned correctly, and inspected regularly for signs of wear or deterioration.

3   Note that the removal of the *inner* section of the timing belt cover is described as part of the cylinder head removal procedure; see Section 11.

Removal

4   Before starting work, immobilize the engine and vehicle as follows:

  a) Disconnect the cable from the negative terminal of the battery. **Caution:** *If the stereo in your vehicle is equipped with an anti-theft system, make sure you have the correct activation code before disconnecting the battery.*

  b) Prevent any vehicle movement by applying the parking brake and chocking the rear wheels.

5   Access to the timing belt covers can be improved by removing the air cleaner housing-to-throttle body ducting the crankcase breather hose.

6   Release the uppermost part of the timing belt cover by prying open the metal spring clips and where applicable, removing the retaining screws. Lift the cover away from the engine.

7   With reference to Section 6, remove the V-belt (if equipped), then remove the ribbed drivebelt.

8   Refer to Section 2 and using the engine alignment markings, set the engine to TDC on No. 1 cylinder.

9   Loosen and withdraw the retaining screws, then remove the pulley for the ribbed drivebelt (together with the V-belt pulley, if equipped) from the crankshaft sprocket. On completion, check that the engine is still set to TDC.

10   Remove the water pump pulley to allow removal of the timing belt lower cover.

11   Remove the retaining screws and clips, and lift off the timing belt lower cover.

12   Refer to Section 5 and relieve the tension on the timing belt by loosening the tensioner mounting nut slightly, allowing it to pivot away from the belt **(see illustration)**.

13   Examine the timing belt for manufacturer's markings that indicate the direction of rotation. If none are present, make your own using typist's correction fluid. **Caution:** *If the belt appears to be in good condition and can be re-used, it is essential that it is installed the same way, otherwise accelerated wear will result, leading to premature failure.*

14   Slide the belt off the sprockets, taking care to avoid twisting or

4.12 Relieve the tension on the timing belt by loosening the tensioner mounting nut (arrow)

4.23 Tension the belt by turning the tensioner clockwise using snap-ring pliers

5.5 Slide the tensioner off its mounting stud

2A

kinking it excessively. Ensure that the sprockets remain aligned with their respective timing markings once the timing belt has been removed. **Caution:** *It is potentially damaging to allow the camshaft to turn with the timing belt removed and the engine set at TDC, as piston-to-valve contact may occur.*

15 Examine the belt for evidence of contamination by coolant or lubricant. If this is the case, identify the source of the contamination before progressing any further. Check the belt for signs of wear or damage, particularly around the leading edges of the belt teeth. Replace the belt if its condition is in doubt; the cost of belt replacement is negligible compared with potential cost of the engine repairs, should the belt fail in service. Similarly, if the belt is known to have covered more than 36, 000 miles, it is prudent to replace it regardless of condition, as a precautionary measure.

16 If the timing belt is not going to be installed for some time, it is a wise precaution to hang a warning label on the steering wheel, to remind yourself (and others) not to attempt starting the engine.

Installation

17 Ensure that the crankshaft, camshaft and intermediate shaft timing marks are still correctly aligned in the TDC on No. 1 cylinder position, as described in Section 2.

18 Loop the timing belt under the crankshaft sprocket loosely, observing the direction of rotation markings.

19 Temporarily install the pulley for the ribbed drivebelt to the crankshaft sprocket, using two of the retaining screws - note that the offset mounting holes allow only one installation position.

20 Verify that the timing marks on the crankshaft pulley and the intermediate shaft sprocket are still correctly aligned; refer to Section 2 for details.

21 Engage the timing belt teeth with the crankshaft sprocket, then maneuver it into position over the intermediate shaft and camshaft sprockets. Observe the direction of rotation markings on the belt.

22 Pass the flat side of the belt over the tensioner roller - avoid bending the belt back on itself or twisting it excessively as you do this. Ensure that the 'front run' of the belt is taut - all the slack should be in the section of the belt that passes over the tensioner roller.

23 Tension the belt by turning the eccentrically mounted tensioner clockwise; two holes are provided in the side of the tensioner hub for this purpose - a pair of sturdy right-angled snap-ring pliers is a suitable substitute for the correct VW tool **(see illustration)**.

24 Test the timing belt tension by grasping it between the fingers at a point mid-way between the intermediate shaft and camshaft sprockets and twisting it; the belt tension is correct when it can just be twisted 90-degrees (quarter of a turn) and no further.

25 When the correct belt tension has been achieved, tighten the tensioner locknut to the specified torque.

26 Using a wrench or wrench and socket on the crankshaft pulley center bolt, rotate the crankshaft two complete revolutions, bringing the timing marks back into alignment. Re-check the timing belt tension and adjust it, if necessary.

27 Remove the pulley for the ribbed drivebelt from the crankshaft sprocket to allow the lower section of the outer timing belt cover to be installed, then install the pulley, noting that the offset of the mounting holes allows only one installation position. Finally, insert and tighten the retaining bolts to the specified torque.

28 Refer to Chapter 3 and install the water pump pulley.

29 Working from Section 6, install and tension the drivebelt(s).

30 Reconnect the negative battery cable.

5 Timing belt tensioner and sprockets - removal, inspection and installation

1 Before starting work, immobilize the engine and vehicle as follows:

a) *Disconnect the cable from the negative terminal of the battery.* **Caution:** *If the stereo in your vehicle is equipped with an anti-theft system, make sure you have the correct activation code before disconnecting the battery.*

b) *Prevent any vehicle movement by applying the parking brake and chocking the rear wheels.*

2 To gain access to the components detailed in this Section, carry out the following:

a) *Refer to Section 6 and remove the drivebelt(s)*

b) *Refer to Chapter 3 and remove the water pump pulley.*

Timing belt tensioner

Refer to illustration 5.5

Removal

3 With reference to the relevant Steps of Sections 2 and 4, set the engine to TDC on No. 1 cylinder, then remove the timing belt upper and lower covers.

4 Loosen the retaining nut at the hub of the tensioner pulley, and allow the assembly to rotate counterclockwise, relieving the tension on the timing belt. Remove the nut and recover the washer.

5 Slide the tensioner off its mounting stud **(see illustration)**.

Inspection

6 Wipe the tensioner clean, but do not use solvents that may contaminate the bearings. Spin the tensioner pulley on its hub by hand. Stiff movement or excessive freeplay is an indication of severe wear; the tensioner is not a serviceable component, and should be replaced.

5.11 To make a camshaft sprocket holding tool, obtain two lengths of steel strap, then bolt the two straps together to form a forked end, leaving the bolt loose so the shorter strip can pivot freely. At the end of each 'prong' of the fork, secure a bolt with a nut and a locknut, which will engage with the cut-outs in the sprocket

Installation

7   Slide the tensioner pulley over the mounting stud, then install the washer and retaining nut - do not fully tighten the nut at this stage.
8   With reference to Section 4, tension the timing belt and install the timing belt covers.
9   Reconnect the cable to the negative battery terminal.

Camshaft timing belt sprocket

Refer to illustration 5.11

Removal

10   With reference to Section 4, remove the timing belt covers and set the engine to TDC on No. 1 cylinder. Loosen the tensioner center nut and rotate it counterclockwise to relieve the tension on the timing belt. Carefully slide the timing belt off the camshaft sprocket.
11   The camshaft sprocket must be held stationary while its retaining bolt is loosened; if access to the correct VW special tool is not possible, a simple home-made tool using basic materials may be fabricated **(see illustration)**.
12   Using the homemade tool, brace the camshaft sprocket. Loosen and remove the retaining bolt; recover the washer (if equipped).
13   Slide the camshaft sprocket from the end of the camshaft. Where applicable, recover the Woodruff key from the keyway.
14   With the sprocket removed, examine the camshaft oil seal for signs of leaking. If necessary, refer to Section 8 and replace it.
15   Wipe the sprocket and camshaft mating surfaces clean.

Installation

16   Where applicable, install the Woodruff key into the keyway. Install the sprocket to the camshaft, engaging the slot in the sprocket with the Woodruff key.
17   Working from Section 2, check that the engine is still set to TDC on No. 1 cylinder, then install and tension the timing belt. Install the timing belt covers.
18   Install the crankshaft and water pump drivebelt pulleys, then insert the bolts and tighten them to the specified torque.
19   With reference to Section 6, install and tension the drivebelt(s).

Crankshaft timing belt sprocket

Refer to illustration 5.22

Removal

20   With reference to Sections 2 and 4, remove the timing belt covers

5.22 Removing the crankshaft sprocket

and set the engine to TDC on No. 1 cylinder. Loosen the tensioner center nut and rotate it counterclockwise to relieve the tension on the timing belt. Carefully slide the timing belt off the crankshaft sprocket.
21   The crankshaft sprocket must be held stationary while its retaining bolt is loosened. If access to the correct VW flywheel locking tool is not available, lock the crankshaft in position by removing the starter motor, as described in Chapter 5A, to expose the flywheel ring gear. Then get an assistant to insert a prybar between the gear teeth and the transaxle bellhousing while the sprocket retaining bolt is loosened.
22   Withdraw the bolt, recover the washer and lift off the sprocket **(see illustration)**.
23   With the sprocket removed, examine the crankshaft oil seal for signs of leaking. If necessary, refer to Section 10 and replace it.
24   Wipe the sprocket and crankshaft mating surfaces clean.

Installation

25   Install the sprocket, engaging the lug on the inside of the sprocket with the recess in the end of the crankshaft. Insert the bolt and tighten it to the specified torque.
26   Working from Section 4, check that the engine is still set to TDC on No. 1 cylinder, then install and tension the timing belt. Install the timing belt covers.
27   Install the crankshaft (and where applicable, water pump) drive-belt pulley(s), then insert the retaining bolts and tighten them to the specified torque.
28   With reference to Section 6, install and tension the drivebelt(s).

Intermediate shaft sprocket

Removal

29   With reference to Section 4, remove the timing belt covers, and set the engine to TDC on No. 1 cylinder. Loosen the tensioner center nut, and rotate it counterclockwise to relieve the tension on the timing belt. Carefully slide the timing belt off the intermediate shaft sprocket.
30   The intermediate shaft sprocket must be held stationary while its retaining bolt is loosened; if access to the correct VW special tool is not possible, a simple home-made tool using basic materials made be fabricated as described in the camshaft sprocket removal sub-Section.
31   Using the home-made tool, brace the intermediate shaft sprocket and loosen and remove the retaining bolt; recover the washer where installed.
32   Slide the sprocket from the end of the intermediate shaft. Where applicable, recover the Woodruff key from the keyway.
33   With the sprocket removed, examine the intermediate shaft oil seal for signs of leaking. If necessary, refer to Section 9 and replace it.
34   Wipe the sprocket and shaft mating surfaces clean.

Installation

35   Install the Woodruff key into the keyway, with the plain surface

facing upwards. Install the sprocket to the intermediate shaft, engaging the slot in the sprocket with the Woodruff key.

36 With reference to Section 2, check that the engine is still set to TDC on No. 1 cylinder. Where applicable, align the intermediate shaft sprocket with the crankshaft pulley timing marks.

37 Tighten the sprocket retaining bolt to the specified torque; hold the sprocket using the method employed during removal.

38 With reference to Section 4, install and tension the timing belt, then install the timing belt covers.

39 Install the crankshaft drivebelt pulley(s), then insert the retaining bolts and tighten them to the specified torque.

40 With reference to Section 6, install and tension the drivebelt(s).

6 Drivebelts - removal and installation

General information

1 Depending on the vehicle specification and engine type, one or two drivebelts may be installed. Both are driven from pulleys mounted on the crankshaft, and provide drive for the alternator, water pump, power steering pump and on vehicles with air conditioning, the refrigerant compressor.

2 The run of the belts and the components they drive are also dependent on vehicle specification and engine type, and because of this, the water pump and power steering pump may be equipped with pulleys to suit either a ribbed belt or a V-belt.

3 The ribbed drivebelt on 2.0L engines (code ABA) is equipped with an automatic tensioning device. On 1.8L engines (code ACC), the belt is tensioned by the alternator mounting, which can be moved on its mounting brackets. The V-belt is tensioned by pivoting the power steering pump on its mount brackets.

4 On installation, the drivebelt must be tensioned correctly, to ensure correct operation under all conditions and prolonged service life.

V-belt

Removal

5 Park the vehicle on a level surface, and apply the parking brake. Disconnect the cable from the negative terminal of the battery. **Caution:** *If the stereo in your vehicle is equipped with an anti-theft system, make sure you have the correct activation code before disconnecting the battery.* Jack up the front of the vehicle and rest it securely on axle stands.

6 Turn the steering to full right lock, then refer to Chapter 11 and remove the plastic air ducting from underneath the right-hand front fender.

7 With reference to Chapter 10, loosen the power steering pump mounting bolts and allow the pump body to pivot around its uppermost mounting towards the engine.

8 Guide the V-belt off the power steering pump pulley and where applicable, the water pump pulley.

9 Examine the belt for signs or wear or damage, and replace it if necessary.

Installation and tensioning

10 Install the belt by reversing the removal procedure, ensuring that it seats evenly in the pulleys.

11 Set the belt tension by grasping the underside of the power steering pump and drawing it towards the front of the vehicle. The tension is correct when the midpoint of the belt's longest run can be deflected by no more than 13/64-inch under moderate pressure. Tighten the power steering pump mounting bolts securely.

12 Rotate the crankshaft in its normal direction of rotation through two turns, then re-check and if necessary adjust the tension.

V-ribbed belt

Removal

13 Park the vehicle on a level surface, and apply the parking brake.

Disconnect the cable from the negative terminal of the battery. **Caution:** *If the stereo in your vehicle is equipped with an anti-theft system, make sure you have the correct activation code before disconnecting the battery.* Jack up the front of the vehicle and rest it securely on axle stands.

14 Turn the steering to full right lock, then refer to Chapter 11 and remove the plastic air ducting from underneath the right-hand front fender.

15 Where applicable, remove the V-belt as described in the previous sub-Section.

16 Examine the ribbed belt for manufacturer's markings, indicating the direction of rotation. If none are present, make some using typist's correction fluid or a dab of paint - do not cut or score the belt in any way.

17 Rotate the tensioner roller arm clockwise against its spring tension, so that the roller is forced away from the belt.

18 Pull the belt off the alternator pulley, then release it from the remaining pulleys.

Installation and tensioning

Caution: *Observe the manufacturer's direction of rotation markings on the belt, when installing.*

19 Pass the ribbed belt underneath the crankshaft pulley, ensuring that the ribs seat in the channels on the surface of the pulley.

20 Rotate the tensioner roller arm clockwise against its spring tension - use an adjustable wrench as a lever.

21 Pass the belt around the water pump pulley or air conditioning compressor pulley (as applicable), then install it over the alternator pulley.

22 Release the tensioner pulley arm and allow the roller to bear against the flat surface of the belt.

23 Refer to Chapter 11 and install the plastic air ducts to the underside of the fender.

24 Where applicable, refer to the previous sub-Section and install the V-belt.

25 Lower the vehicle to the ground, then reconnect the negative battery cable.

7 Camshaft cover - removal and installation

Removal

1 Disconnect the cable from the negative terminal of the battery. **Caution:** *If the stereo in your vehicle is equipped with an anti-theft system, make sure you have the correct activation code before disconnecting the battery.*

2.0L engine

2 Detach the air intake duct from the throttle body.

3 Remove the upper intake manifold (plenum) (see Chapter 4B).

4 Working around the edge of the camshaft cover, progressively loosen and remove the retaining nuts.

1.8L engine

5 Remove crankcase breather hose from the cover; cut off the crimp-type clamp - use an equivalent sized worm-drive clamp on installation.

6 To gain greater working space, refer to Chapter 4A and disconnect the throttle cable from the throttle housing.

7 Working around the edge of the camshaft cover, progressively loosen and remove the retaining nuts.

All engines

8 Lift the cover away from the cylinder head; if it sticks, do not attempt to lever it off with an implement - instead free it by working around the cover and tapping it lightly with a soft-faced mallet.

9 Where applicable, lift the oil baffle plate off the camshaft bearing cap studs, noting its orientation.

2A

10   Recover the camshaft cover gasket; note that the gasket may be made up of several pieces, depending on the engine. Rubber gaskets can be reused if they are in good condition (not cracked or hardened). Other gaskets *must* be replaced.
11   Clean the mating surfaces of the cylinder head and camshaft cover thoroughly, removing all traces of oil and old gasket - take care to avoid damaging the surfaces as you do this.

Installation

12   Install the camshaft cover by following the removal procedure in reverse, noting the following points:

a) *Ensure that all sections of the gasket are correctly seated on the cylinder head, and take care to avoid displacing it as the camshaft cover is lowered into position.*

b) *Tighten the camshaft cover retaining screws/nuts to the specified torque.*

c) *When installing hoses that were originally secured with crimp-type clamps, use standard worm-drive clamps in their place on installation.*

13   Reconnect the cable to the negative terminal of the battery.

8   Camshaft oil seal - replacement

Refer to illustration 8.8

1   Disconnect the cable from the negative terminal of the battery. **Caution:** *If the stereo in your vehicle is equipped with an anti-theft system, make sure you have the correct activation code before disconnecting the battery.*
2   Refer to Section 6 and remove the drivebelt(s).
3   With reference to Sections 2, 4 and 5 of this Chapter, remove the drivebelt pulleys and timing belt cover, then set the engine to TDC on No. 1 cylinder and remove the timing belt, timing belt tensioner and camshaft sprocket.
4   After removing the retaining screws, lift the inner timing belt cover away from the engine block - this will expose the oil seal.
5   Drill two small holes into the existing oil seal, diagonally opposite each other. Thread two self-tapping screws into the holes, and using two pairs of pliers, pull on the heads of the screws to extract oil seal. Take great care to avoid drilling through into the seal housing or camshaft sealing surface.
6   Clean out the seal housing and sealing surface of the camshaft by wiping it with a lint-free cloth - avoid using solvents that may enter the cylinder head and affect component lubrication. Remove any debris or burrs that may cause the seal to leak.
7   Lubricate the lip of the new oil seal with clean engine oil, and push it over the camshaft until it is positioned above its housing.
8   Using a hammer and a socket of suitable diameter, drive the seal squarely into its housing **(see illustration)**. **Note:** *Select a socket that bears only on the hard outer surface of the seal, not the inner lip which can easily be damaged.*
9   With reference to Sections 2, 4 and 5 of this Chapter, install the inner timing belt cover and the timing sprockets, then install and tension the timing belt. On completion, install the timing belt outer cover.
10   With reference to Section 6, install and tension the drivebelt(s).

9   Intermediate shaft oil seal - replacement

1   Disconnect the cable from the negative terminal of the battery. **Caution:** *If the stereo in your vehicle is equipped with an anti-theft system, make sure you have the correct activation code before disconnecting the battery.*
2   Refer to Section 6 and remove the drivebelt(s).
3   With reference to Sections 4 and 5 of this Chapter, remove the drivebelt pulleys, timing belt outer cover, timing belt, tensioner and intermediate shaft sprocket.
4   After removing the retaining screws, lift the inner timing belt cover

8.8 Drive the camshaft oil seal squarely into its housing

away from the engine block - this will expose the intermediate shaft sealing flange.
5   With reference to Section 7 of Chapter 2C, remove the intermediate shaft flange and replace the shaft and flange oil seals.
6   Referring to Sections 4 and 5 of this Chapter, carry out the following:

a) *Install the inner timing belt cover.*

b) *Install the intermediate shaft timing belt sprocket.*

c) *Install and tension the timing belt.*

d) *Install the timing belt outer cover.*

7   With reference to Section 6 of this Chapter, install and tension the drivebelt(s).

10   Crankshaft oil seals - replacement

Crankshaft front oil seal

Refer to illustrations 10.9 and 10.10

1   Disconnect the cable from the negative terminal of the battery. **Caution:** *If the stereo in your vehicle is equipped with an anti-theft system, make sure you have the correct activation code before disconnecting the battery.*
2   Drain the engine oil - see Chapter 1.
3   Raise the front of the vehicle and rest it securely on axle stands.
4   Remove the screws and detach the plastic air ducting from underneath the right-hand front fender.
5   Remove the drivebelt(s) (see Section 6).
6   With reference to Sections 4 and 5 of this Chapter, remove the drivebelt pulleys, timing belt outer covers, timing belt and crankshaft sprocket.
7   Remove the oil seal, using the same method as that described for the camshaft oil seal removal, in Section 8.
8   Clean out the seal housing and sealing surface of the crankshaft by wiping it with a lint-free cloth - avoid using solvents that may enter the crankcase and affect component lubrication. Remove any debris or burrs that could cause the seal to leak.
9   Lubricate the lip of the new oil seal with clean engine oil, and position it over the housing **(see illustration)**.
10   Using a hammer and a socket of suitable diameter, drive the seal squarely into its housing **(see illustration)**. **Note:** *Select a socket that bears only on the hard outer surface of the seal, not the inner lip which can easily be damaged.*
11   With reference to Sections 2, 4 and 5 of this Chapter, install the crankshaft timing belt sprocket, then install and tension the timing belt.

10.9 Lubricate the new crankshaft oil seal, and position it over the housing

10.10 Using a hammer and a socket, drive the seal squarely into its housing

10.22 Tighten the front oil seal housing bolts to the specified torque

2A

On completion, install the timing belt outer cover, and drivebelt pulley(s).

12 The remainder of the installation procedure is a reversal of removal, as follows:

a) With reference to Section 6, install and tension the drivebelt(s).
b) Install the plastic air ducting to the underside of the fender, working from Chapter 11.
c) Refer to Chapter 1 and refill the engine with the correct grade and quantity of oil.
d) Reconnect the cable to the negative terminal of the battery.

Crankshaft front oil seal housing - gasket replacement

Refer to illustration 10.22

13 Proceed as described in Steps 1 to 6 above, then refer to Section 15 and remove the oil pan.

14 Progressively loosen and then remove the oil seal housing retaining bolts.

15 Lift the housing away from the cylinder block, together with the crankshaft oil seal, using a twisting motion to ease the seal along the shaft.

16 Recover the old gasket from the seal housing on the cylinder block. If it has disintegrated, scrape the remains off with a knife blade. Take care to avoid damaging the mating surfaces.

17 Pry the old oil seal from the housing using a screwdriver.

18 Wipe the oil seal housing clean, and check it visually for signs of distortion or cracking. Lay the housing on a work surface, with the mating surface face down. Press in a new oil seal, using a block of wood as a press to ensure that the seal enters the housing squarely.

19 Smear the crankcase mating surface with multi-purpose grease, and lay the new gasket in position.

20 Pad the end of the crankshaft with a layer of electrical tape; this will protect the oil seal as it is being installed.

21 Lubricate the inner lip of the crankshaft oil seal with clean engine oil, then install the seal and its housing to the end of the crankshaft. Ease the seal along the shaft using a twisting motion, until the housing is flush with the crankcase.

22 Insert the retaining bolts and tighten them progressively to the specified torque (see illustration). Caution: *The housing is fabricated from a light alloy, and may be distorted if the bolts are not tightened progressively.*

23 Refer to Section 15 and install the oil pan.

24 With reference to Sections 2, 4 and 5 of this Chapter, install the crankshaft timing belt sprocket, then install and tension the timing belt. On completion, install the timing belt outer cover, and drivebelt pulley(s).

25 The remainder of the installation procedure is a reversal of removal, as follows:

a) With reference to Section 6, install and tension the drivebelt(s).
b) Install the plastic air ducting to the underside of the fender.
c) Refer to Chapter 1 and refill the engine with the correct grade and quantity of oil.
d) Reconnect the cable to the negative terminal of the battery.

Crankshaft rear oil seal (flywheel end)

Refer to illustration 10.40

26 Proceed as described in Steps 1 to 3 above, then refer to Section 15 and remove the oil pan.

27 Remove the screws and detach the plastic air ducting from underneath the left-hand front fender.

28 Refer to Chapter 7A or B as applicable, and remove the transaxle from the engine.

29 On vehicles with manual transaxle, refer to Section 13 of this Chapter and remove the flywheel, then refer to Chapter 6 and remove the clutch friction plate and pressure plate.

30 On vehicles with automatic transaxle, refer to Section 13 of this Chapter and remove the driveplate from the crankshaft.

31 Where applicable, remove the retaining bolts and lift the intermediate plate away from the cylinder block.

32 Progressively loosen and then remove the oil seal housing retaining bolts.

33 Lift the housing away from the cylinder block, together with the crankshaft oil seal, using a twisting motion to ease the seal along the shaft.

34 Recover the old gasket from the seal housing on the cylinder block. If it has disintegrated, scrape the remains off with a knife blade. Take care to avoid damaging the mating surfaces.

35 Pry the old oil seal from the housing using a screwdriver.

36 Wipe the oil seal housing clean, and check it visually for signs of distortion or cracking. Lay the housing on a work surface, with the mating surface face down. Press in a new oil seal, using a block of wood as a press to ensure that the seal enters the housing squarely.

37 Smear the crankcase mating surface with multi-purpose grease, and lay the new gasket in position.

38 A protective plastic cap is supplied with genuine VW crankshaft oil seals; when placed over the end of the crankshaft, the cap prevents damage to the inner lip of the oil seal as it is being installed. Use electrical tape to pad the end of the crankshaft if a cap is not available.

39 Lubricate the inner lip of the crankshaft oil seal with clean engine oil, then install the seal and its housing to the end of the crankshaft. Ease the seal along the shaft using a twisting motion, until the housing is flush with the crankcase.

40 Insert the retaining bolts and tighten them progressively to the

specified torque **(see illustration)**. **Caution:** *The housing is fabricated from a light alloy, and may be distorted if the bolts are not tightened progressively.*

41   Refer to Section 15 and install the oil pan.

42   Install the intermediate plate to the cylinder block, then insert and tighten the bolts.

43   On vehicles with automatic transaxle, work from Section 13 of this Chapter and install the driveplate to the crankshaft.

44   On vehicles with manual transaxle, refer to Section 13 of this Chapter and install the flywheel, then refer to Chapter 6 and install the clutch friction plate and pressure plate.

45   Referring to Chapter 7A or B as applicable, install the transaxle to the engine.

46   The remainder of the installation procedure is a reversal of removal, as follows:

a) *Install the plastic air ducting to the underside of the fender.*

b) *Refer to Chapter 1 and refill the engine with the correct grade and quantity of oil.*

c) *Reconnect the negative battery cable.*

11   Cylinder head and manifolds - removal, separation and installation

Removal

1   Select a solid, level surface to park the vehicle upon. Give yourself enough space to move around it easily.

2   Refer to Chapter 11 and remove the hood from its hinges.

3   Disconnect the cable from the negative terminal of the battery. **Caution:** If *the stereo in your vehicle is equipped with an anti-theft system, make sure you have the correct activation code before disconnecting the battery.*

4   Referring to Chapter 1, carry out the following :

a) *Drain the engine oil.*

b) *Drain the cooling system.*

5   Refer to Section 6 and remove the drivebelt(s).

6   With reference to Section 2, set the engine to TDC on No. 1 cylinder.

7   Refer to Chapter 3 and perform the following:

a) *Loosen the clamps and disconnect the radiator top and bottom hoses from the ports on the cylinder head and water pump/thermostat housing (as applicable).*

b) *Loosen the clamps and disconnect the expansion tank and heater inlet and outlet coolant hoses from the ports on the cylinder head.*

8   The "lock carrier" is a panel assembly comprising the front bumper moulding, radiator and grille, cooling fan(s) headlight units, front valance and hood lock mechanism. Although its removal is not essential, its does give greatly-improved access to the engine. Its removal is described at the beginning of the engine removal procedure - refer to Chapter 2C for details.

9   With reference to Chapter 6, unplug the oxygen sensor cabling from the main harness at the multiway connector (where applicable).

10   Remove the spark plug wires from the spark plugs and the distributor.

11   On multi-point fuel-injected models, refer to Chapter 4B and remove the throttle body, the upper section of the intake manifold (2.0L engine only), the fuel rail and the fuel injectors.

12   On 1.8L engines, refer to Chapter 4A, remove the throttle body airbox, then remove the throttle body.

13   With reference to Sections 2, 4 and 7, carry out the following:

a) *Remove the camshaft cover.*

b) *Remove the timing belt outer covers, and disengage the timing belt from the camshaft sprocket.*

14   Loosen and withdraw the retaining screws and lift off the inner timing belt cover(s).

15   With reference to Chapter 4A or B as applicable, unplug the wiring harness from the coolant temperature sensor at the connector.

16   Refer to Chapter 6 and separate the exhaust downpipe from the

10.40 Tighten the rear oil seal housing bolts to the specified torque

exhaust manifold flange.

17   Where applicable, detach the warm-air intake hose from the exhaust manifold heat shield.

18   Loosen and remove the bolt securing the engine oil dipstick tube to the cylinder head.

19   Remove the retaining screw and detach the engine harness connector bracket from the cylinder head.

20   Following the *reverse* of the tightening sequence shown in **illustration 11.36**, progressively loosen the cylinder head bolts, by half a turn at a time, until all bolts can be unscrewed by hand.

21   Check that nothing remains connected to the cylinder head, then lift the head away from the cylinder block; seek assistance if possible, as it is a heavy assembly, especially if it is being removed complete with the manifolds.

22   Remove the gasket from the top of the block, noting the locating dowels. If the dowels are a loose fit, remove them and store them with the head for safe-keeping. Do not discard the gasket - on some models it will be needed for identification purposes.

23   If the cylinder head is to be disassembled for overhaul refer to Chapter 2C.

Manifold separation

24   Intake manifold removal and installation is described in Chapter 4A or B as applicable.

25   Progressively loosen and remove the exhaust manifold retaining nuts. Lift the manifold away from the cylinder head and recover the gaskets. Where applicable, loosen the union and detach the CO sampling pipe from the manifold.

26   Ensure that the mating surfaces are completely clean, then install the exhaust manifold, using new gaskets. Tighten the retaining nuts to the specified torque.

Preparation for installation

27   The mating faces of the cylinder head and cylinder block/crankcase must be perfectly clean before installing the head. Use a hard plastic or wood scraper to remove all traces of gasket and carbon; also clean the piston crowns. Take particular care during the cleaning operations, as aluminum alloy is easily damaged. Also, make sure that the carbon is not allowed to enter the oil and water passages - this is particularly important for the lubrication system, as carbon could block the oil supply to the engine's components. Using adhesive tape and paper, seal the water, oil and bolt holes in the cylinder block/crankcase.

28   Check the mating surfaces of the cylinder block/crankcase and the cylinder head for nicks, deep scratches and other damage. If slight, they may be removed carefully with a file, but if excessive, machining may be the only alternative to replacement.

29   If warpage of the cylinder head gasket surface is suspected, use a straightedge to check it for distortion. Refer to Part C of this Chapter if necessary.

11.32a Lay a new head gasket on the block, engaging it with the locating dowels

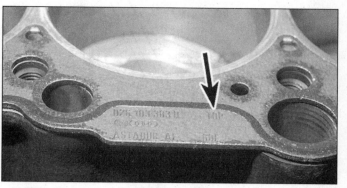

11.32b Ensure that the manufacturer's "TOP" mark and part number are face up

30   Check the condition of the cylinder head bolts, and particularly their threads, whenever they are removed. Wash the bolts in suitable solvent, and wipe them dry. Check each for any sign of visible wear or damage, replacing any bolt if necessary. Measure the length of each bolt, to check for stretching (although this is not a conclusive test, if all bolts have stretched by the same amount). The manufacturer does not actually specify that the bolts must be replaced, however, it is strongly recommended that the bolts be replaced as a complete set whenever they are removed.

31   It is possible for the piston crowns to strike and damage the valve heads, if the camshaft is rotated with the timing belt removed and the crankshaft set to TDC. For this reason, the crankshaft must be set to a position other than TDC on No. 1 cylinder before the cylinder head is installed. Use a wrench and socket on the crankshaft pulley center bolt to turn the crankshaft in its normal direction of rotation, until all four pistons are positioned halfway down their bores, with No. 1 piston on its upstroke (90-degrees before TDC).

Installation

Refer to illustrations 11.32a, 11.32b and 11.36

32   Lay a new head gasket on the cylinder block, engaging it with the locating dowels. Ensure that the manufacturer's "TOP" and part number markings are face up **(see illustrations)**.

33   With the help of an assistant, place the cylinder head and manifolds centrally on the cylinder block, ensuring that the locating dowels engage with the holes in the cylinder head. Check that the head gasket is correctly seated before allowing the weight the full weight of the cylinder head to rest upon it.

34   Apply a smear of grease to the thread and to the underside of the head of the cylinder head bolts; use a good-quality high-melting point grease.

35   Carefully enter each bolt into its relevant hole (*do not drop them in*) and screw them in by hand only until finger-tight.

36   Working progressively and in the sequence shown, tighten the cylinder head bolts to their Stage 1 torque setting, using a torque wrench and suitable socket **(see illustration)**. Repeat the exercise in the same sequence for the Stage 2 torque setting.

37   Once all the bolts have been tightened to their Stage 2 settings, working again in the given sequence, angle-tighten the bolts through the specified Stage 3 angle, using a socket and extension bar. It is recommended that an angle-measuring gauge is used during this stage of the tightening, to ensure accuracy. If a gauge is not available, use white paint to make alignment marks between the bolt head and cylinder head prior to tightening; the marks can then be used to check that the bolt has been rotated through the correct angle during tightening. Repeat the exercise for the Stage 4 setting.

38   Install the timing belt inner cover, tightening the retaining screws securely.

39   Refer to Section 2 and follow the procedure for setting the engine

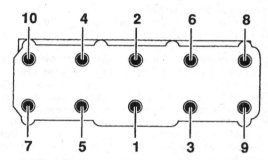

11.36 Cylinder head bolt tightening sequence

to TDC on No. 1 cylinder with the timing belt removed. On completion, refer to Section 4 and install the camshaft timing belt.

40   The remainder of the installation sequence is a reversal of the removal procedure, as follows:

  a) *Bolt the engine dipstick tube to the cylinder head, where applicable.*
  b) *Refer to Chapter 6 and reconnect the exhaust downpipe to the exhaust manifold.*
  c) *On multi-point fuel-injected systems, refer to Chapter 4B and install the fuel injectors, fuel rail, upper section of the intake manifold (where applicable) and the throttle body.*
  d) *On single-point fuel-injected models, refer to Chapter 4A and install the throttle body and air box.*
  e) *Connect the spark plug wires.*
  f) *Refer to Section 7 and install the camshaft cover.*
  g) *With reference to the information in Chapter 2C, install the lock carrier assembly, if it was removed for greater access.*
  h) *Reconnect the radiator, expansion tank and heater coolant hoses, referring to Chapter 3 for guidance. Reconnect the coolant temperature sensor wiring.*
  i) *Refer to Section 6 and install the drivebelt(s).*
  j) *Restore the battery connection.*
  k) *Refer to Chapter 11 and install the hood.*

41   On completion, refer to Chapter 1 and carry out the following:

  a) *Refill the engine cooling system with the correct quantity of new coolant.*
  b) *Refill the engine lubrication system with the correct grade and quantity of oil.*

12 Hydraulic lifters - operation check

Refer to illustration 12.6

1   The hydraulic lifters are self-adjusting, and require no attention while in service.

2   If the hydraulic lifters become excessively noisy, their operation

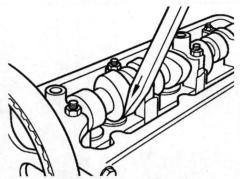

12.6 Press down on the lifter, until it contacts the top of the valve stem

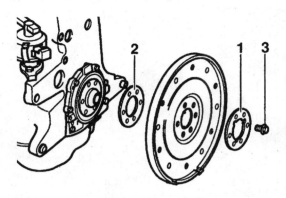

13.6 Driveplate components

1	*Plate*	*3*	*Mounting bolt*
2	*Shim*		

can be checked as described below.

3   Run the engine until it reaches its normal operating temperature. Switch off the engine, then refer to Section 7 and remove the camshaft cover.

4   Rotate the camshaft by turning the crankshaft with a socket and wrench, until the first cam lobe over No. 1 cylinder is pointing upwards.

5   Using a feeler gauge, measure the clearance between the base of the cam lobe and the top of the tappet. If the clearance is greater than 0.003-inch, then the tappet is defective and must be replaced.

6   If the clearance is less than 0.003-inch, press down on the top of the tappet, until it is felt to contact the top of the valve stem **(see illustration)**. Use a wooden or plastic implement that will not damage the surface of the tappet.

7   If the tappet travels more than 0.003-inch before making contact, then it is defective and must be replaced.

8   Hydraulic tappet removal and installation is described as part of the cylinder head overhaul sequence - see Chapter 2C for details.

13 Flywheel/driveplate - removal, inspection and installation

General information

Manual transaxle models

1   The mounting arrangement of the flywheel and clutch components depends on the type of transaxle installed.

2   The clutch pressure plate is bolted directly to the end of the crankshaft. The flywheel is then bolted to the pressure plate. Removal of these components is therefore described in Chapter 8. Inspect the flywheel as described below.

Automatic transaxle models

3   The torque converter driveplate is bolted directly to the end of the crankshaft; removal is as described below. Removal of the automatic transaxle and torque converter is described in Chapter 7B.

Driveplate

Removal

Refer to illustration 13.6

4   Remove the transaxle as described in Chapter 7B.

5   Lock the driveplate in position by bolting a piece of scrap metal between the driveplate and one of the transaxle bellhousing mounting holes. Mark the position of the driveplate with respect to the crankshaft using a dab of paint.

6   Loosen and withdraw the driveplate mounting bolts, then lift off the driveplate. Recover the packing plate and the shim (where applicable) **(see illustration)**.

Installation

7   Installation is a reversal of removal, using the alignment marks

made during removal. Install new mounting bolts and tighten them to the specified torque. Remove the locking tool, and install the transaxle as described in Chapter 7B.

Flywheel

Inspection

Note: *Removal of the flywheel is covered in Chapter 8.*

8   If the flywheel's clutch mating surface is deeply scored, cracked or otherwise damaged, the flywheel must be replaced. However, it may be possible to have it surface-ground; seek the advice of a VW dealer or engine rebuilding specialist.

9   If the ring gear is badly worn or has missing teeth, the flywheel must be replaced.

Installation

10   Flywheel installation is covered in Chapter 8.

14 Engine mounts - inspection and replacement

Inspection

1   If improved access is required, raise the front of the car and support it securely on axle stands.

2   Check the mounts to see if they are cracked, hardened or separated from the metal at any point; replace the mounting if any such damage or deterioration is evident.

3   Check that all fasteners are securely tightened.

4   Using a large screwdriver or a crowbar, check for wear in the mounting by carefully levering against it to check for free play. Where this is not possible, enlist the aid of an assistant to move the engine/transaxle back and forth, or from side to side, while you watch the mounting. While some freeplay is to be expected even from new components, excessive wear should be obvious. If excessive free play is found, check first that the fasteners are correctly secured, then replace any worn components as described below.

Replacement

Front engine mount

5   Disconnect the battery negative cable. **Caution:** *If the stereo in your vehicle is equipped with an anti-theft system, make sure you have the correct activation code before disconnecting the battery.*

6   Position a floor jack underneath the engine and position it such that the jack head is directly underneath the engine/bellhousing mating surface.

7   Raise the jack until it just takes the weight of the engine off the front engine mount.

8   Unscrew and withdraw the through-bolt.

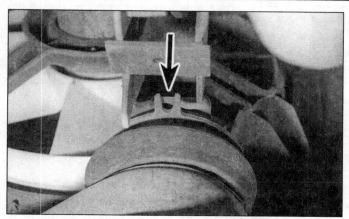

14.13 Lug (arrow) on top of the mounting block engages in the recess in the bracket

14.25 Removing the rear left-hand engine mount block

9   Remove the starter motor.
10   Loosen and withdraw the engine mount-to-transaxle bellhousing bolts and remove the bracket.
11   Working under the engine mount front crossmember, remove the engine mount block retaining screw.
12   Lift the engine mount block out of the crossmember cup.
13   Installation is a reversal of removal, noting the following points:

a) Ensure that the orientation lug that protrudes from the top of surface of the engine mounting block engages with the recess in the mounting bracket (see illustration).
b) Tighten all bolts securely.

Rear right-hand engine mount

14   Disconnect the battery negative cable. **Caution:** If the stereo in your vehicle is equipped with an anti-theft system, make sure you have the correct activation code before disconnecting the battery.
15   Mount an engine lifting beam across the engine bay, and attach the jib to the engine lifting eyes on the cylinder head. Alternatively, an engine hoist can be used. Raise the hoist/lifting beam jib to take the weight of the engine off the engine mount.
16   Loosen and withdraw the engine mount through-bolt.
17   Unbolt the engine mounting bracket from the cylinder block.
18   Unbolt the engine mounting block from the body, and remove it from the engine bay.
19   Installation is a reversal of removal, noting the following points:

a) Ensure that the orientation lug that protrudes from the top of surface of the engine mounting block engages with the recess in the mounting bracket.
b) Tighten all bolts securely.

Rear left-hand mount

Refer to illustration 14.25
20   Disconnect the battery negative cable. **Caution:** If the stereo in your vehicle is equipped with an anti-theft system, make sure you have the correct activation code before disconnecting the battery.
21   Position a floor jack below the engine, with the jack head directly underneath the engine/bellhousing mating surface.
22   Raise the jack until it just takes the weight of the engine off the rear right-hand engine mounting.
23   Loosen and withdraw the engine mounting through-bolt.
24   Unbolt the engine mounting bracket from the end of the transaxle casing.
25   Unbolt the engine mounting block from the body, and remove it from the engine bay (see illustration).
26   Installation is a reversal of removal, noting the following points:

a) Ensure that the orientation lug that protrudes from the top of surface of the engine mounting block engages with the recess in the mounting bracket.
b) Tighten all bolts securely.

15 Oil pan - removal and installation

Refer to illustration 15.5

Removal

1   Disconnect the battery negative cable. **Caution:** If the stereo in your vehicle is equipped with an anti-theft system, make sure you have the correct activation code before disconnecting the battery. Refer to Chapter 1 and drain the engine oil. Where applicable, remove the screws and lower the engine undertray away from the vehicle.
2   Park the vehicle on a level surface, apply the parking brake and chock the rear wheels.
3   Raise the front of the vehicle and support it securely on axle stands.
4   To improve access to the oil pan, refer to Chapter 8 and disconnect the right-hand driveaxle from the transaxle output flange.
5   Working around the outside of the oil pan, progressively loosen and withdraw the oil pan retaining bolts (see illustration). Where applicable, unbolt and remove the flywheel cover plate from the transaxle to gain access to the left-hand oil pan fixings.
6   Break the joint by striking the oil pan with the palm of your hand, then lower the oil pan and withdraw it from underneath the vehicle. Recover and discard the oil pan gasket. Where a baffle plate is installed, note that it can only be removed once the oil pump has been unbolted (see Section 16).
7   While the oil pan is removed, take the opportunity to check the oil pump pick-up/strainer for signs of clogging or disintegration. If necessary, remove the pump as described in Section 16, and clean or replace the strainer.

15.5 Removing the oil pan bolts (engine removed and inverted for clarity)

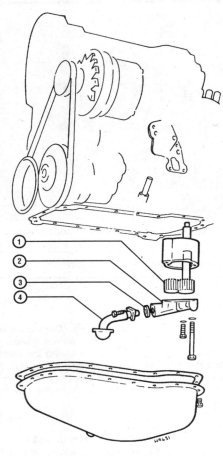

16.3 Oil pump components

1	*Oil pump gears*	*3*	*O-ring seal*
2	*Oil pump cover*	*4*	*Pickup tube*

Installation

8   Clean all traces of sealant from the mating surfaces of the cylinder block/crankcase and oil pan, then use a clean rag to wipe out the oil pan.

9   Ensure that the oil pan and cylinder block/crankcase mating surfaces are clean and dry, then apply a coating of suitable sealant to the oil pan and crankcase mating surfaces.

10   Lay a new oil pan gasket in position on the oil pan mating surface, then install the oil pan and install the retaining bolts. Tighten the nuts and bolts evenly and progressively to the specified torque.

11   Install the driveaxle and engine bay undertray.

12   Refer to Chapter 1 and refill the engine with the specified grade and quantity of oil.

13   Reconnect the negative battery cable.

16   Oil pump and pickup - removal, inspection and installation

Refer to illustrations 16.3, 16.7 and 16.8

1   The oil pump and pickup are both mounted in the oil pan. Drive is taken from the intermediate shaft, which rotates at half crankshaft speed.

Removal

2   Refer to Section 15 and remove the oil pan from the crankcase.

3   Loosen and remove the bolts securing the oil pump to the base of

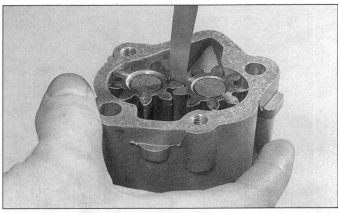

16.7 Checking the oil pump backlash

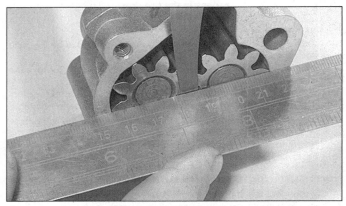

16.8 Checking the oil pump axial clearance

the crankcase **(see illustration)**.

4   Lower the oil pump and pickup away from the crankcase. Where applicable, recover the baffle plate.

Inspection

5   Remove the screws from the mating flange, and lift off the pickup tube. Recover the O-ring seal. Loosen and withdraw the screws, then remove the oil pump cover.

6   Clean the pump thoroughly, and inspect the gear teeth for signs of damage or wear.

7   Check the pump backlash by inserting a feeler gauge between the meshed gear teeth; rotate the gears against each other slightly, to give the maximum clearance **(see illustration)**. Compare the measurement with the limit quoted in Specifications.

8   Check the pump axial clearance as follows. Lay a straightedge across the oil pump casing, then using a feeler gauge, measure the clearance between the straightedge and the pump gears **(see illustration)**. Compare the measurement with the limit listed in the Specifications.

9   If either measurement is outside of the specified limit, this indicates that the pump is worn and must be replaced.

Installation

10   Install the oil pump cover, then install and tighten the screws to the specified torque.

12   Reassemble the oil pickup to the oil pump, using a new O-ring seal. Tighten the retaining screws to the specified torque.

13   Where applicable, install the crankcase baffle plate.

14   Install the oil pump to the crankcase, then install the mounting bolts and tighten them to the specified torque.

15   Refer to Section 15 and install the oil pan.

Chapter 2 Part B
Diesel engine

Contents

Specifications

General

Engine code*
  1896cc, mechanical fuel injection, turbocharged, 75 h.p. .... AAZ

*Note: See "Buying Spare Parts and Vehicle Identification" for the location of the code marking on the engine.

Bore	3.13 inches
Stroke	3.76 inches
Compression ratio	22.5:1
Compression pressures (wear limit)	377 psi
Firing order	1 - 3 - 4 - 2
Cylinder No. 1 location	Timing belt end
Timing belt tension (measured using Volkswagen tool VW 210)	Scale reading of 12 to 13 units

Lubrication system

Oil pump type	Oil pan-mounted, driven indirectly from intermediate shaft
Normal operating oil pressure	29 psi minimum (at 2000 rpm, oil temperature 176°F)
Oil pump backlash	0.0079 inch (wear limit)
Oil pump axial clearance	0.0059 inch (wear limit)

Torque specifications

	Ft-lbs
Alternator mounting bolts	18
Camshaft cover screws	7
Camshaft bearing cap nuts	15
Camshaft sprocket bolt	33
Crankshaft drivebelt pulley screws	18
Crankshaft front oil seal housing bolts	18
Crankshaft rear oil seal housing bolts	7
Crankshaft sprocket bolt	
Stage 1	66
Stage 2	Angle-tighten a further 90°
Cylinder head bolts:	
Stage 1	430
Stage 2	44
Stage 3	Angle-tighten a further 90°
Stage 4	Angle-tighten a further 90°
Engine mountings	
Through-bolts	37
Front mounting block bolt	37
Front mounting bracket bolts	41
Left rear mounting bracket bolts	18
Rear mounting block-to-body bolts	18
Right rear mounting bracket bolts	18
Exhaust manifold nuts	18
Intake manifold bolts	18
Intermediate shaft sprocket bolt	33
Oil pump cover screws	7
Oil pump mounting bolts	18
Oil pump pickup tube screws	7
Power steering pump mounting bolts	18
Oil pan retaining bolts	15
Timing belt tensioner locknut	15

---

1   General information

Using this Chapter

Chapter 2 is divided into three parts; A, B and C. Repair operations that can be carried out with the engine in the vehicle are described in Parts A (gasoline engines) and B (diesel engines). Part C covers the removal of the engine/transaxle as a unit and describes the engine disassembly and overhaul procedures.

In Parts A and B, the assumption is made that the engine is installed in the vehicle, with all ancillaries connected. If the engine has been removed for overhaul, the preliminary disassembly information which precedes each operation may be ignored.

Access to the engine bay can be improved by removing the hood and the front lock carrier assembly; these procedures are described in Chapter 11 and Chapter 2C respectively.

Engine description

The diesel engine is a water-cooled, single overhead camshaft, in-line four cylinder unit with cast-iron cylinder blocks and aluminum-alloy cylinder head. It's mounted transversely at the front of the vehicle, with the transaxle bolted to the left-hand side of the engine.

The cylinder head carries the camshaft, which are driven by a toothed timing belt. It also houses the intake and exhaust valves, which are closed by single or double coil springs, and which run in guides pressed into the cylinder head. The camshaft actuates the valves directly via hydraulic lifters, mounted in the cylinder head. The cylinder head contains integral oilways which supply and lubricate the lifters.

The crankshaft is supported by five main bearings, and endplay is controlled by a thrust bearing installed between cylinders Nos. 2 and 3.

The diesel engine is fitted with a timing belt-driven intermediate shaft, which provides drive for the brake booster vacuum pump and the oil pump.

Engine coolant is circulated by a pump, driven by the auxiliary drivebelt. For details of the cooling system, refer to Chapter 3.

Lubricant is circulated under pressure by a pump, driven either by the crankshaft or by the intermediate shaft, depending on engine type. Oil is drawn from the oil pan through a strainer, and then forced through an externally-mounted, replaceable screw-on filter. From there, it is distributed to the cylinder head, where it lubricates the camshaft journals and hydraulic lifters, and also to the crankcase, where it lubricates the main bearings, connecting rod big- and small-ends, wrist pins and cylinder bores. Oil jets are fitted to the base of each cylinder - these spray oil onto the underside of the pistons, to improve cooling. An oil cooler, supplied with engine coolant, reduces the temperature of the oil before it re-enters the engine.

Repairs possible with the engine installed in the vehicle

The following operations can be performed without removing the engine:

a) Auxiliary drivebelts - removal and installation.
b) Camshaft - removal and installation.*
c) Camshaft oil seal - replacement.
d) Camshaft sprocket - removal and installation.
e) Coolant pump - removal and installation (refer to Chapter 3).
f) Crankshaft oil seals - replacement.
g) Crankshaft sprocket - removal and installation.
h) Cylinder head - removal and installation.*
i) Engine mountings - inspection and replacement.
j) Intermediate shaft oil seal - replacement.
k) Oil pump and pickup assembly - removal and installation.
l) Oil pan - removal and installation.
m) Timing belt, sprockets and cover - removal, inspection and installation.

*Cylinder head disassembly procedures are in Chapter 2C, and also contain details of camshaft and hydraulic tappet removal.

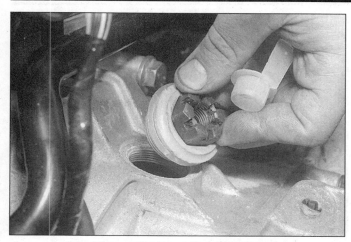

2.2a Remove the inspection cover from the transaxle bellhousing

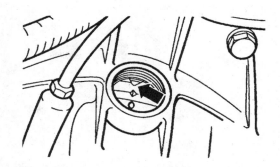

2.2b Timing mark on the edge of the flywheel (arrow) lined up with pointer on bellhousing casting

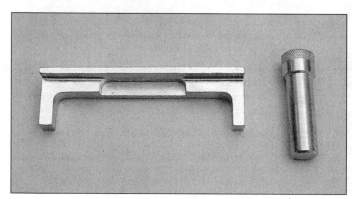

2.3 Engine locking tools

Note: *It is possible to remove the pistons and connecting rods (after removing the cylinder head and oil pan) without removing the engine from the vehicle. However, this procedure is not recommended. Work of this nature is more easily and thoroughly completed with the engine on the bench - refer to Chapter 2C.*

2   Top Dead Center (TDC) for number one piston - locating

Refer to illustrations 2.2a, 2.2b, 2.3, 2.4, 2.6 and 2.7

1   Remove camshaft cover, auxiliary drivebelts and timing belt outer covers as described in Sections 7, 6 and 4 respectively.

2   Remove the inspection cover from the transaxle bellhousing. Rotate the crankshaft clockwise with a wrench and socket, or a wrench, until the timing mark machined onto the edge of the flywheel lines up with pointer on the bellhousing casting **(see illustrations)**.

3   To lock the engine in the TDC position, the camshaft (not the sprocket) and fuel injection pump sprocket must be secured in a reference position, using special locking tools. Improvised tools may be fabricated, but due to the exact measurements and machining involved, it is strongly recommended that a kit of locking tools is either borrowed or rented from a VW dealer, or purchased from a reputable tool manufacturer **(see illustration)**.

4   Engage the edge of the locking bar with the slot in the end of the camshaft **(see illustration)**.

5   With the locking bar still inserted, turn the camshaft slightly (by

turning the crankshaft clockwise, as before), so that the locking bar rocks to one side, allowing one end of the bar to contact the cylinder head surface. At the other side of the locking bar, measure the gap between the end of the bar and the cylinder head using a feeler gauge.

6   Turn the camshaft back slightly, then pull out the feeler gauge. The idea now is to level the locking bar by inserting two feeler gauges, each with a thickness equal to half the originally measured gap, on either side of the camshaft between each end of the locking bar and the cylinder head. This centers the camshaft, and sets the valve timing in reference condition **(see illustration)**.

7   Insert the locking pin through the fuel injection pump sprocket alignment hole, and thread it into the support bracket behind the sprocket. This locks the fuel injection pump in a reference condition **(see illustration)**.

8   The engine is now set to TDC on No. 1 cylinder.

2.4 Engage the locking bar with the slot in the camshaft

2.6 Camshaft centered and locked using locking bar and feeler gauges

2.7 Injection pump sprocket locked using locking pin (arrow)

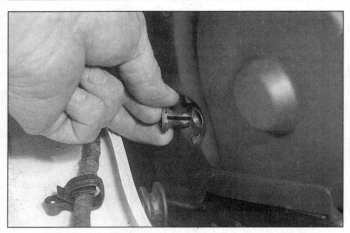

4.6 Removing the press-stud fixings from the timing belt upper cover

4.9 Removing the crankshaft drivebelt pulleys

3   Cylinder compression check

Compression test

Note: *A compression tester specifically designed for diesel engines must be used for this test.*

1   When engine performance is down, or if misfiring occurs, a compression test can provide diagnostic clues as to the engine's condition. If the test is performed regularly, it can give warning of trouble before any other symptoms become apparent.

2   A compression tester specifically intended for diesel engines must be used, because of the higher pressures involved. The tester is connected to an adapter which screws into the glow plug or injector hole. It is unlikely to be worthwhile buying such a tester for occasional use, but it may be possible to borrow or hire one - if not, have the test performed by a garage.

3   Unless specific instructions to the contrary are supplied with the tester, observe the following points:

   a) *The battery must be in a good state of charge, the air filter must be clean, and the engine should be at normal operating temperature.*

   b) *All the injectors or glow plugs should be removed before starting the test. If removing the injectors, also remove the flame shield washers, otherwise they may be blown out.*

   c) *The stop solenoid must be disconnected, to prevent the engine from running or fuel from being discharged.*

4   There is no need to hold the accelerator pedal down during the test, because the diesel engine air intake is not throttled.

5   VW specify wear limits for compression pressures - refer to the Specifications. Seek the advice of a VW dealer or other diesel specialist if in doubt as to whether a particular pressure reading is acceptable.

6   The cause of poor compression is less easy to establish on a diesel engine than on a gasoline one. The effect of introducing oil into the cylinders ("wet" testing) is not conclusive, because there is a risk that the oil will sit in the swirl chamber or in the recess on the piston crown, instead of passing to the rings. However, the following can be used as a rough guide to diagnosis.

7   All cylinders should produce very similar pressures; a difference of more than 73 psi between any two cylinders indicates the existence of a fault. Note that the compression should build up quickly in a healthy engine; low compression on the first stroke, followed by gradually-increasing pressure on successive strokes, indicates worn piston rings. A low compression reading on the first stroke, which does not build up during successive strokes, indicates leaking valves or a blown head gasket (a cracked head could also be the cause).

8   A low reading from two adjacent cylinders is almost certainly due to the head gasket having blown between them; the presence of coolant in the engine oil will confirm this.

9   If the compression reading is unusually high, the cylinder head surfaces, valves and pistons are probably coated with carbon deposits. If this is the case, the cylinder head should be removed and decarbonized (refer to Part C of this Chapter).

Leakdown test

10   A leakdown test measures the rate at which compressed air fed into the cylinder is lost. It is an alternative to a compression test, and in many ways it is better, since the escaping air provides easy identification of where pressure loss is occurring (piston rings, valves or head gasket).

11   The equipment needed for leakdown testing is unlikely to be available to the home mechanic. If poor compression is suspected, have the test performed by a suitably-equipped garage.

4   Timing belt and outer covers - removal and installation

General information

1   The primary function of the toothed timing belt is to drive the camshaft, but it is also used to drive the fuel injection pump and intermediate shaft. Should the belt slip or break in service, the valve timing will be disturbed and piston-to-valve contact may occur, resulting in serious engine damage.

2   For this reason, it is important that the timing belt is tensioned correctly, and inspected regularly for signs of wear or deterioration.

3   Note that the removal of the *inner* section of the timing belt cover is described as part of the cylinder head removal procedure; see Section 11 later in this Chapter.

Removal

Refer to illustrations 4.6 and 4.9

4   Before starting work, immobilize the engine by disconnecting the fuel cut-off solenoid cable (see Chapter 4C). Prevent any vehicle movement by applying the parking brake and blocking the rear wheels.

5   Access to the timing belt covers can be improved by removing the air cleaner housing - refer to Chapter 4C.

6   Release the uppermost part of the timing belt outer cover by prying open the metal spring clips and where applicable, removing the press-stud fixings **(see illustration)**. Lift the cover away from the engine.

7   With reference to Section 6, remove the auxiliary drivebelt(s). Loosen and withdraw the screws, and lift off the coolant pump pulley.

8   Refer to Section 2, and using the engine alignment markings, set the engine to TDC on No. 1 cylinder.

9   Loosen and withdraw the retaining screws, then remove the pul-

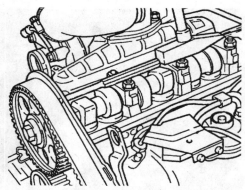

4.19 Releasing the camshaft sprocket from the taper using a pin punch

ley for the ribbed drivebelt (together with the V-belt pulley, where installed) from the crankshaft sprocket **(see illustration)**. On completion, check that the engine is still set to TDC. Note: *To prevent the drivebelt pulley from rotating while the mounting bolts are being loosened, grip the sprocket by wrapping a length of old rubber hose or inner tube around it.*

10 Remove the retaining screws and clips, and lift off the timing belt lower cover.

11 Ensure that the fuel injection pump sprocket, sprocket locking pin is firmly in position (see Section 2). **Caution:** *Do not loosen the sprocket center bolt, as this will alter the fuel injection pump's basic timing setting.*

12 With reference to Section 5, relieve the tension on the timing belt by slackening the tensioner mounting nut slightly, allowing it to pivot away from the belt.

13 Examine the timing belt for manufacturer's markings that indicate the direction of rotation. If none are present, make your own using typist's correction fluid or a dab of paint - do not cut or score the belt in any way. **Caution:** *If the belt appears to be in good condition and can be re-used, it is essential that it is reinstalled the same way around, otherwise accelerated wear will result, leading to premature failure.*

14 Slide the belt off the sprockets, taking care to avoid twisting or kinking it excessively.

15 Examine the belt for evidence of contamination by coolant or lubricant. If this is the case, find the source of the contamination before progressing any further. Check the belt for signs of wear or damage, particularly around the leading edges of the belt teeth.

16 Replace the belt if its condition is in doubt; the cost of belt replacement is negligible compared with potential cost of the engine

repairs, should the belt fail in service. Similarly, if the belt is known to have covered more than 36 000 miles, it is prudent to replace it regardless of condition, as a precautionary measure.

17 If the timing belt is not going to be reinstalled for some time, it is a wise precaution to hang a warning label on the steering wheel, to remind yourself (and others) not to start the engine.

Installation

Refer to illustrations 4.19, 4.25a and 4.25b

18 Ensure that the crankshaft is still set to TDC on No. 1 cylinder, as described in Section 2.

19 Refer to Section 5 and loosen the camshaft sprocket bolt by half a turn. Release the sprocket from the camshaft taper mounting by carefully tapping it with a pin punch, inserted through the hole provided in the timing belt inner cover **(see illustration)**.

20 Loop the timing belt loosely under the crankshaft sprocket. **Caution:** *Observe the direction of rotation markings on the belt.*

21 Engage the timing belt teeth with the crankshaft sprocket, then maneuver it into position over the camshaft and injection pump sprockets. Ensure the belt teeth seat correctly on the sprockets. **Note:** *Slight adjustments to the position of the camshaft sprocket may be necessary to achieve this.*

22 Pass the flat side of the belt over the intermediate shaft pulley and tensioner roller - avoid bending the belt back on itself or twisting it excessively as you do this.

23 Remove the locking pin from the fuel injection pump sprocket (see Section 2).

24 Ensure that the 'front run' of the belt is taut - i.e. all the slack should be in the section of the belt that passes over the tensioner roller.

25 Tension the belt by turning the eccentrically-mounted tensioner clockwise; two holes are provided in the side of the tensioner hub for this purpose - a pair of sturdy right-angled snap-ring pliers is a suitable substitute for the correct VW tool. **(see illustrations)**.

26 Test the timing belt tension by grasping it between the fingers at a point mid-way between the intermediate shaft and camshaft sprockets, and twisting it. The belt tension is correct when the belt can just be twisted through 90° (quarter of a turn) and no further.

27 When the correct belt tension has been achieved, tighten the tensioner locknut to the specified torque.

28 The belt tension must be accurately checked, and if necessary adjusted - this involves the use of dedicated belt tension measuring device (Volkswagen tool VW 210), and it is advisable to have this operation carried out by a VW dealer.

29 At this point, check the crankshaft is still set to TDC on No. 1 cylinder (see Section 2).

30 Refer to Section 5 and tighten the camshaft sprocket bolt to the specified torque.

2B

4.25a Tensioning the timing belt using a pair of snap-ring pliers in the belt tensioner

4.25b Timing belt correctly installed

5.4 Remove the tensioner nut and recover the washer

5.5 Slide the tensioner off its mounting stud

31   Remove the sprocket locking pin.

32   With reference to Section 2, remove the camshaft locking bar.

33   Using a wrench or wrench and socket on the crankshaft pulley center bolt, rotate the crankshaft through two complete revolutions. Reset the engine to TDC on No. 1 cylinder, with reference to Section 2 and check that the fuel injection pump sprocket locking pin can be inserted. Re-check the timing belt tension and adjust it, if necessary.

34   Install the upper and lower sections of the timing belt outer cover, tightening the retaining screws securely.

35   Where applicable, install the coolant pump pulley and tighten the retaining screws to the specified torque.

36   Install the crankshaft drivebelt pulley and tighten the retaining screws to the specified torque, using the method employed during removal. Note that the offset of the pulley mounting holes allows only one fitting position.

37   Working from Section 6, install and tension the auxiliary drive-belt(s).

38   Restore the fueling system by reconnecting the fuel cut-off sole-noid wiring (see Chapter 4C).

39   On completion, refer to Chapter 4C and check the fuel injection pump timing.

---

5   Timing belt tensioner and sprockets - removal and installation

---

1   Before starting work, disable the fuel system by disconnecting the wiring from the fuel cut-off solenoid (see Chapter 4C). Prevent any vehicle movement by applying the parking brake and blocking the rear wheels.

2   To gain access to the components detailed in this Section, refer to Section 6 and remove the auxiliary drivebelt(s).

Timing belt tensioner

Removal

Refer to illustrations 5.4 and 5.5

3   With reference to the relevant paragraphs of Sections 2 and 4, set the engine to TDC on No. 1 cylinder, then remove the upper and lower sections of the timing belt outer cover.

4   Loosen the retaining nut at the hub of the tensioner pulley and allow the assembly to rotate counterclockwise, relieving the tension on the timing belt. Remove the nut and recover the washer **(see illustration)**.

5   Slide the tensioner off its mounting stud **(see illustration)**.

6   Wipe the tensioner clean, but do not use solvents that may conta-minate the bearings. Spin the tensioner pulley on its hub by hand. Stiff movement or excessive freeplay is an indication of severe wear; the tensioner is not a serviceable component, and should be replaced.

Installation

7   Slide the tensioner pulley over the mounting stud.

8   Install the tensioner washer and retaining nut - do not fully tighten the nut at this stage.

9   With reference to Section 4, install and tension the timing belt.

10   Restore the fuel system by reconnecting the fuel cut-off solenoid wiring.

11   Refer to Section 4 and install the timing belt covers.

Camshaft timing belt sprocket

Removal

Refer to illustrations 5.14 and 5.16

12   Refer to Section 2 and 4, set the engine to TDC on No. 1 cylinder, then remove the timing belt outer covers.

13   With reference to the previous sub-Section, loosen the tensioner center nut and allow it to rotate counterclockwise, to relieve the ten-sion on the timing belt. Carefully slide the timing belt off the camshaft sprocket.

14   The camshaft sprocket must be held stationary while its retaining

5.14 To make a camshaft sprocket holding tool, obtain two lengths of steel strip about 1/4-inch thick by 1-inch wide or similar, one 24-inches long, the other 8-inches long (all dimensions approximate). Bolt the two strips together to form a forked end, leaving the bolt slack so that the shorter strip can pivot freely. At the end of each 'prong' of the fork, secure a bolt with a nut and a locknut, to act as the fulcrums; these will engage with the cut-outs in the sprocket, and should protrude by about 1-1/4 inch

5.16 Removing the camshaft sprocket

5.28a Insert the crankshaft sprocket bolt . . .

5.28b . . . tighten it to the Stage 1 torque . . .

2B

bolt is slackened; if access to the correct VW special tool is not possible, a simple home-made tool using basic materials may be fabricated **(see illustration)**.

15 Using the home-made tool, brace the camshaft sprocket and loosen and remove the retaining bolt; recover the washer where fitted.

16 Slide the camshaft sprocket from the end of the camshaft **(see illustration)**. Where applicable, recover the Woodruff key from the keyway.

17 With the sprocket removed, examine the camshaft oil seal for signs of leaking. If necessary, refer to Section 8 and replace it.

18 Wipe the sprocket and camshaft mating surfaces clean.

Installation

19 Where applicable, install the Woodruff key into the keyway with the plain surface facing upwards. Install the sprocket to the camshaft, engaging the slot in the sprocket with the Woodruff key. Where a key is not used, ensure the lug in the sprocket hub engages with recess in the end of the camshaft.

20 Working from Sections 2 and 4, check that the engine is still set to TDC on No. 1 cylinder, then install and tension the timing belt. Install the timing belt covers.

21 Install the crankshaft drivebelt pulley(s), then insert the retaining screws and tighten them to the specified torque.

22 With reference to Section 6, install and tension the auxiliary drivebelt(s).

Crankshaft timing belt sprocket

Removal

23 Refer to Section 2 and 4, set the engine to TDC on No. 1 cylinder, then remove the timing belt outer covers. With reference to the previous sub-Section, loosen the tensioner center hut and allow it to rotate counterclockwise, to relieve the tension on the timing belt. Carefully slide the timing belt off the camshaft sprocket.

24 The crankshaft sprocket must be held stationary while its retaining bolt is slackened. If access to the correct VW flywheel locking tool is not available, lock the crankshaft in position by removing the starter motor, as described in Chapter 5A, to expose the flywheel ring gear. Get an assistant insert a stout lever between the ring gear teeth and the transaxle bellhousing while the sprocket retaining bolt is slackened.

25 Withdraw the bolt, recover the washer and lift off the sprocket.

26 With the sprocket removed, examine the crankshaft oil seal for signs of leaking. If necessary, refer to Section 10 and replace it.

27 Wipe the sprocket and crankshaft mating surfaces clean.

5.28c . . . then through the Stage 2 angle

Installation

Refer to illustrations 5.28a, 5.28b, 5.28c and 5.34

28 Refit the sprocket to the crankshaft, engaging the lug on the inside of the sprocket with the recess in the end of the crankshaft. Insert the retaining bolt and tighten it to the specified torque **(see illustrations)**.

29 Working from Sections 2 and 4, check that the engine is still set to TDC on No. 1 cylinder, then install and tension the timing belt. Install the timing belt covers.

30 Install the crankshaft drivebelt pulley(s).

31 With reference to Section 6, install and tension the auxiliary drivebelt(s).

Intermediate shaft sprocket

Removal

32 With reference to Sections 2 and 4, remove the timing belt covers and set the engine to TDC on No. 1 cylinder. Loosen the tensioner center nut and rotate it counterclockwise to relieve the tension on the timing belt. Carefully slide the timing belt off the camshaft sprocket.

33 The intermediate shaft sprocket must be held stationary while its retaining bolt is slackened; if access to the correct VW special tool is not possible, a simple home-made tool using basic materials may be fabricated as described in the camshaft sprocket removal sub-Section.

5.34 Brace the intermediate shaft sprocket, then remove the retaining bolt

6.8 Removing the auxiliary V-belt

34   Using a socket and extension bar, brace the intermediate shaft sprocket. Loosen and remove the retaining bolt; recover the washer, where fitted **(see illustration)**.

35   Slide the sprocket from the end of the intermediate shaft. Where applicable, recover the Woodruff key from the keyway.

36   With the sprocket removed, examine the intermediate shaft oil seal for signs of leaking. If necessary, refer to Section 9 and replace it.

37   Wipe the sprocket and shaft mating surfaces clean.

Installation

38   Where applicable, install the Woodruff key into the keyway with the plain surface facing upwards. Refit the sprocket to the intermediate shaft, engaging the slot in the sprocket with the Woodruff key.

39   Tighten the sprocket retaining bolt to the specified torque; hold the sprocket using the method employed during removal.

40   With reference to Section 2, check that the engine is still set to TDC on No. 1 cylinder. Working from Section 4, install and tension the timing belt, then install the timing belt covers.

41   Install the crankshaft drivebelt pulley(s), then insert the retaining screws and tighten them to the specified torque.

42   With reference to Section 6, install and tension the auxiliary drivebelt(s).

Fuel injection pump sprocket

43   Refer to Chapter 4C.

---

6   Drivebelts - removal and installation

General information

1   Depending on the vehicle specification and engine type, one or two auxiliary drivebelts may be installed. Both are driven from pulleys mounted on the crankshaft, and provide drive for the alternator, coolant pump, power steering pump and on vehicles with air conditioning, the refrigerant compressor.

2   The run of the belts and the components they drive are also dependent on vehicle specification and engine type. Because of this, the coolant pump and power steering pump may have pulleys to suit either a ribbed belt or a V-belt.

3   The ribbed drivebelt may have an automatic tensioner, depending on its run (and hence the number of components it is driving). Otherwise, the belt is tensioned by the alternator mountings, which have an in-built tensioning spring. The V-belt is tensioned by pivoting the power steering pump on its mounting.

4   On installation, the drivebelt must be tensioned correctly to ensure correct operation and prolonged service life.

Auxiliary V-belt

Refer to illustration 6.8

Removal

5   Park the vehicle on a level surface and apply the parking brake. Jack up the front of the vehicle and rest it on axle stands - refer to *"Jacking and Vehicle Support"*. Disable the starting system by unplugging the starter solenoid at the connector; see Chapter 5A.

6   Turn the steering to full right lock, then refer to Chapter 11 and remove the plastic air ducting from below the right-hand front fender.

7   With reference to Chapter 10, loosen the power steering pump mounting bolts and allow the pump body to pivot around its uppermost mounting towards the engine.

8   Guide the V-belt off the power steering pump pulley and where applicable, the coolant pump pulley **(see illustration)**.

9   Examine the belt for signs or wear or damage, and replace it if necessary.

Installation and tensioning

Refer to illustration 6.11

10   Install the belt by reversing the removal procedure. Ensure it seats evenly in the pulleys.

11   Set the belt tension by grasping the underside of the power steering pump and drawing it towards the front of the vehicle. The tension is correct when the midpoint of the belt's longest run can be deflected by

6.11 Tightening the steering pump bolt

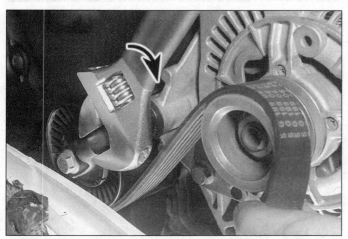

6.17 Rotate the tensioner roller arm clockwise - use an adjustable wrench - and remove the belt

7.2 Crankcase breather regulator valve

no more than 1/4-inch. Tighten the power steering pump bolt to the correct torque **(see illustration)**

12   Rotate the crankshaft in its normal direction of rotation through two turns, then re-check and if necessary adjust the tension.

Auxiliary ribbed belt

Removal

Refer to illustration 6.17

13   Park the vehicle on a level surface and apply the parking brake. Jack up the front of the vehicle and rest it securely on axle stands - refer to *"Jacking and Vehicle Support"*. Disable the starting system by unplugging the starter solenoid at the connector; refer to Chapter 5A.

14   Turn the steering to full right lock, then refer to Chapter 11 and remove the plastic air ducting from underneath the right-hand front fender.

15   Where applicable, remove the auxiliary V-belt as described in the previous sub-Section.

16   Examine the ribbed belt for manufacturer's markings, indicating the direction of rotation. If none are present, make some using typist's correction fluid or a dab of paint - do not cut or score the belt in any way.

Vehicles with a roller-arm automatic tensioning device

17   Rotate the tensioner roller arm clockwise against its spring tension so that the roller is forced away from the belt - use an adjustable wrench as a lever **(see illustration)**.

Vehicles with rotary automatic tensioning device

18   Install an end-wrench to the tensioner center nut and rotate the assembly counterclockwise, against its spring tension.

Vehicles without an automatic tensioning device

19   Loosen the alternator upper and lower mounting bolts by between one and two turns.

20   Push the alternator down to its stop against the spring tension, so that it rotates around its uppermost mounting.

All vehicles

21   Pull the belt off the alternator pulley, then release it from the remaining pulleys.

Installation and tensioning

Caution: *Observe the manufacturer's direction of rotation markings on the belt, when installing.*

22   Pass the ribbed belt underneath the crankshaft pulley, ensuring that the ribs seat securely in the channels on the surface of the pulley.

Vehicles with roller-arm automatic tensioning device

23   Rotate the tensioner roller arm clockwise against its spring tension - use an adjustable wrench as a lever **(see to illustration 6.17)**.

24   Pass the belt around the coolant pump pulley or air conditioning refrigerant pump pulley (as applicable), then install it over the alternator pulley.

25   Release the tensioner pulley arm and allow the roller to bear against the flat surface of the belt.

Vehicles with rotary automatic tensioning device

26   Install an end-wrench to the tensioner center nut and rotate the assembly counterclockwise, against its spring tension.

27   Pass the flat side of the belt underneath the tensioner roller, then place it over the power steering pump and alternator pulleys.

28   Release the wrench, and let the tensioner roller bear against the flat side of the belt.

Vehicles without an automatic tensioning device

29   Repeatedly push the alternator down to its stop against the spring tension, so that it rotates around its uppermost mounting and check that it moves back freely when released. If necessary, loosen the alternator mounting bolts by a further half a turn.

30   Keep the alternator pushed down against its stop, pass the belt over the alternator pulley, then release the alternator and allow it to tension the belt.

31   Restore the starting system, then start the engine and allow it to idle for approximately ten seconds.

32   Switch the engine off, then tighten first the lower, then the alternator upper mounting bolts to the specified torque.

All vehicles

33   Refer to Chapter 11 and install the plastic air ducts to the underside of the fender.

34   Where applicable, refer to the previous sub-Section and install the auxiliary V-belt.

35   Lower the vehicle to the ground, then (if not already carried out) restore the starting system with reference to Chapter 5A.

7   Camshaft cover - removal and installation

Removal

Refer to illustrations 7.2, 7.3, 7.4 and 7.5

1   Immobilize the engine by unplugging the electrical wiring from the fuel cut-off solenoid at the connector; refer to Chapter 4C for guidance.

2   Disconnect the crankcase breather hose and regulator valve from the camshaft cover **(see illustration)**.

7.3 Camshaft cover retaining nut

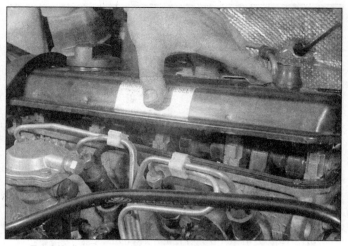

7.4 Lift the camshaft cover away from the cylinder head

3   Loosen and withdraw the three camshaft cover retaining nuts - recover the washers and seals **(see illustration)**.
4   Lift the cover away from the cylinder head **(see illustration)**; if it sticks, do not attempt to lever it off - instead free it by working around the cover and tapping it lightly with a soft-faced mallet.
5   Recover the camshaft cover gasket **(see illustration)**. Inspect the gasket carefully, and replace it if damage or deterioration is evident.
6   Clean the mating surfaces of the cylinder head and camshaft cover thoroughly, removing all traces of oil and old gasket - take care to avoid damaging the surfaces as you do this.

Installation

Refer to illustration 7.7

7   Install the camshaft cover by following the removal procedure in reverse, noting the following points:

a) *Ensure that the gasket is correctly seated on the cylinder head, and take care to avoid displacing it as the camshaft cover is lowered into position* **(see illustration)**.
b) *Tighten the camshaft cover retaining screws/nuts to the specified torque.*
c) *When refitting hoses that were originally secured with crimp-type clips, use standard worm-drive clips in their place on refitting.*

8   On completion, restore the fuel system by reconnecting the fuel cut-off solenoid lead.

8   Camshaft oil seal - replacement

Refer to illustration 8.5

1   Immobilize the engine by unplugging the electrical wiring from the fuel cut-off solenoid at the connector; refer to Chapter 4C for guidance.
2   Refer to Section 6 and remove the drivebelt(s).
3   With reference to Sections 2, 4 and 5 of this Chapter, remove the drivebelt pulleys and timing belt cover; set the engine to TDC on No. 1 cylinder and remove the timing belt, timing belt tensioner and camshaft sprocket.
4   After removing the retaining screws, lift the timing belt inner cover away from the engine block.
5   Working from the relevant Section of Chapter 2C, carry out the following:

a) *Unbolt the camshaft No. 1 bearing cap, and slide off the camshaft oil seal.*
b) *Lubricate the surface of a new camshaft oil seal with clean engine oil, and place it over the end of the camshaft.*
c) *Apply a suitable sealant to the mating surface of the bearing cap, then install it and tighten its mounting nuts progressively to the specified torque* **(see illustration)**.

6   Refer to Section 7 and install the camshaft cover.

7.5 Recover the camshaft cover gasket

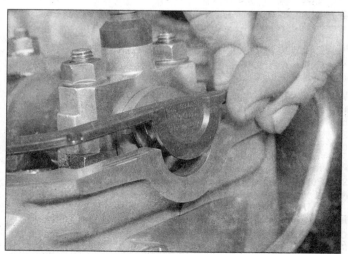

7.7 Ensure that the camshaft cover gasket is correctly seated on the cylinder head

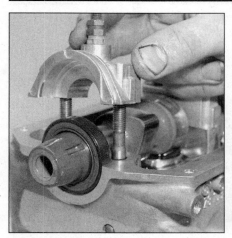

8.5 Installing the camshaft bearing cap

10.7 Removing the crankshaft front oil seal using self-tapping screws

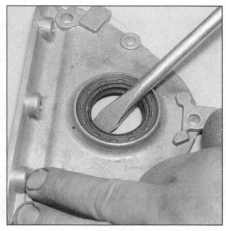

10.17 Pry the old oil seal from the housing

7   With reference to Sections 2, 4 and 5 of this Chapter, install the timing belt inner cover and timing sprockets, then install and tension the timing belt. On completion, install the timing belt outer cover.

8   With reference to Section 6, install and tension the drivebelt(s).

9   Intermediate shaft oil seal - replacement

1   Immobilize the engine by unplugging the electrical wiring from the fuel cut-off solenoid at the connector; refer to Chapter 4C for guidance.

2   Refer to Section 6 and remove the drivebelt(s).

3   With reference to Sections 4 and 5 of this Chapter, remove the drivebelt pulleys, timing belt outer cover, timing belt, tensioner and intermediate shaft sprocket.

4   After removing the retaining screws, lift the inner timing belt cover away from the engine block - this will expose the intermediate shaft sealing flange.

5   With reference to Section 7 of Chapter 2C, remove the intermediate shaft flange and replace the shaft and flange oil seals.

6   Refer to Sections 4 and 5 of this Chapter, carry out the following:

   a) Install the timing belt inner cover.
   b) Install the intermediate shaft timing belt sprocket.
   c) Install and tension the timing belt.
   d) Install the timing belt outer cover.

7   With reference to Section 6 of this Chapter, install and tension the drivebelt(s).

10   Crankshaft oil seals - replacement

Crankshaft front oil seal

Refer to illustration 10.7

1   Immobilize the engine by unplugging the electrical wiring from the fuel cut-off solenoid at the connector; refer to Chapter 4C for guidance.

2   Refer to Chapter 1 and drain the engine oil.

3   With reference to "*Jacking and Vehicle Support*", raise the front of the vehicle and rest it securely on axle stands.

4   Working from Chapter 11, remove the screws and detach the plastic air ducting from underneath the right-hand front fender.

5   Refer to Section 6 and remove the drivebelt(s).

6   With reference to Sections 4 and 5 of this Chapter, remove the drivebelt pulleys, timing belt outer covers, timing belt and crankshaft sprocket.

7   Drill two small holes into the existing oil seal, diagonally opposite

each other. Thread two self-tapping screws into the holes and using two pairs of pliers, pull on the heads of the screws to extract the oil seal (**see illustration**). Take great care to avoid drilling through into the seal housing or crankshaft sealing surface.

8   Clean out the seal housing and sealing surface of the crankshaft by wiping it with a lint-free cloth - avoid using solvents that may enter the crankcase and affect component lubrication. Remove any debris or burrs that could cause the seal to leak.

9   Smear the lip of the new oil seal with clean engine oil, and position it over the housing.

10   Using a hammer and a socket of suitable diameter, drive the seal squarely into its housing. **Note:** *Select a socket that bears only on the hard outer surface of the seal, not the inner lip, which can easily be damaged.*

11   With reference to Sections 2, 4 and 5 of this Chapter, install the crankshaft timing belt sprocket, then install and tension the timing belt. On completion, install the timing belt outer cover, and drivebelt pulley(s).

12   The remainder of the refitting procedure is a reversal of removal, as follows:

   a) With reference to Section 6, install and tension the drivebelt(s).
   b) Install the plastic air ducting to the underside of the fender - see Chapter 11.
   c) Refer to Chapter 1 and refill the engine with the correct grade and quantity of oil.
   d) Restore the fuel system.

Crankshaft front oil seal housing - gasket replacement

Refer to illustrations 10.17, 10.19 and 10.21

13   Proceed as described in paragraphs 1 to 6 above, then refer to Section 15 and remove the oil pan.

14   Progressively loosen and then remove the oil seal housing retaining bolts.

15   Lift the housing away from the cylinder block, together with the crankshaft oil seal, using a twisting motion to ease the seal along the shaft.

16   Recover the old gasket from the seal housing on the cylinder block. If it has disintegrated, scrape the remains off with a gasket scraper. Take care to avoid damaging the mating surfaces.

17   Pry the old oil seal from the housing using a stout screwdriver (**see illustration**).

18   Wipe the oil seal housing clean, and check it visually for signs of distortion or cracking. Lay the housing on a work surface, with the mating surface face down. Press in a new oil seal, using a block of wood as a press to ensure that the seal enters the housing squarely.

19   Smear the crankcase mating surface with multi-purpose grease,

**10.19 Locate the new crankshaft front oil seal housing gasket
in position**

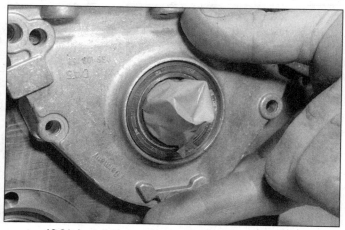

**10.21 Install the seal and its housing to the end of
the crankshaft**

and lay the new gasket in position **(see illustration)**.
20   Pad the end of the crankshaft with a layer of PVC tape; this will
protect the oil seal as it is being installed.
21   Lubricate the inner lip of the crankshaft oil seal with clean engine
oil, then install the seal and its housing to the end of the crankshaft.
Ease the seal along the shaft using a twisting motion, until the housing
is flush with the crankcase **(see illustration)**.
22   Insert the bolts and tighten them to the specified torque. **Caution:**
*The housing is light alloy, and may be distorted if the bolts are not tight-
ened progressively.*
23   Refer to Section 15 and install the oil pan.
24   With reference to Sections 2, 4 and 5 of this Chapter, install the
crankshaft timing belt sprocket, then install and tension the timing belt.
On completion, install the timing belt outer cover, and drivebelt pul-
ley(s).
25   The remainder of the refitting procedure is a reversal of removal,
as follows:

a) *With reference to Section 6, install and tension the drivebelt(s).*
b) *Install the plastic air ducting to the underside of the fender, working from Chapter 11.*
c) *Refer to Chapter 1 and refill the engine with the correct grade and
   quantity of oil.*
d) *Restore the fuel system.*

Crankshaft rear oil seal (flywheel end)

Refer to illustrations 10.35, 10.36, 10.37, 10.38, 10.39 and 10.40
26   Proceed as described in paragraphs 1 to 3 above, then refer to

Section 15 and remove the oil pan.
27   Working from Chapter 11, remove the screws and detach the
plastic air ducting from underneath the left-hand front fender.
28   Refer to Chapter 7A or B as applicable, and remove the transaxle
from the engine.
29   On vehicles with manual transaxle, refer to Section 13 of this
Chapter and remove the flywheel; refer to Chapter 8 and remove the
clutch friction plate and pressure plate.
30   On vehicles with automatic transaxle, refer to Section 13 of this
Chapter and remove the driveplate from the crankshaft.
31   Where applicable, remove the retaining bolts and lift the intermediate plate away from the cylinder block.
32   Progressively loosen and then remove the oil seal housing retaining bolts.
33   Lift the housing away from the cylinder block, together with the
crankshaft oil seal, using a twisting motion to ease the seal along the
shaft.
34   Recover the old gasket from the seal housing and cylinder block.
If it has disintegrated, scrape the remains off with a gasket scraper.
Take care to avoid damaging the mating surfaces.
35   Pry the old oil seal from the housing using a stout screwdriver
(see illustration).
36   Wipe the oil seal housing clean, and check it visually for signs of
distortion or cracking. Lay the housing on a work surface, with the
mating surface face down. Press in a new oil seal, using a block of
wood as a press to ensure that the seal enters the housing squarely
(see illustration).

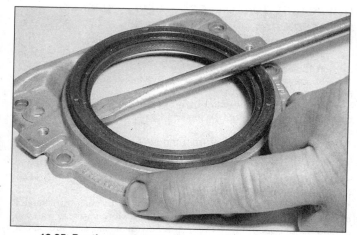

10.35 Pry the crankshaft rear oil seal from the housing

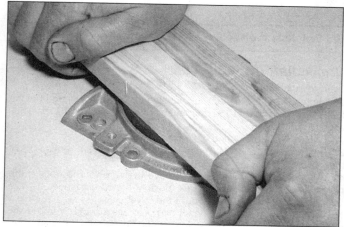

10.36 Press in a new oil seal, using a block of wood

10.37 Locate the new crankshaft rear oil seal housing gasket in position

10.38 A protective plastic cap is supplied with genuine VW crankshaft oil seals

37   Smear the crankcase mating surface with multi-purpose grease, and lay the new gasket in position **(see illustration)**.

38   A protective plastic cap is supplied with genuine VW crankshaft oil seals; when installed over the end of the crankshaft, the cap prevents damage to the inner lip of the oil seal as it is being installed **(see illustration)**. Use PVC tape to pad the end of the crankshaft if a cap is not available.

39   Lubricate the inner lip of the crankshaft oil seal with clean engine oil, then install the seal and its housing to the end of the crankshaft. Ease the seal along the shaft using a twisting motion, until the housing is flush with the crankcase **(see illustration)**.

40   Insert the retaining bolts and tighten them progressively to the specified torque **(see illustration)**. **Caution:** *The housing is light alloy, and may be distorted if the bolts are not tightened progressively.*

41   Refer to Section 15 and install the oil pan.

42   Install the intermediate plate to the cylinder block, then insert and tighten the retaining bolts.

43   On vehicles with automatic transaxle, work from Section 13 of this Chapter and install the driveplate to the crankshaft.

44   On vehicles with manual transaxle, refer to Chapter 8 and install the flywheel, pressure plate and clutch friction plate.

45   With reference to Chapter 7A or B as applicable, install the transaxle to the engine.

46   The remainder of the refitting procedure is a reversal of removal, as follows:

a) *Install the plastic air ducting to the underside of the fender, working from Chapter 11.*

b) *Refer to Chapter 1 and refill the engine with the correct grade and quantity of oil.*

c) *Restore the fuel systems.*

11   Cylinder head, intake and exhaust manifolds - removal, separation and installation

Removal

Refer to illustrations 11.7, 11.11a, 11.11b, 11.12, 11.13, 11.14 and 11.16

1   Select a level surface to park the vehicle upon. Give yourself enough space to move around it easily.

2   Refer to Chapter 11 and remove the hood from its hinges.

3   Disconnect the battery negative cable, and position It away from the terminal. **Caution:** *If the stereo in your vehicle is equipped with an anti-theft system, make sure you have the correct activation code before disconnecting the battery.*

4   With reference to Chapter 1, carry out the following:

a) *Drain the engine oil.*

b) *Drain the cooling system.*

5   Refer to Section 6 and remove the drivebelt(s).

6   With reference to Section 2, set the engine to TDC on No. 1 cylinder.

7   Refer to Chapter 3 and perform the following:

10.39 Fitting the crankshaft rear oil seal and housing

10.40 Tightening the crankshaft rear oil seal housing retaining bolts

11.7 Disconnect the heater coolant hoses from the ports on the cylinder head

11.11a Loosen and withdraw the retaining screws . . .

11.11b . . . and lift off the timing belt inner covers

a) *Loosen the clips and disconnect the radiator hoses from the ports on the cylinder head.*
b) *Loosen the clips and disconnect the expansion tank hose, and the heater intake and outlet coolant hoses, from the ports on the cylinder head* **(see illustration)**.

8   The "lock carrier" is a panel assembly comprising the front bumper molding, radiator and grille, cooling fan(s) headlight units, front valence and hood lock mechanism. Although its removal is not essential, its does give greatly-improved access to the engine; refer to the beginning of the engine removal procedure, in Chapter 2C.

9   Refer to Chapter 4C and carry out the following:

a) *Disconnect and remove the injector fuel supply hoses from the injectors and the injection pump head.*
b) *Disconnect the injector bleed hose from the injection pump fuel return port.*
c) *Unplug all fuel system electrical cabling at the relevant connectors, labeling each cable to aid refitting later.*

10   With reference to Sections 2, 4 and 7, carry out the following:

a) *Remove the camshaft cover.*
b) *Remove the timing belt outer covers, and disengage the timing belt from the camshaft sprocket.*
c) *Remove the timing belt tensioner, camshaft sprocket and fuel injection pump sprocket.*

11   Loosen and withdraw the retaining screws and lift off the timing belt inner covers **(see illustrations)**.

12   With reference to Chapter 3 and 4C (as applicable), disconnect the wiring plug from the coolant temperature sensor **(see illustration)**.

13   Refer to Chapter 6 and carry out the following:

11.12 Disconnect the wiring plug from the coolant temperature sensor

a) *Remove the bolts and separate the exhaust downpipe from the exhaust manifold flange.*
b) *Where applicable, remove the turbocharger from the exhaust manifold.*
c) *Where applicable, remove the EGR valve and its connecting pipework from the intake and exhaust manifolds.*
d) *Unbolt the supply cable from the glow plug in cylinder No. 4* **(see illustration)**.

14   Remove the retaining screw and detach the engine harness connector bracket from the cylinder head **(see illustration)**.

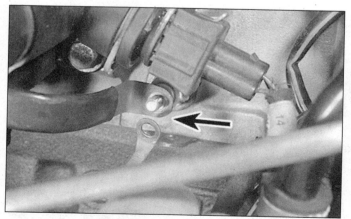

11.13 Unbolt the electrical supply cable from No 4 cylinder glow plug

11.14 Removing the engine harness connector bracket from the cylinder head

11.16 Lifting the cylinder head away from the engine

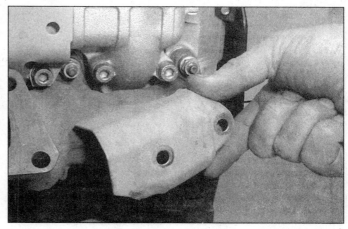

11.20 Unbolt and remove the exhaust manifold heat shield

2B

11.21a Install the exhaust manifold gaskets . . .

11.21b . . . then install the exhaust manifold. Tighten the nuts to the specified torque

15   Working in the reverse of the sequence shown in **illustration 11.37a**, progressively loosen the cylinder head bolts, by half a turn at a time, until all bolts can be unscrewed by hand. Discard the bolts - new ones must be installed on reassembly.

16   Check that nothing remains connected to the cylinder head, then lift the head away from the cylinder block; seek assistance if possible, as it is a heavy assembly, especially if it is being removed complete with the manifolds **(see illustration)**.

17   Remove the gasket from the top of the block, noting the locating dowels. If the dowels are a loose fit, remove them and store them with the head for safe-keeping. Do not discard the gasket - it will be needed for identification purposes.

18   If the cylinder head is to be disassembled for overhaul refer to Chapter 2C.

Manifold separation

Refer to illustrations 11.20, 11.21a, 11.21b, 11.23a, 11.23b and 11.23c

19   With the cylinder head on a work surface, loosen and withdraw the intake manifold securing bolts. Lift the manifold away, and recover the gasket.

20   Unbolt the heat shield **(see illustration)**, then progressively loosen and remove the exhaust manifold retaining nuts. Lift the manifold away from the cylinder head, and recover the gaskets.

21   Ensure that the intake and exhaust manifold mating surfaces are completely clean. Install the exhaust manifold, using new gaskets. Ensure that the gaskets are installed the correct way around, otherwise they will obstruct the intake manifold gasket. Tighten the exhaust manifold retaining nuts to the specified torque **(see illustrations)**.

22   Install the heat shield to the studs on the exhaust manifold, then install and tighten the retaining nuts.

23   Install a new intake manifold gasket on the cylinder head, then lift the intake manifold into position. Insert the retaining bolts and tighten them to the specified toque **(see illustrations)**.

11.23a Install a new intake manifold gasket on the cylinder head . . .

11.23b . . . then lift the intake manifold into position

11.23c Insert the retaining bolts and tighten them to the specified torque

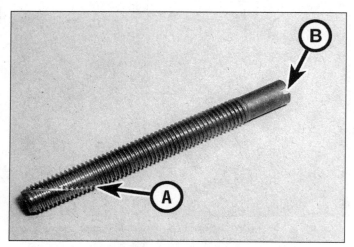

11.27 If a tap is not available, make a home-made substitute by cutting a slot (A) down the threads of one of the old cylinder head bolts. After use, the bolt head can be cut off, and the shank can then be used as an alignment dowel to assist cylinder head refitting. Cut a screwdriver slot (B) in the top of the bolt, to allow it to be unscrewed

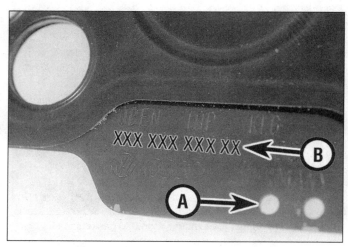

11.29 Cylinder head gasket punched holes (A) and part number (B)

Preparation for refitting

Refer to illustration 11.27

24   The mating faces of the cylinder head and cylinder block/crankcase must be perfectly clean before installing the head. Use a hard plastic or wood scraper to remove all traces of gasket and carbon; also clean the piston crowns. Take particular care during the cleaning operations, as aluminum alloy is easily damaged. Also, make sure that the carbon is not allowed to enter the oil and water passages - this is particularly important for the lubrication system, as carbon could block the oil supply to the engine's components. Using adhesive tape and paper, seal the water, oil and bolt holes in the cylinder block/crankcase.

25   Check the mating surfaces of the cylinder block/crankcase and the cylinder head for nicks, deep scratches and other damage. If slight, they may be removed carefully with abrasive paper, but note that head machining will not be possible - refer to Chapter 2C.

26   If warpage of the cylinder head gasket surface is suspected, use a straight-edge to check it for distortion. Refer to Part C of this Chapter if necessary.

27   Clean out the cylinder head bolt drillings using a suitable tap. If a tap is not available, make a home-made substitute (see illustration).

28   On the engine covered in this Chapter, it is possible for the piston

crowns to strike and damage the valve heads, if the camshaft is rotated with the timing belt removed and the crankshaft set to TDC. For this reason, the crankshaft must be set to a position other than TDC on No. 1 cylinder, before the cylinder head is reinstalled. Use a wrench and socket on the crankshaft pulley center bolt to turn the crankshaft in its normal direction of rotation, until all four pistons are positioned halfway down their bores, with No. 1 piston on its upstroke - approximately 90-degrees before TDC.

Installation

Refer to illustrations 11.29, 11.32, 11.36, 11.37a, 11.37b and 11.38

29   Examine the old cylinder head gasket for manufacturer's identification markings. These will either be in the form of punched holes or a part number, on the edge of the gasket (see illustration). Unless new pistons have been installed, the new cylinder head gasket must be the same type as the old one.

30   If new piston assemblies have been installed as part of an engine overhaul, before purchasing the new cylinder head gasket, refer to Section 13 of Chapter 2C and measure the piston projection. Purchase a new gasket according to the results of the measurement (see Chapter 2C Specifications).

31   Lay the new head gasket on the cylinder block, engaging it with the locating dowels. Ensure that the manufacturer's "TOP" and part number markings are face up.

32   Cut the heads from two of the old cylinder head bolts. Cut a slot, big enough for a screw-driver blade, in the end of each bolt. These can

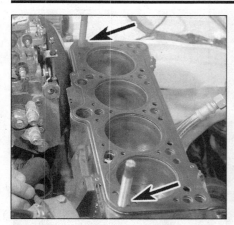

11.32 Two of the old head bolts (arrows) used as cylinder head alignment dowels

11.36 Oil the cylinder head bolt threads, then place each bolt into its relevant hole

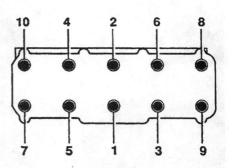

11.37a Cylinder head bolt tightening sequence

11.37b Tightening the cylinder head bolts using a torque wrench and socket

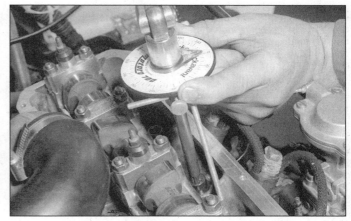

11.38 Angle-tightening a cylinder head bolt

be used as alignment dowels to assist in cylinder head installation (see illustration).

33   With the help of an assistant, place the cylinder head and manifolds centrally on the cylinder block, ensuring that the locating dowels engage with the recesses in the cylinder head. Check that the head gasket is correctly seated before allowing the full weight of the cylinder head to rest upon it.

34   Unscrew the home-made alignment dowels, using a flat bladed screwdriver.

35   Apply a smear of grease to the threads, and to the underside of the heads, of the new cylinder head bolts.

36   Oil the bolt threads, then carefully enter each bolt into its relevant hole (do not drop them in) and screw in, by hand only, until finger-tight (see illustration).

37   Working progressively and in the sequence shown, tighten the cylinder head bolts to their Stage 1 torque setting, using a torque wrench and suitable socket (see illustrations). Repeat the exercise in the same sequence for the Stage 2 torque setting.

38   Once all the bolts have been tightened to their Stage 2 settings, working again in the given sequence, angle-tighten the bolts through the specified Stage 3 angle, using a socket and extension bar. It is recommended that an angle-measuring gauge is used during this stage of the tightening, to ensure accuracy. If a gauge is not available, use white paint to make alignment marks between the bolt head and cylinder head prior to tightening; the marks can then be used to check the bolt has been rotated through the correct angle during tightening. Repeat for the Stage 4 setting (see illustration).

39   Install the timing belt inner cover, tightening the retaining screws securely.

40   With reference to Sections 2 and 5, install the timing belt tensioner and sprockets.

41   Refer to Section 2 and set the engine to TDC on No. 1 cylinder. On completion, refer to Section 4 and install the camshaft timing belt and outer covers.

42   The remainder of installation is a reversal of the removal procedure, as follows:

a) Refer to Chapter 6 and install the turbocharger (where applicable), the exhaust downpipe, the EGR valve (where applicable) and the glow plug cabling.

b) Refer to Chapter 4C and install the injector fuel supply hoses to the injectors and the injection pump head. Reconnect all fuel system electrical cabling. Install the injector bleed hose to the injection pump fuel return port.

c) Install the engine harness connector bracket to the cylinder head.

d) Install the camshaft cover (see Section 7).

e) With reference to the information in Chapter 2C, install the lock carrier assembly, if it was removed for greater access.

f) Reconnect the radiator, expansion tank and heater coolant hoses, referring to Chapter 3 for guidance. Reconnect the coolant temperature sensor wiring.

g) Refer to Section 6 and install the drivebelt(s).

h) Restore the battery connection.

i) Refer to Chapter 11 and install the hood.

43   On completion, refer to Chapter 1 and carry out the following:

a) Refill the engine cooling system with the correct quantity of new coolant.

b) Refill the engine lubrication system with the correct grade and quantity of oil.

14.10a Remove the engine mounting-to-transaxle bellhousing bolts . . .

14.10b . . . and remove the bracket

44   After completing a preliminary inspection, start the engine and allow it to reach operating temperature. Stop the engine and remove the camshaft cover. With the engine hot, tighten the cylinder head bolts, following the recommended sequence, an additional 90-degrees (1/4-turn). Refit the camshaft cover.

12 Hydraulic lifters - check

1   The hydraulic lifters are self-adjusting, and require no attention while in service.
2   If the hydraulic lifters become excessively noisy, their operation can be checked as described below.
3   Run the engine until it reaches its normal operating temperature. Switch off the engine, then refer to Section 7 and remove the camshaft cover.
4   Rotate the camshaft by turning the crankshaft with a socket and wrench, until the first cam lobe over No. 1 cylinder is pointing upwards.
5   Using a feeler gauge, measure the clearance between the base of the cam lobe and the top of the tappet. If the clearance is greater than 0.004-inch, then the tappet is defective and must be replaced.
6   If the clearance is less than 0.004-inch, press down on the top of the lifter, until it is felt to contact the top of the valve stem. Use a wooden or plastic implement that will not damage the surface of the tappet.
7   If the tappet travels more than 0.04-inch before making contact, then it is defective and must be replaced.
8   Hydraulic lifter removal and installation is described as part of the cylinder head overhaul sequence - see Chapter 2C for details.

13 Flywheel/driveplate - removal, inspection and installation

General information

1   The mounting arrangement of the flywheel and clutch components depends on the type of transaxle installed.
2   On vehicles with the 020 (5-speed) transaxle, the clutch pressure plate is bolted directly to the end of the crankshaft. The flywheel is then bolted to the pressure plate. Removal of these components is therefore described in Chapter 8.
3   On vehicles with the 084 (4-speed) transaxle, the layout is more conventional; the flywheel is mounted on the crankshaft, with the pressure plate bolted to it. Removal of the flywheel is as described in Section 13 of Chapter 2A.

14 Engine mountings - inspection and replacement

Inspection

1   If improved access is required, raise the front of the car and support it securely on axle stands.
2   Check the mounting rubbers to see if they are cracked, hardened or separated from the metal at any point; replace the mounting if any such damage or deterioration is evident.
3   Check that all the mounting's fasteners are securely tightened; use a torque wrench to check if possible.
4   Using a large screwdriver or a crowbar, check for wear in the mounting by carefully levering against it to check for free play. Where this is not possible, enlist the aid of an assistant to move the engine/transaxle back and forth, or from side to side, while you watch the mounting. While some free play is to be expected even from new components, excessive wear should be obvious. If excessive free play is found, check first that the fasteners are correctly secured, then replace any worn components as described below.

Replacement

Front engine mounting

Refer to illustrations 14.10a, 14.10b, 14.11 and 14.12

5   Disconnect the battery negative cable, and position it away from the terminal.
6   Position a floor jack underneath the engine, and position it such that the jack head is directly underneath the engine/bellhousing mating surface.
7   Raise the jack until it just takes the weight of the engine off the front engine mounting.
8   Loosen and withdraw the engine mounting through-bolt.
9   Refer to Chapter 5A and remove the starter motor.
10   Loosen and withdraw the engine mounting-to-transaxle bellhousing bolts, and remove the bracket **(see illustrations)**.
11   Working under the engine mounting front crossmember, remove the engine mounting block retaining screw **(see illustration)**.
12   Lift the engine mounting block out of the crossmember cup **(see illustration)**.
13   Installation is a reversal of removal, noting the following points:
   a) *Ensure that the orientation lug that protrudes from the top of surface of the engine mounting block engages with the recess in the mounting bracket* **(see illustration)**.
   b) *Tighten all bolts to the specified torque.*

Rear right-hand engine mounting

Refer to illustration 14.13

14   Disconnect the battery negative cable, and position it away from

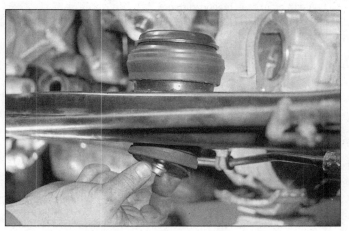

14.11 Remove the engine mounting block retaining screw

14.12 Lift the engine mounting block out of the crossmember cup

2B

the terminal.

15 Mount an engine lifting beam across the engine bay, and attach the jib to the engine lifting eyes on the cylinder head. Alternatively, an engine hoist can be used. Raise the hoist/lifting beam jib to take the weight of the engine off the engine mounting.

16 Loosen and withdraw the engine mounting through-bolt.

17 Unbolt the engine mounting bracket from the cylinder block.

18 Unbolt the engine mounting block from the body, and remove it from the engine bay.

19 Installation is a reversal of removal, noting the following points:

a) Ensure that the orientation lug that protrudes from the top of surface of the engine mounting block engages with the recess in the mounting bracket.

b) Tighten all bolts to the specified torque.

Rear left-hand mounting

Refer to illustration 14.25

20 Disconnect the battery negative cable, and position it away from the terminal.

21 Position a floor jack underneath the engine, and position it such that the jack head is directly underneath the engine/bellhousing mating surface.

22 Raise the jack until it just takes the weight of the engine off the rear right-hand engine mounting.

23 Loosen and withdraw the engine mounting through-bolt.

24 Unbolt the engine mounting bracket from the end of the transaxle casing.

25 Unbolt the engine mounting block from the body, and remove it from the engine bay **(see illustration)**.

26 Installation is a reversal of removal, noting the following points:

a) Ensure that the orientation lug that protrudes from the top of surface of the engine mounting block engages with the recess in the mounting bracket.

b) Tighten all bolts to the specified torque.

15 Oil pan - removal, inspection and installation

Caution: *If the stereo in your vehicle is equipped with an anti-theft system, make sure you have the correct activation code before disconnecting the battery.*

Removal

1 Disconnect the battery negative cable, and position it away from the terminal. Refer to Chapter 1 and drain the engine oil. Where applicable, remove the screws and lower the engine undertray away from the vehicle.

2 Park the vehicle on a level surface, apply the parking brake and chock the rear wheels.

3 Raise the front of the vehicle, rest it securely on axle stands or wheel ramps; refer to *"Jacking and Vehicle Support"*.

4 To improve access to the oil pan, refer to Chapter 8 and disconnect the right-hand driveshaft from the transaxle output flange.

14.13 Lug (arrow) on top of the mounting engages with the recess in the bracket

14.25 Removing the rear left-hand engine mounting

5   Working around the outside of the oil pan, progressively loosen and withdraw the oil pan retaining bolts. Where applicable, unbolt and remove the flywheel cover plate from the transaxle to gain access to the left-hand oil pan fixings.

6   Break the joint by striking the oil pan with the palm of your hand, then lower the oil pan and withdraw it from underneath the vehicle. Recover and discard the oil pan gasket. Where a baffle plate is installed, note that it can only be removed once the oil pump has been unbolted (see Section 16).

7   While the oil pan is removed, take the opportunity to check the oil pump pick-up/strainer for signs of clogging or disintegration. If necessary, remove the pump as described in Section 16, and clean or replace the strainer.

Installation

8   Clean all traces of sealant from the mating surfaces of the cylinder block/crankcase and oil pan, then use a piece of clean rag to wipe out the oil pan.

9   Ensure that the oil pan and cylinder block/crankcase mating surfaces are clean and dry, then apply a coating of suitable sealant to the oil pan and crankcase mating surfaces.

10   Lay a new oil pan gasket in position on the oil pan mating surface, then refit the oil pan and install the retaining bolts. Tighten the nuts and bolts evenly and progressively to the specified torque.

11   Where applicable, install the driveshaft and engine undertray.

12   Refer to Chapter 1 and refill the engine with the specified grade and quantity of oil.

13   Restore the battery connection.

16   Oil pump and pickup - removal and installation

General information

1   The oil pump and pickup are both mounted in the oil pan. Drive is taken from the intermediate shaft, which rotates at half crankshaft speed.

2   The oil pump arrangement is identical to that described for gasoline engines. Refer to Chapter 2A for further details.

Chapter 2 Part C
General engine overhaul procedures

Contents

2C

Specifications

Engine codes
1.8L gasoline engine	ACC
2.0L gasoline engine..	ABA
1.9L diesel engine..	AAZ

Cylinder head
Cylinder head gasket surface, maximum distortion	0.004 inch
Minimum cylinder head height:	
Gasoline engines...	5.220 inches
Diesel engine..	Head reworking not possible.
Maximum swirl chamber projection (diesel engine)	0.003 inch

Cylinder head gasket
Identification markings (punched holes), diesel engine only*:
Piston projection	
0.0260 to 0.0339 inch	1 hole
0.0343 to 0.0354 inch	2 holes
0.0358 to 0.0402 inch	3 holes

*Note: See text in Chapter 2B and Sections 4 and 13 of this Chapter for details.

Valve assembly
Valve stem diameter:	
Intake	
1.8L gasoline engine, 1.9L diesel engine.....................................	0.3138 inch
2.0L gasoline engine...	0.2746 inch
Exhaust:	
1.8L gasoline engine, 1.9L diesel engine.....................................	0.3130 inch
2.0L gasoline engine...	0.2738 inch
Maximum valve head deflection (end of valve stem flush with top of guide)	
Intake:	
Diesel engine..	0.051 inch
Gasoline engines	0.039 inch
Exhaust (all engines)..	0.051 inch
Valve spring free length	N/A
Valve spring squareness limit	N/A

Camshaft

Maximum shaft runout .... 0.0004 inch
Cam lobe height:
    Intake .... N/A
    Exhaust... N/A
Endplay, all engine codes.. 0.0059 inch
Maximum running clearance .... 0.004 inch
Camshaft identification codes:
    1.8L gasoline engine .... Q 026
    2.0L gasoline engine .... B 037
    1.9L diesel engine .... N/A

Intermediate shaft (gasoline engines)

Maximum endplay .... 0.0098 inch

Cylinder block

Bore diameter:
    2.0L gasoline engine
        Standard .... 3.2484 inches
        1st oversize... 3.2583 inches
        2nd oversize.. 3.2681 inches
        Maximum bore wear .... 0.003 inch
    1.8L gasoline engine
        Standard .... 3.1894 inches
        1st oversize... 3.1992 inches
        2nd oversize.. 3.2090 inches
        Maximum bore wear .... 0.003 inch
    1.9L diesel engine
        Standard .... 3.1303 inches
        1st oversize... 3.1402 inches
        2nd oversize.. 3.1500 inches
        Maximum bore wear .... 0.004 inch

Pistons and piston rings

Piston diameter
    2.0L gasoline engine
        Standard .... 3.2474 inches
        1st oversize... 3.2572 inches
        2nd oversize.. 3.2671 inches
        Maximum deviation.................................... 0.0015 inch
    1.8L gasoline engine
        Standard .... 3.1883 inches
        1st oversize... 3.1982 inches
        2nd oversize.. 3.2080 inches
        Maximum deviation.................................... 0.0015 inch
    1.9L diesel engine
        Standard .... 3.1291 inches
        1st oversize... 3.1390 inches
        2nd oversize.. 3.1488 inches
        Maximum deviation.................................... 0.0015 inch
Wrist pin external diameter:
    1.9L diesel engine .... 26.0 mm
    Gasoline engines... N/A
Piston ring-to-groove clearance
    Gasoline engines
        Top compression ring
            Standard.. 0.0008 to 0.0020 inch
            Service limit .... 0.0059 inch
        2nd compression ring
            Standard.. 0.0008 to 0.0020 inch
            Service limit .... 0.0059 inch
        Oil scraper ring:
            Standard.. 0.0008 to 0.0020 inch
            Service limit .... 0.0059 inch
    Diesel engine
        Top compression ring
            Standard.. 0.0035 to 0.0047 inch
            Service limit .... 0.0098 inch

Diesel engine
 2nd compression ring
  Standard .... 0.0020 to 0.0031 inch
  Service limit .... 0.0098 inch
 Oil scraper ring
  Standard .... 0.0012 to 0.0024 inch
  Service limit .... 0.0059 inch
Piston ring end gap
 Gasoline engines
  Top compression ring
   Standard .... 0.0078 to 0.0157 inch
   Service limit .... 0.039 inch
  2nd compression ring
   Standard .... 0.0078 to 0.0157 inch
   Service limit .... 0.039 inch
  Oil scraper ring
   Standard .... 0.0098 to 0.0197 inch
   Service limit .... 0.039 inch
 Diesel engine
  Top compression ring
   Standard .... 0.0079 to 0.0157 inch
   Service limit .... 0.0472 inch
  2nd compression ring
   Standard .... 0.0079 to 0.0157 inch
   Service limit .... 0.0236 inch
  Oil scraper ring
   Standard .... 0.0098 to 0.0197 inch
   Service limit .... 0.0472 inch

Connecting rods

Length:
 Diesel engine .... 5.67 inches
 Gasoline engines .... N/A
Big-end thrust (side) clearance
 Gasoline engines
  Standard .... 0.0020 to 0.0122 inch
  Service limit .... 0.0146 inch
 Diesel engine
  Standard .... N/A
  Service limit .... 0.0146 inch

Crankshaft

Maximum runout .... N/A
Endplay
 Gasoline engines
  Standard .... 0.0028 to 0.0067 inch
  Service limit .... 0.0098 inch
 Diesel engine
  Standard .... 0.0028 to 0.0067 inch
  Service limit .... 0.0146 inch
Main bearing journal diameters
 Standard .... 2.1260 inches
 1st undersize .... 2.1161 inches
 2nd undersize .... 2.1063 inches
 3rd undersize .... 2.0965 inches
 Tolerance .... -0.00087 to -0.00165 inch
Main bearing running clearances
 Gasoline engines
  Standard .... 0.0008 to 0.0024 inch
  Service limit .... 0.0067 inch
 Diesel engine
  Standard .... 0.0012 to 0.0031 inch
  Service limit .... 0.0067 inch
Crankpin journal diameters
 Standard .... 1.8819 inches
 1st undersize .... 1.8720 inches
 2nd undersize .... 1.8622 inches
 3rd undersize .... 1.8524 inches
 Tolerance .... -0.00087 to -0.00165 inch

2C

Crankshaft (continued)

Connecting rod bearing oil clearance
    Gasoline engines
        Standard .... 0.0004 to 0.0024 inch
        Service limit.. 0.0047 inch
    Diesel engine
        Standard .... N/A
        Service limit.. 0.0031 inch

Oil pump

Gear backlash
    Standard.. 0.0020 inch
    Service limit .... 0.0079 inch
Gear axial clearance (service limit) .... 0.0059 inch

Torque specifications

Ft-lbs (unless otherwise indicated)

Connecting rod bearing cap bolts/nuts
    Stage 1 22
    Stage 2 .... Angle-tighten a further 90-degrees
Camshaft bearing cap nuts/bolts 15
Crankshaft main bearing cap bolts.. 48
Engine crossmember-to-body bolts.. 37
Intermediate shaft flange bolts 18
Intermediate shaft sprocket bolt.. 59
Lock carrier-to-chassis bolts.. 17
Lock carrier-to-fender screws .... 48 in-lbs
Piston oil jet/pressure relief valve.. 19
Transaxle bellhousing to engine:
    M10 bolts 44
    M12 bolts 59

---

1 Engine and transaxle removal - preparation and precautions

If you have decided that the engine must be removed for overhaul or major repair work, several preliminary steps should be taken.

Locating a suitable place to work is extremely important. Adequate work space, along with storage space for the vehicle, will be needed. If a workshop or garage is not available, at the very least a solid, level, clean work surface is required.

If possible, clear some shelving close to the work area and use it to store the engine components and ancillaries as they are removed and disassembled. In this manner, the components stand a better chance of staying clean and undamaged during the overhaul. Laying out components in groups together with their bolts, screws etc. will save time and avoid confusion when the engine is reinstalled.

Clean the engine compartment and engine/transaxle before beginning the removal procedure; this will help visibility and help to keep tools clean.

The help of an assistant should be available; there are certain instances when one person cannot safely perform all of the operations required to remove the engine from the vehicle. Safety is of primary importance, considering the potential hazards involved in this kind of operation. A second person should always be in attendance to offer help in an emergency. If this is the first time you have removed an engine, advice and aid from someone more experienced would also be beneficial.

Plan the operation ahead of time. Before starting work, obtain (or arrange for the hire of) all of the tools and equipment you will need. Access to the following items will allow the task of removing and installing the engine/transaxle to be completed safely and with relative ease: a heavy-duty floor jack - rated in excess of the combined weight of the engine and transaxle, complete sets of spanners and sockets as described in the front of this manual, wooden blocks, and plenty of rags and cleaning solvent for mopping up spilled oil, coolant and fuel. A selection of different sized plastic storage bins will also prove useful for keeping disassembled components grouped together. If any of the equipment must be hired, make sure that you arrange for it in advance, and perform all of the operations possible without it beforehand; this may save you time and money.

Plan on the vehicle being out of use for quite a while, especially if you intend to carry out an engine overhaul. Read through the whole of this Section and work out a strategy based on your own experience and the tools, time and workspace available to you. Some of the overhaul processes may have to carried out by a VW dealer or other qualified repair shop - these establishments often have busy schedules, so it would be prudent to consult them before removing or disassembling the engine, to get an idea of the amount of time required to carry out the work.

When removing the engine from the vehicle, be methodical about the disconnection of external components. Labeling cables and hoses as they are removed will greatly assist the installation process.

Always be extremely careful when lifting the engine/transaxle assembly from the engine bay. Serious injury can result from careless actions. If help is required, it is better to wait until it is available rather than risk personal injury and/or damage to components by continuing alone. By planning ahead and taking your time, a job of this nature, although major, can be accomplished successfully and without incident.

On all models described in this manual, the engine and transaxle are removed as a complete assembly through the front of the vehicle. This involves the removal of the lock carrier, which is the panel assembly that forms the front of the engine bay. Although the lock carrier is a large assembly, its removal is not difficult, and the benefits in terms of ease of access are well worth the effort involved.

Note that the engine and transaxle should ideally be removed with the vehicle standing on all four wheels, but access to the driveaxles and exhaust system downpipe will be improved if the vehicle can be temporarily raised onto axle stands.

2   Engine and transaxle - removal, separation and installation

Removal

All models

1   Select a solid, level surface to park the vehicle upon. Give yourself enough space to move around it easily.

2   Refer to Chapter 11 and remove the hood from its hinges.

3   Disconnect the battery negative cable, and position It away from the terminal. **Caution:** *If the stereo in your vehicle is equipped with an anti-theft system, make sure you have the correct activation code before disconnecting the battery.*

4   With reference to Chapter 1 as applicable, carry out the following:

a) *If the engine is to be disassembled, drain the engine oil.*
b) *Drain the cooling system.*
c) *Where applicable, remove the drive V-belt.*
d) *Remove the ribbed drivebelt.*

5   Refer to Chapter 3 and carry out the following:

a) *Loosen the clamps and disconnect the radiator top and bottom hoses from the ports on the cylinder head, and from the thermostat housing/water pump (as applicable).*
b) *Loosen the clamps and disconnect the coolant hoses from the expansion tank and heater intake and outlet ports at the firewall.*

6   On vehicles with air conditioning, refer to Chapter 3 and carry out the following additional operations:

a) *Remove the retaining bolts from the clamps that secure the refrigerant condenser supply and return pipes to the engine crossmember.*
b) *Unbolt the air conditioning compressor from the engine, and allow it to rest on the engine front crossmember.*

7   Make reference to Chapter 11 and carry out the following:

a) *Remove the undertray from the underside of the engine.*
b) *Remove the screws and clamps that secure the plastic inner wheel arch liners to the front valance.*

8   The "lock carrier" is a panel assembly comprising the front bumper molding, radiator and grille, cooling fan(s) headlight units, front valance and hood lock mechanism. Its removal gives greatly-improved access to the engine and transaxle, and allows them to be lifted out of the vehicle via the front of the engine bay. To remove the lock carrier, carry out the following:

a) *Unplug the wiring harness at the multi-way connector. The connector is a bayonet fit; twist the housing to unlock the two halves, then pull them apart. Recover the internal seal if it has become loose and cover the connector housings with a plastic bag to prevent the ingress of dirt. Release the harness from all the metal retaining clamps.*
b) *Refer to Chapter 11 and detach the hood lock release cable form the lock mechanism.*
c) *Remove the lock carrier fasteners at the following locations: two flange screws on the uppermost edge above the headlight units, four bolts (two on each side) behind fog lamp units/reflector panels, threaded into the ends of the chassis rails.*
d) *Lift the lock carrier assembly away from the front of the vehicle and rest it on a dust sheet; tilt the assembly forward as you remove it, to avoid spilling any coolant that may remain in the radiator. Note: On vehicles with air conditioning, the compressor remains connected to the refrigerant condenser by the supply and return hoses and is removed together with the lock carrier assembly. Caution: Take care to avoid kinking the air conditioning refrigerant hoses.*

Gasoline models

9   With reference to Chapter 6, unplug the oxygen sensor cabling from the main harness at the multi-way connector.

10   Disconnect the ignition coil wire from the center terminal of the distributor cap, and tie it back away from the engine.

11   Refer to Chapter 9 and disconnect the brake booster vacuum hose from the port on the intake manifold.

12   On vehicles with an Exhaust Gas Recirculation (EGR) system, refer to Chapter 6 and disconnect the vacuum hoses from the connection points on the EGR valve, brake booster vacuum hose, air intake hose and where applicable, fuel injection pump. Make a careful note of the order of connection to ensure correct installation.

13   Refer to Chapter 6 and disconnect the vacuum hose from the port on the throttle body. Make a careful note of the point of connection to ensure correct installation.

Single-point injection models

14   With reference to Chapter 4A, carry out the following operations:

a) *Depressurize the fuel system*
b) *Remove the exhaust manifold-to-air cleaner and throttle body air-box-to-air cleaner ducting from the engine bay.*
c) *Remove the airbox from the top of the throttle body; make a note of the vacuum hose connections to ensure correct refitting later.*
d) *Disconnect the accelerator cable from the throttle spindle lever.*
e) *Disconnect the fuel supply and return hoses from the throttle body - observe the precautions at the beginning of Chapter 4A.*

Multi-point injection models

15   With reference to Chapter 4B, carry out the following operations:

a) *Depressurize the fuel system.*
b) *Loosen the clamps and remove the exhaust manifold-to-air cleaner and throttle body-to-airflow meter ducting from the engine bay.*
c) *Disconnect the accelerator cable from the throttle spindle lever.*
d) *Disconnect the fuel supply and return hoses from the throttle body - observe the precautions at the beginning of Chapter 4B.*

16   On vehicles with automatic transaxle, extra clearance is required when removing the engine and transaxle as one assembly. Removing the ribbed drivebelt pulleys from the crankshaft achieves this - refer to Chapter 2A for details.

Diesel models

Caution: *When disassembling any part of the air intake system on a turbocharged vehicle, ensure that no foreign material can get into the turbo air intake port; cover the opening with a sheet of plastic, secured with an elastic band. The turbocharger compressor blades could be severely damaged if debris is allowed to enter.*

17   Refer to Chapter 9 and disconnect the brake booster vacuum hose from the vacuum pump on the cylinder block.

18   On vehicles with an Exhaust Gas Recirculation (EGR) system, refer to Chapter 6 and disconnect the vacuum hoses from the connection points on the EGR valve, brake booster vacuum hose, air intake hose and where applicable, fuel injection pump. Make a careful note of the order of connection to ensure correct refitting.

19   Refer to Chapter 4C and carry out the following operations:

a) *Loosen and withdraw the banjo bolts, then disconnect the fuel supply and return hoses from the fuel injection pump.*
b) *Release the clamp, then disconnect the injector bleed hose from the port on the fuel return fitting.*
c) *Loosen the clamps and remove the intake air hose from the air cleaner, crankcase ventilation hose and turbocharger inlet.*
d) *Disconnect the accelerator cable from the fuel injection pump.*
e) *Where applicable, disconnect the cold start accelerator cable from the fuel injection pump.*

All models

Refer to illustrations 2.20, 2.32 and 2.33

20   Isolate the main engine harness from the vehicle at the connector, which is mounted on a bracket at transaxle end of the cylinder block. The connector is a bayonet fit; twist the housing to unlock the two halves, then pull them apart. Recover the internal seal if it has become loose. Cover the connector housings with a plastic bag to prevent the

2.20 Main engine harness connector

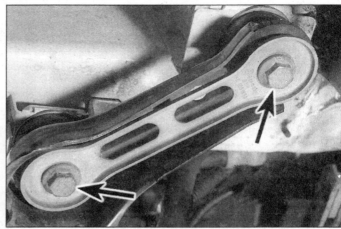

2.32 Front crossmember mounting bolts (left-hand side shown)

ingress of dirt **(see illustration)**.

21   Refer to Chapter 5A and disconnect the wiring from the alternator, starter motor and solenoid.

22   With reference to Chapter 5B and Chapter 4A, B or C as applicable, identify those sections of the ignition and fuel system electrical harness that remain connected to sensors and actuators on the engine, that are not part of the main engine harness. These sections will not be isolated at the engine harness connector, and must be disconnected individually. Label each connector carefully to ensure correct installation.

23   On vehicles with power steering, refer to Chapter 10 and carry out the following:

   a) *Loosen the retaining screws and release the clamps that secure the power steering supply and return pipes to the engine front crossmember.*

   b) *Remove the power steering fluid reservoir retaining screws and lower it away from the battery tray, allowing it to rest on the engine crossmember.*

   c) *Unbolt the power steering pump, together with its mounting brackets from the engine and hang it from the engine crossmember using wire or a large cable-tie.* **Note:** *The power steering hoses can remain connected to the pump and reservoir, so there is no need to drain the hydraulic fluid from the system.*

24   On vehicles with manual transaxle, refer to Chapter 7A and carry out the following:

   a) *At the top of the transaxle casing, disconnect the wiring from the speedometer drive transducer and reversing light switch.*

   b) *Disconnect the gear selection mechanism from the transaxle.*

   c) *Refer to Chapter 8 and disconnect the clutch cable from the release mechanism at the front of the transaxle casing.*

25   On vehicles with automatic transaxle, refer to Chapter 7B and carry out the following:

   a) *Release the selector cable from the selector lever at the top of the transaxle casing.*

   b) *Clamp the coolant hoses leading to the transaxle fluid cooler, then release the clamps and disconnect the hoses from the cooler ports.*

   c) *Unplug the wiring harness from the transaxle at the connectors; label each connector to aid installation later.*

26   Refer to Chapter 8 and separate the driveaxles from the transaxle differential output shafts.

27   With reference to Chapter 6, unbolt the exhaust down pipe from the exhaust manifold (or turbocharger on diesel engine models). Recover and discard the gasket.

28   Unbolt the engine and transaxle ground straps from the bodywork.

29   With reference to Chapter 2A, loosen and withdraw the through-bolts from all three engine mountings.

2.33 Car supported on stands, with engine and transaxle removed

30   Attach the jib of an engine lifting beam or hoist to the lifting eyelets at the front of the cylinder head. Raise the jib slightly and tilt the engine towards the rear of the engine bay, so that the front engine mounting is raised above the front crossmember.

31   Unscrew the retaining bolts and withdraw the starter motor; refer to Chapter 5A for greater detail. Remove the front engine mounting bracket.

32   Working under the front of the vehicle, loosen and withdraw the retaining bolts, then lower the front crossmember to the ground, together with the power steering pump. Remove the retaining bolt from underneath, and extract the engine mounting rubber block **(see illustration)**.

33   Carry out a final check to ensure that nothing else remains connected to the engine, then slowly wheel the hoist back from the vehicle, and maneuver the engine and transaxle out through the front of the engine bay. Rotate the assembly slightly as it is withdrawn, so that the timing belt end of the engine emerges first. Guide the pulleys past the inner fender to avoid damaging the paint **(see illustration)**.

Separation

34   Rest the engine and transaxle assembly on a firm, flat surface, and use wooden blocks as wedges to keep the unit steady.

Manual transaxle

35   The transaxle is secured to the engine by a combination of machine screws and studs, threaded into the cylinder block and bell-

housing - the total number of fixings depends on the type of transaxle and vehicle specification. Note that two of these fixings also serve as mountings for the starter motor and the front engine mounting.

36   Starting at the bottom, remove all the screws and nuts then carefully draw the transaxle away from the engine, resting it securely on wooden blocks. Collect the locating dowels if they are loose enough to be extracted. **Caution:** *Take care to prevent the transaxle from tilting, until the input shaft is fully disengaged from the clutch friction plate.*

37   With reference to Chapter 8, remove the clutch release mechanism, pressure plate and friction plate.

Automatic transaxle

38   Unbolt the skid plate from the underside of the transaxle oil pan.

39   Unbolt the protection plate from the bottom of the transaxle bellhousing, this will expose the rear face of the driveplate.

40   Mark the position of the torque converter with respect to the driveplate, using chalk or a marker pen. Remove the three nuts that secure the driveplate to the torque converter; turn the engine over using a socket and wrench on the crankshaft sprocket to rotate the driveplate and expose each nut in turn.

41   The transaxle is secured to the engine by a combination of bolts and studs with nuts, threaded into the cylinder block and bellhousing - the total number of fasteners depends on the type of transaxle and vehicle specification. Note that two of these fasteners also serve as mountings for the starter motor.

42   Starting at the bottom, remove all the bolts and nuts then carefully draw the transaxle away from the engine, resting it securely on wooden blocks. Collect the locating dowels if they are loose enough to be extracted. **Caution:** *Take care to prevent the torque converter from sliding off the transaxle input shaft - hold it in place as the transaxle is withdrawn.*

43   Place a length of batten across the open face of the bellhousing, fastening it with cable-ties, to keep the torque converter in place in its housing.

Installation

44   If the engine and transaxle have not been separated, proceed from Step 50 onward.

Manual transaxle

45   Smear a quantity of high-melting-point grease on the splines of the transaxle input shaft. Do not use an excessive amount as there is the risk of contaminating the clutch friction plate. Carefully install the transaxle to the cylinder block, guiding the dowels into the mounting holes in cylinder block.

46   Install the bellhousing bolts and nuts, hand tightening them to secure the transaxle in position. **Note:** *Do not tighten them to force the engine and transaxle together.* Ensure that the bellhousing and cylinder block mating faces will butt together evenly without obstruction, before tightening the bolts and nuts to their specified torque.

Automatic transaxle

47   Remove the torque converter restraint from the face of the bellhousing. Check that the drive lugs on the torque converter hub are correctly engaged with the recesses in the inner wheel of the automatic transaxle fluid pump.

48   Carefully mate the transaxle to the cylinder block, guiding the dowels into the mounting holes in cylinder block. Observe the markings made during the removal, to ensure correct alignment between the torque converter and the driveplate.

49   Install the bellhousing bolts and nuts, hand-tightening them to secure the transaxle in position. **Note:** *Do not tighten them to force the engine and transaxle together.* Ensure that the bellhousing and cylinder block mating faces will butt together evenly without obstruction, before tightening the bolts and nuts to their specified torque.

All models

50   With reference to Chapter 5A, install the starter motor, together with the front engine mounting bracket and tighten the retaining bolts to the specified torque.

51   Attach the jib of an engine hoist to the lifting eyelets on the cylinder head, and raise the engine and transaxle from the ground.

52   Wheel the hoist up to the front of the vehicle and with the help of an assistant, guide the engine and transaxle in through the front of the engine bay. Rotate the assembly slightly so that the transaxle casing enters first, then guide the drivebelt pulleys past the bodywork.

53   Align the rear engine mounting brackets with the mounting points on the body. Note that alignment lugs protrude from the metal discs that are bonded to the top of each of each engine mounting; these must engage with the recesses on the underside of the engine mounting brackets - refer to Chapter 2A or B for details.

54   Lift the front engine mounting crossmember into position. Apply a little clean engine oil to the threads of the retaining bolts, then insert and tighten them to the correct torque. **Caution:** *Ensure that no weight bears on the crossmember until the bolts are tightened.*

55   Install the front engine mounting rubber block into the cup in the crossmember, then insert the retaining bolt through the underside of the crossmember and tighten it to the specified torque.

56   Lower the engine and transaxle into position, ensuring that the locating lugs on the front engine mounting engage with the recess in the mounting bracket. Insert the front and rear engine mounting through-bolts, tightening them by hand initially.

57   Detach the engine hoist jib from the lifting eyelets.

58   Settle the engine and transaxle assembly on its mountings by rocking it backwards and forwards, then tighten the mounting through-bolts to the specified torque.

59   Refer to Chapter 8 and reconnect the driveaxles to the transaxle.

60   The remainder of the installation sequence is the direct reverse of the removal procedure, noting the following points:

a)   *Ensure that all sections of the wiring harness follow their original routing; use new cable-ties to secure the harness in position, keeping it away from sources of heat and abrasion.*

b)   *On vehicles with manual transaxle, refer to Chapter 7A and reconnect the gear shift mechanism to the transaxle, then check the overall operation of the gear shift mechanism. If necessary, adjust the gear shift rod/cables.*

c)   *Refer to Chapter 8 and reconnect the cable to the transaxle, then check the operation of the automatic adjustment mechanism.*

d)   *On vehicles with automatic transaxle, refer to Chapter 7B and reconnect the selector cable to the transaxle, then check (and if necessary adjust) the overall operation of the gear selection mechanism.*

e)   *Refer to Chapter 11 and install the lock carrier assembly to the front of the vehicle; ensure that all wiring harness connections are remade correctly and tighten the retaining fasteners to the specified torque.*

f)   *Ensure that all hoses are correctly routed and are secured with the correct hose clamps, where applicable. If the hose clamps originally installed were of the crimp variety, they cannot be used again; worm-drive clamps must be installed in their place, unless otherwise specified.*

g)   *Refill the cooling system as described in Chapter 1.*

h)   *Refill the engine with appropriate grades and quantities of oil, as detailed in Chapter 1.*

Diesel models

i)   *With reference to Chapter 6, after reconnecting the cold start accelerator cable to fuel injection pump, check and if necessary adjust the operation of the cold start acceleration system.*

Gasoline models

j)   *With reference to Chapter 4A or B as applicable, reconnect the throttle cable and adjust it as necessary.*

k)   *With reference to Chapter 5B, check and adjust the engine idle speed.*

All models

61   When the engine is started for the first time, check for air, coolant, lubricant and fuel leaks from manifolds, hoses etc. If the engine has been overhauled, read the cautionary notes in Section 14 before attempting to start it.

2C

3   Engine overhaul - general information

It is much easier to disassemble and work on the engine if it is mounted on a portable engine stand. These stands can often be hired from a tool hire shop. Before the engine is mounted on a stand, the flywheel should be removed, so that the stand bolts can be tightened into the end of the cylinder block/crankcase.

If a stand is not available, it is possible to disassemble the engine with it blocked up on a sturdy workbench, or on the floor. Be very careful not to tip or drop the engine when working without a stand.

If you intend to obtain a reconditioned engine, all ancillaries must be removed first, to be transferred to the replacement engine (just as they will if you are doing a complete engine overhaul yourself). These components include the following:

Gasoline engines
a) *Power steering pump (Chapter 10) - where applicable.*
b) *Air conditioning compressor (Chapter 3) - where applicable.*
c) *Alternator (including mounting brackets) and starter motor (Chapter 5A).*
d) *The ignition system and engine management components including all sensors, distributor, spark plug wires and spark plugs (Chapters 1 and 5).*
e) *The fuel injection system components (Chapters 4A and B).*
f) *All electrical switches, actuators and sensors, and the engine wiring harness (Chapters 4A and B, Chapter 5B).*
g) *Intake and exhaust manifolds (Chapter 2C).*
h) *The engine oil level dipstick and its tube (Chapter 2C).*
i) *Engine mount (Chapter 2A and B).*
j) *Flywheel/driveplate (Chapter 2C).*
k) *Clutch components (Chapter 8) - manual transaxle.*

Diesel engines
a) *Power steering pump (Chapter 10) - where applicable.*
b) *Air conditioning compressor (Chapter 3) - where applicable.*
c) *Alternator (including mounting brackets) and starter motor (Chapter 5A).*
d) *The glow plug/pre-heating system components (Chapter 5C)*
e) *All fuel system components, including the fuel injection pump, all sensors and actuators (Chapter 4C).*
f) *The vacuum pump (Chapter 2B).*
g) *All electrical switches, actuators and sensors, and the engine wiring harness (Chapters 4A and B, Chapter 5B).*
h) *Intake and exhaust manifolds and the turbocharger (Chapter 2C).*
i) *The engine oil level dipstick and its tube (Chapter 2C).*
j) *Engine mount (Chapters 2A and B).*
k) *Flywheel/driveplate (Chapter 2C).*
l) *Clutch components (Chapter 8) - manual transaxle.*

Note: *When removing the external components from the engine, pay close attention to details that may be helpful or important during refitting. Note the installed position of gaskets, seals, spacers, pins, washers, bolts, and other small components.*

If you are obtaining a "short block" (which consists of the engine cylinder block/crankcase, crankshaft, pistons and connecting rods, all fully assembled), then the cylinder head, oil pan and baffle plate, oil pump, timing belt (together with its tensioner and covers), drivebelt (together with its tensioner), water pump, thermostat housing, coolant outlet elbows, oil filter housing and where applicable oil cooler will also have to be removed.

If you are planning a complete overhaul, the engine can be disassembled in the order given below:
a) *Intake and exhaust manifolds.*
b) *Timing belt, sprockets and tensioner.*
c) *Cylinder head.*
d) *Flywheel/driveplate.*
e) *Oil pan.*
f) *Oil pump.*
g) *Piston/connecting rod assemblies.*
h) *Crankshaft.*

4.6 Keep groups of components together in labeled bags or boxes

Before beginning the disassembling and overhaul procedures, make sure that you have all of the tools necessary. Refer to *"Maintenance techniques, tools and working facilities"* for further information.

4   Cylinder head - disassembly, cleaning, inspection and reassembly

Note: *New and reconditioned cylinder heads will be available from the original manufacturer, and from engine overhaul specialists. It should be noted that some specialist tools are required for the disassembly and inspection procedures, and new components may not be readily available. It may, therefore, be more practical and economical for the home mechanic to purchase a reconditioned head, rather than to disassemble, inspect and recondition the original head.*

Disassembly
Refer to illustration 4.6
1   Remove the cylinder head from the engine block, and separate the intake and exhaust manifolds from it, as described in Part A or B of this Chapter.
2   On diesel models, remove the injectors and glow plugs, as described Chapter 4C and Chapter 5C.
3   Refer to Chapter 3 and remove the coolant outlet elbow together with its gasket/O-ring.
4   Where applicable, unscrew the coolant sensor and oil pressure switch from the cylinder head.
5   Refer to Chapter 2A or B as applicable and remove the timing belt sprocket from camshaft.
6   It is important that groups of components are kept together when they are removed and, if they are still serviceable, refitted in the same groups. If they are refitted randomly, accelerated wear leading to early failure will occur. Stowing groups of components in plastic bags or storage bins will help to keep everything in the right order - label them according to their installed location, e.g. 'No. 1 exhaust', 'No. 2 intake', etc. **(see illustration)**. (Note that No.1 cylinder is nearest the timing belt end of the engine.)
7   Check that the manufacturer's identification markings are visible on camshaft bearing caps; if none can be found, make your own using a scriber or center-punch.
8   The camshaft bearing cap retaining nuts must be removed progressively and in sequence to avoid stressing the camshaft, as follows:

Diesel engine
9   Loosen the retaining nuts from bearing caps Nos. 5, 1 and 3 first, then at bearing caps 2 and 4. Loosen the nuts alternately and diagonally half a turn at a time until they can be removed by hand. **Note:** *Camshaft bearing caps are numbered 1 to 5 from the timing belt end.*

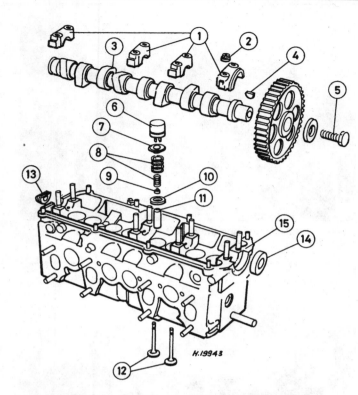

4.10 Cylinder head components - gasoline engines

1	Camshaft bearing cap	9	Valve stem seals
2	Nut	10	Valve spring lower seat
3	Camshaft	11	Valve guides
4	Woodruff key	12	Valves
5	Camshaft sprocket bolt	13	Plug
6	Hydraulic tappet	14	Camshaft oil seal
7	Valve spring retainer	15	Cylinder head casting
8	Valve springs		

Gasoline engines

10   Loosen and remove the retaining nuts from bearing caps Nos. 1 and 3 first, then at bearing caps 2 and 5. Loosen the nuts alternately and diagonally half a turn at a time until they can be removed by hand. **Note:** *Camshaft bearing caps are numbered 1 to 5 from the timing belt end - there is no bearing cap installed at cylinder No. 4* **(see illustration).**

All engines

Refer to illustrations 4.15a, 4.15b and 4.18

11   Slide the oil seal from the timing sprocket end of the camshaft and discard it; a new one must be used on reassembly.
12   Carefully lift the camshaft from the cylinder head; do not tilt it and support both ends as it is removed so that the journals and lobes are not damaged.
13   Lift the hydraulic lifters from their bores and store them with the valve contact surface facing downwards, to prevent the oil from draining out.
14   Make a note of the position of each tappet, as they must be fitted to the same valves on reassembly - accelerated wear leading to early failure will result if they are interchanged.
15   Turn the cylinder head over, and rest it on one side. Using a valve spring compressor, compress each valve spring in turn, extracting the keepers when the valve spring retainer has been pushed far enough down the valve stem to free them. If the retainer sticks, lightly tap the upper jaw of the spring compressor with a hammer to free it **(see illustrations).**
16   Release the valve spring compressor and remove the retainer, valve spring(s) and lower spring seat. **Note:** *Some engines may have concentric double valve springs, or single valve springs with no lower spring seat.*
17   Use a pair of pliers to extract the valve stem oil seal. Withdraw the valve itself from the head gasket side of the cylinder head. If the valve sticks in the guide, carefully deburr the end face with fine abrasive paper. Repeat this process for the remaining valves.
18   On diesel models, if the swirl chambers are badly coated or burned and are in need of replacement, insert a pin punch through each injector hole, and carefully drive out the swirl chambers using a mallet **(see illustration).**

Cleaning

19   Using a suitable degreasing agent, remove all traces of oil deposits from the cylinder head, paying particular attention to the journal bearings, hydraulic tappet bores, valve guides and oilways. Scrape off any traces of old gasket from the mating surfaces, taking care not to score or gouge them. If using emery paper, do not use a grade of less than 100. Turn the head over and using a blunt blade, scrape any carbon deposits from the combustion chambers and ports. **Caution:** *Do not erode the sealing surface of the valve seat. Finally, wash the entire head casting with a suitable solvent to remove the remaining debris.*
20   Clean the valve heads and stems using a fine wire brush. If the valve is heavily coated, scrape off the majority of the deposits with a blunt blade first, then use the wire brush. **Caution:** *Do not erode the sealing surface of the valve face.*
21   Thoroughly clean the remainder of the components using solvent and allow them to dry completely. Discard the oil seals, as new items must be installed when the cylinder head is reassembled.

2C

4.15a Valve spring compressor jaws correctly located on the retainer . . .

4.15b . . . and on the valve head

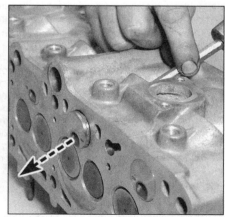

4.18 Swirl chamber removal (diesel models)

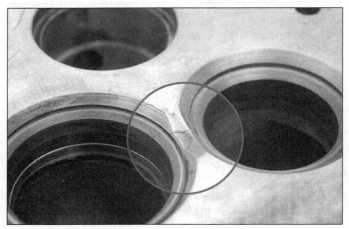

4.22 Look for cracking between the valve seats

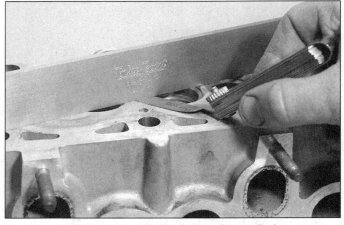

4.24 Measuring the distortion of the cylinder head gasket surface

Inspection

Refer to illustrations 4.22 and 4.24

Cylinder head casting

Note: *On diesel engines the cylinder heads cannot be milled; new or exchange units must be obtained.*

22   Examine the head casting closely to identify any damage sustained or cracks that may have developed **(see illustration)**. Pay particular attention to the areas around the mounting holes, valve seats and spark plug holes. If cracking is discovered between the valve seats, Volkswagen state that the cylinder head may be re-used, provided the cracks are no larger than 0.020-inch wide. More serious damage will mean the replacement of the cylinder head casting.

23   Moderately pitted and scorched valve seats can be repaired by lapping the valves in during reassembly, as described later in this Chapter. Badly worn or damaged valve seats may be restored by recutting; this is a highly specialized operation involving precision machining and accurate angle measurement and as such should be entrusted to a professional cylinder head re-builder.

24   Measure any distortion of the gasketed surfaces using a straight edge and a set of feeler gauges. Take one measurement longitudinally on both the intake and exhaust manifold mating surfaces. Take several measurements across the head gasket surface, to assess the level of distortion in all planes **(see illustration)**. Compare the measurements with the figures in the Specifications. On gasoline engines, if the head is distorted out of specification, it may be possible to repair it by smoothing down any high-spots on the surface with fine abrasive paper.

25   Minimum cylinder head heights (measured between the cylinder head gasket surface and the valve cover gasket surface), where quoted by the manufacturer, are listed in Specifications. If the cylinder head is to be professionally machined, bear in mind the following:

a) *The minimum cylinder head height dimension (where specified) must be adhered to.*

b) *The valve seats will need to be recut to suit the new height of the cylinder head, otherwise valve to piston crown contact may occur.*

c) *Before the valve seats can be recut, check that there is enough material left on the cylinder head to allow repair; if too much material is removed, the valve stem may protrude too far above the top of the valve guide and this would prevent the hydraulic lifters from operating correctly. Refer to a professional head rebuilder or machine shop for advice.* **Note:** *Depending on engine type, it may be possible to obtain new valves with shorter valve stems - refer to your VW dealer for advice.*

Camshaft

Refer to illustrations 4.26 and 4.30

26   The camshaft is identified by means of markings stamped onto the side of the shaft, between the intake and exhaust lobes - refer to

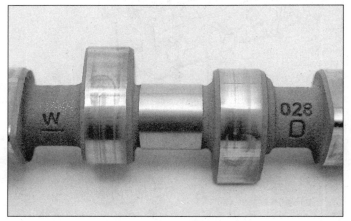

4.26 Camshaft identification markings

Specifications for details **(see illustration)**.

27   Visually inspect the camshaft for evidence of wear on the surfaces of the lobes and journals. Normally their surfaces should be smooth and have a dull shine; look for scoring, erosion or pitting and areas that appear highly polished - these are signs that wear has begun to occur. Accelerated wear will occur once the hardened exterior of the camshaft has been damaged, so always replace worn items. **Note:** *If these symptoms are visible on the tips of the camshaft lobes, check the corresponding tappet, as it will probably be worn as well.*

28   Where applicable, examine the distributor drive gear for signs of wear or damage. Slack in the drive caused by worn gear teeth will affect ignition timing.

29   If the machined surfaces of the camshaft appear discolored or "blued", it is likely that it has been overheated at some point, probably due to inadequate lubrication. This may have distorted the shaft, so check the runout as follows: place the camshaft between two V-blocks and using a dial indicator, measure the runout at the center journal. Compare your reading with the figure listed in the Specifications. If it exceeds this figure, camshaft replacement should be considered.

30   To measure the camshaft endplay, temporarily install the camshaft to the cylinder head, then install the first and last bearing caps and tighten the retaining nuts to the specified first stage torque setting - refer to *"Reassembly"* for details. Anchor a dial indicator to the timing pulley end of the cylinder head and align the gauge probe with the camshaft axis. Push the camshaft to one end of the cylinder head as far as it will travel, then rest the probe on the end of the camshaft, and zero the gauge display. Push the camshaft as far as it will go to the other end of the cylinder head, and record the gauge reading. Verify the reading by pushing the camshaft back to its original position and

4.30 Checking camshaft endplay using a dial indicator

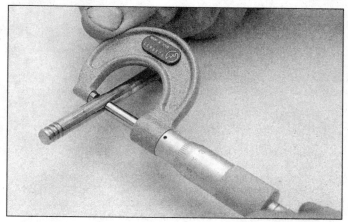

4.44 Measure the diameter of a valve stem with a micrometer

checking that the gauge indicates zero again **(see illustration)**. **Note:** *The hydraulic lifters must not be installed while this measurement is being taken.*

31 Check that the camshaft endplay measurement is within the limit listed in the Specifications. Wear outside of this limit is unlikely to be confined to any one component, so replacement of the camshaft, cylinder head and bearing caps must be considered; seek the advice of a cylinder head rebuilding specialist.

32 The difference between the outside diameters of the camshaft bearing surfaces and the internal diameters formed by the bearing caps and the cylinder head must now be measured, this dimension is known as the camshaft "running clearance" or oil clearance.

33 The dimensions of the camshaft bearing journals are not quoted by the manufacturer, so running clearance measurement by means of a micrometer and a bore gauge or internal vernier calipers cannot be recommended in this case.

34 Another (more accurate) method of measuring the running clearance involves the use of Plastigage. This is a soft, plastic material supplied in thin "sticks" of about the same diameter as a sewing needle. Lengths of Plastigage are cut to length as required, laid on the camshaft bearing journals and crushed as the bearing caps are temporarily installed and tightened. The Plastigage spreads widthwise as it is crushed; the running clearance can then be determined by measuring the increase in width using the card gauge supplied with the Plastigage kit.

35 The following Steps describe this measurement procedure step by step, but note that a similar method is used to measure the crankshaft running clearances; refer to the illustrations in Section 11 for further guidance.

36 Ensure that the cylinder head, bearing cap and camshaft bearing surfaces are completely clean and dry. Lay the camshaft in position in the cylinder head.

37 Lay a length of Plastigage on top of each of the camshaft bearing journals.

38 Lubricate each bearing cap with a little silicone release agent, then place them in position over the camshaft and tighten the retaining nuts down to the specified torque - refer to *Reassembly* later in this Section for guidance. **Note:** *Do not rotate the camshaft while the bearing caps are in place, as the measurements will be affected.*

39 Carefully remove the bearing caps again, lifting them vertically away from the camshaft to avoid disturbing the Plastigage. The Plastigage should remain on the camshaft bearing surface, squashed into a uniform sausage shape. If it disintegrates as the bearing caps are removed, re-clean the components and repeat the exercise, using a little more release agent on the bearing cap.

40 Hold the scale card supplied with the kit against each bearing journal, and match the width of the crushed Plastigage with the graduated markings on the card, use this to determine the running clearances.

41 Compare the camshaft running clearance measurements with

those listed in the Specifications; if any are outside the specified tolerance, the camshaft and cylinder head should be replaced.

42 Note that undersize camshafts with bearing shells may be obtained from VW dealers, but only as part of an exchange cylinder head package.

43 On completion, remove the bearing caps and camshaft, and clean of all remaining traces of Plastigage and silicone release agent.

Valves and associated components

Refer to illustrations 4.44, 4.49, 4.50 and 4.51
Note: *On all engines, the valve heads cannot be re-cut (although they may be lapped in); new or exchange units must be obtained.*

44 Examine each valve closely for signs of wear. Inspect the valve stems for wear ridges, scoring or variations in diameter; measure their diameters at several points along their lengths with a micrometer **(see illustration)**.

45 Check the overall length of each valve, and compare the measurements with the figure in the Specifications.

46 The valve heads should not be cracked, badly pitted or charred. Note that light pitting of the valve head can be rectified by grinding-in the valves during reassembly, as described later in this Section.

47 Check that the valve stem end face is free from excessive pitting or indentation; this would be caused by defective hydraulic lifters.

48 Place the valves in a V-block and using a dial indicator, measure the runout at the valve head. A maximum figure is not quoted by the manufacturer, but the valve should be replaced if the runout appears excessive.

49 Insert each valve into its respective guide in the cylinder head and set up a dial indicator against the edge of the valve head. With the

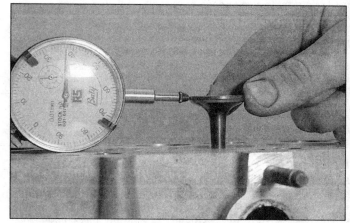

4.49 Measure the maximum deflection of the valve in its guide, using a dial indicator

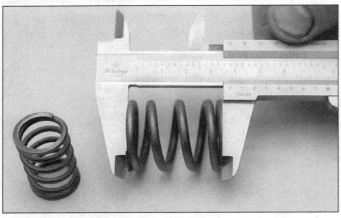

4.50 Measure the free length of each of the valve springs

4.51 Checking the squareness of a valve spring

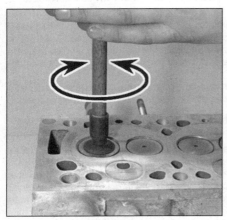

4.54 Grinding-in a valve

4.59a Fitting a swirl chamber (diesel engines)

4.59b Swirl chamber locating recess

valve end face flush with the top of the valve guide, measure the maximum side to side deflection of the valve in its guide **(see illustration)**. If the measurement is out of tolerance, the valve and valve guide should be replaced as a pair. **Note:** *Valve guides are an interference fit in the cylinder head and their removal requires access to a hydraulic press. For this reason, it would be wise to entrust the job to an engineering workshop or head rebuilding specialist.*

50   Using vernier calipers, measure the free length of each of the valve springs. As a manufacturer's figure is not quoted, the only way to check the length of the springs is by comparison with a new component. Note that valve springs are usually replaced during a major engine overhaul **(see illustration)**.

51   Stand each spring on its end on a flat surface, against an engineers square **(see illustration)**. Check the squareness of the spring visually; if it appears distorted, replace the spring.

52   Measuring valve spring pre-load involves compressing the valve by applying a specified weight and measuring the reduction in length. This may be a difficult operation to conduct in the home workshop, so it would be wise to consult your local dealer service department other repair shop for assistance. Weakened valve springs will at best, increase engine running noise and at worst, cause poor compression, so defective items should be replaced.

Reassembly

Refer to illustrations 4.54, 4.59a, 4.59b, 4.60, 4.61, 4.62a, 4.62b, 4.63a, 4.63b, 4.64, 4.65a, 4.65b, 4.67, 4.68a, 4.68b, 4.71 and 4.72

Caution: *Unless all new components are to be used, maintain groups when installing valve train components - do not mix components between cylinders and ensure that components are installed in their original positions.*

53   To achieve a gas-tight seal between the valves and their seats, it will be necessary to grind, or 'lap', the valves in. To complete this process you will need a quantity of fine/coarse grinding paste and a grinding tool - this can either be of the dowel and rubber sucker type, or the automatic type which are driven by a rotary power tool.

54   Smear a small quantity of *fine* grinding paste on the sealing face of the valve head. Turn the cylinder head over so that the combustion chambers are facing upwards and insert the valve into the correct guide. Attach the grinding tool to the valve head and using a backward/forward rotary action, grind the valve head into its seat. Periodically lift the valve and rotate it to redistribute the grinding paste **(see illustration)**.

55   Continue this process until the contact between valve and seat produces an unbroken, matte gray ring of uniform width, on both faces. Repeat the operation for the remaining valves.

56   If the valves and seats are so badly pitted that coarse grinding paste must be used, check first that there is enough material left on both components to make this operation worthwhile - if too little material is left remaining, the valve stems may protrude too far above their guides, impeding the correct operation of the hydraulic lifters. Refer to a machine shop or cylinder head rebuilding specialist for advice.

57   Assuming the repair is feasible, work as described in the previous Step but use the coarse grinding paste initially, to achieve a dull finish on the valve face and seat. Then, wash off coarse paste with solvent and repeat the process using fine grinding paste to obtain the correct finish.

58   When all the valves have been ground in, remove all traces of grinding paste from the cylinder head and valves with solvent, and allow them to dry completely.

59   Where necessary on diesel engines, install new swirl chambers by

4.60 Measuring swirl chamber projection using a dial indicator

4.61 Install the lower spring seat in place, with the convex face facing the cylinder head

4.62a Lubricate the valve stem with clean engine oil and insert it into the guide

2C

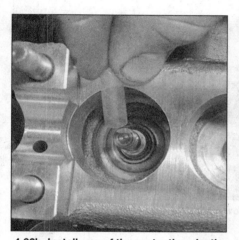

4.62b Install one of the protective plastic sleeves over the valve end face

4.63a Install a new valve stem seal over the valve

4.63b Use a long-reach socket to press on the oil seal

driving them squarely into their housings with a mallet - use a block of wood to protect the face of the swirl chamber. Note the locating recess on the side of the chamber and the corresponding groove in the housing **(see illustrations)**.

60   On completion, the projection of the swirl chamber from the face of the cylinder head must be measured using a dial indicator and compared with the limit quoted in the Specifications **(see illustration)**. If this limit is exceeded, there is a risk that the chamber may be struck by the piston, and in this case the advice of a professional cylinder head rebuilder or machine shop should be sought.

61   Turn the head over and place it on a stand, or wooden blocks. Where applicable, install the first lower spring seat into place, with the convex side facing the cylinder head **(see illustration)**.

62   Working on one valve at a time, lubricate the valve stem with clean engine oil, and insert it into the guide. Install one of the protective plastic sleeves supplied with the new valve stem oil seals over the valve end face - this will protect the oil seal while it is being installed **(see illustrations)**.

63   Dip a new valve stem seal in clean engine oil, and carefully push it over the valve and onto the top of the valve guide - take care not to damage the stem seal as it passes over the valve end face. Use a suitable long reach socket to press it firmly into position **(see illustrations)**.

64   Locate the valve spring(s) over the valve stem **(see illustration)**. Where a lower spring seat is installed, ensure that the springs locate squarely on the stepped surface of the seat. **Note:** *Depending on age*

4.64 Installing a valve spring

and specification, engines may have either concentric double valve springs, or single valve springs with no lower spring seat.

65   Install the retainer over the top of the springs, then using a valve spring compressor, compress the springs until the retainer is pushed beyond the keeper grooves in the valve stem. Install the keepers, using

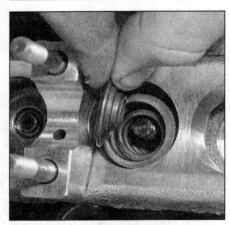

4.65a Install the retainer over the top of the valve spring

4.65b Use grease to hold the keepers in place

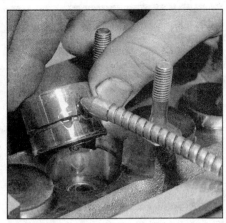

4.67 Install the lifters into their bores in the cylinder head

4.68a Lubricate the camshaft bearings with clean engine oil . . .

4.68b . . . then lower the camshaft into position on the cylinder head

a dab of grease to hold the two halves in the grooves **(see illustrations)**. Gradually release the spring compressor, checking that the keeper remains correctly seated as the spring extends. When correctly seated, the retainer should force the keepers together, and hold them securely in the grooves in the end of the valve.

66   Repeat this process for the remaining sets of valve components. To settle the components after installation, strike the end of each valve stem with a mallet, using a block of wood to protect the stem from damage. Check before progressing any further that the keepers remain firmly held in the end of the valve stem by the retainer.

67   Smear some clean engine oil onto the sides of the hydraulic

lifters, and install them into position in their bores in the cylinder head. Push them down until they contact the valves, then lubricate the camshaft lobe contact surfaces **(see illustration)**.

68   Lubricate the camshaft and cylinder head bearing journals with clean engine oil, then carefully lower the camshaft into position on the cylinder head. Support the ends of the shaft as it is inserted, to avoid damaging the lobes and journals **(see illustrations)**.

69   Turn the camshaft so that the lobes for No 1 cylinder are pointing upwards.

70   On diesel engines, with reference to Chapter 2B, lubricate the lip of a new camshaft oil seal with clean engine oil and locate it over the end of the camshaft.

71   Slide the seal along the camshaft until it locates in the lower half of its housing in the cylinder head **(see illustration)**.

72   Oil the upper surfaces of the camshaft bearing journals, then install the bearing caps in place **(see illustration)**. Ensure that they are

4.71 Installing the camshaft oil seal (diesel engine)

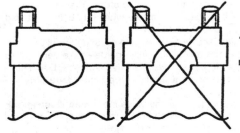

4.72 The camshaft bearing cap mounting holes are drilled off-center

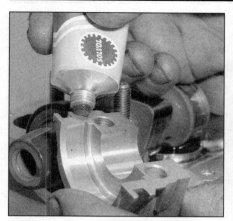

4.75 Smear the mating surfaces of cap Nos. 1 and 5 with sealant (diesel engine)

4.77 Install the coolant outlet elbow, using a new gasket/O-ring as necessary

4.80 Installing a new camshaft oil seal (gasoline engine)

installed the right way around and in the correct locations, then install and tighten the retaining nuts, as follows:

Diesel engine

Refer to illustration 4.75

73   The bearing cap mounting holes are drilled off-center; ensure that they are installed the correct way.

74   Install caps Nos. 2 and 4 over the camshaft, and tighten the retaining nuts alternately and diagonally to the specified torque.

75   Smear the mating surfaces of caps Nos. 1 and 5 with sealant then install them, together with cap No 3, over the camshaft and tighten the nuts to the specified first stage torque **(see illustration)**.

Gasoline engines

Refer to illustration 4.76

76   With reference to Chapter 2A or B as applicable, lubricate the lip of a new camshaft oil seal with clean engine oil, and locate it over the end of the camshaft. Using a mallet and a long-reach socket of an appropriate diameter, drive the seal squarely into its housing until it bears against the inner stop - do not attempt to force it in any further **(see illustration)**.

All engines

Refer to illustration 4.77

77   Install the coolant outlet elbow, using a new gasket/O-ring as necessary **(see illustration)**. Tighten the retaining bolts securely.

78   Install the coolant sensor and oil pressure switch.

79   With reference to Chapter 2A or B as applicable, carry out the following:

a) *Install the timing belt sprocket to the camshaft.*

b) *Install the intake and exhaust manifolds, complete with new gaskets.*

80   On diesel engines, refer to Chapter 4C and 5C, and install the fuel injectors and glow plugs.

81   Refer to Chapter 2A or B as applicable install the cylinder head to the cylinder block.

---

5   Pistons and connecting rods - removal and inspection

Removal

Refer to illustrations 5.5a, 5.5b, 5.6, 5.7, 5.10a and 5.10b

1   Refer to Part A or B of this Chapter (as applicable) and remove the cylinder head, flywheel, oil pan and baffle plate, oil pump and pickup.

2   Inspect the tops of the cylinder bores; any wear ridges found at the point where the pistons reach top dead center must be removed; otherwise the pistons may be damaged when they are pushed out of their bores. This can be accomplished with a scraper or ridge reamer.

3   Scribe the number of each piston on its crown, to allow identification later; note that No 1 is at the timing belt end of the engine.

4   Using a set of feeler gauges, measure the thrust clearance at each connecting rod and record the measurements for later reference.

5   Where applicable, remove the retaining screw and withdraw the piston cooling jets from their mounting holes. On some engines the jet mounting incorporates a pressure relief valve, take care to avoid damaging it during removal **(see illustrations)**.

6   Rotate the crankshaft until pistons No 1 and 4 are at bottom dead

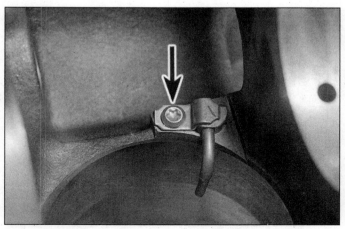

5.5a Remove the piston cooling jet retaining screw (arrow) . . .

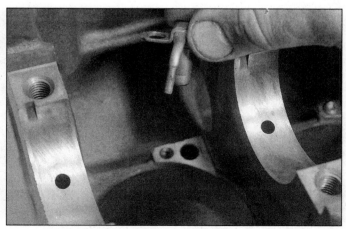

5.5b . . . and withdraw the jet from its mounting hole

2C

5.6 Mark the connecting rod caps and connecting rods with their respective piston numbers (arrows)

5.7 Pad the bolt threads with tape

center. Unless they are already identified, mark the bearing caps and connecting rods with their respective piston numbers, using a center-punch or a scribe **(see illustration)**. Note the orientation of the bearing caps in relation to the connecting rod; it may be difficult to see the manufacturer's markings at this stage, so scribe alignment arrows on them both to ensure correct reassembly. Unbolt the bearing cap bolts/nuts, half a turn at a time, until they can be removed by hand. Recover the bottom shell bearing, and install it in the cap for safe keeping. Note that if the shell bearings are to be re-used, they must be installed to the same connecting rod.

7   On certain engines, the bearing cap bolts will remain in the connecting rod; in this case the threads of the bolts should be padded with insulating tape, to prevent them from scratching the crankpins when the pistons are removed from their bores **(see illustration)**.

8   Drive the pistons out of the top of their bores by pushing on the underside of the piston crown with a piece of dowel or a hammer handle. As the piston and connecting rod emerge, recover the top shell bearing and tape it to the connecting rod for safekeeping.

9   Turn the crankshaft through half a turn and working as described above, remove No 2 and 3 pistons and connecting rods. Remember to maintain the components in their cylinder groups, while they are in a disassembled state.

10   Insert a small flat-bladed screwdriver into the removal slot and pry the wrist pin circlips from each piston. Push out the wrist pin, and separate the piston and connecting rod **(see illustrations)**. Discard the circlips as new items must be installed on reassembly. If the pin proves difficult to remove, heat the piston to 140-degrees F with hot water - the resulting expansion will then allow the two components to be separated.

Inspection

Refer to illustrations 5.11, 5.18, 5.19 and 5.21

11   Before an inspection of the pistons can be carried out, the existing piston rings must be removed, using a removal/installation tool, or an old feeler gauge if such a tool is not available. Always remove the upper piston rings first, expanding them to clear the piston crown. The rings are very brittle and will snap if they are stretched too much - sharp edges are produced when this happens, so protect your eyes and hands. Discard the rings on removal, as new items must be installed when the engine is reassembled **(see illustration)**.

12   Use a section of old piston ring to scrape the carbon deposits out of the ring grooves, taking care not to score or gouge the edges of the groove.

13   Carefully scrape away all traces of carbon from the top of the piston. A hand-held wire brush (or a piece of fine emery cloth) can be used, once the majority of the deposits have been scraped away. Be careful not to remove any metal from the piston, as it is relatively soft. **Note:** *Take care to preserve the piston number markings that were made during removal.*

14   Once the deposits have been removed, clean the pistons and connecting rods with solvent, and dry thoroughly. Make sure that the oil return holes in the ring grooves are clear.

15   Examine the piston for signs of terminal wear or damage. Some normal wear will be apparent, in the form of a vertical 'grain' on the piston thrust surfaces and a slight looseness of the top compression ring in its groove. Abnormal wear should be carefully examined, to assess whether the component is still serviceable and what the cause of the wear might be.

5.10a Insert a small screwdriver into the slot and pry out the circlips

5.10b Push out the wrist pin and separate the piston and connecting rod

5.11 Piston rings can be removed using an old feeler gauge

5.18 Using a micrometer, measure the diameter of all four pistons

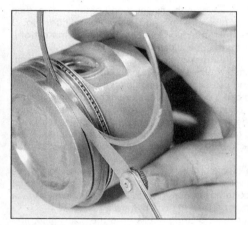

5.19 Measuring the piston ring-to-groove clearance using a feeler gauge

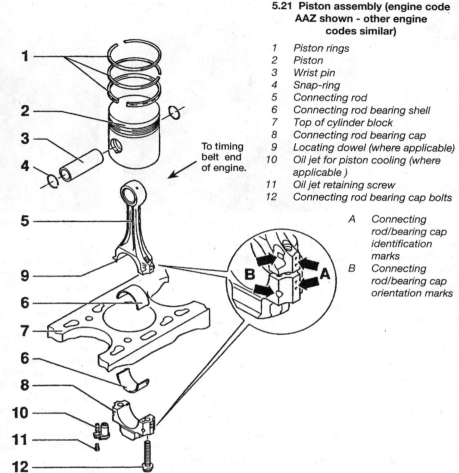

5.21 Piston assembly (engine code AAZ shown - other engine codes similar)

1 Piston rings
2 Piston
3 Wrist pin
4 Snap-ring
5 Connecting rod
6 Connecting rod bearing shell
7 Top of cylinder block
8 Connecting rod bearing cap
9 Locating dowel (where applicable)
10 Oil jet for piston cooling (where applicable)
11 Oil jet retaining screw
12 Connecting rod bearing cap bolts

A Connecting rod/bearing cap identification marks
B Connecting rod/bearing cap orientation marks

To timing belt end of engine.

2C

16 Scuffing or scoring of the piston skirt may indicate that the engine has been overheating, through inadequate cooling, lubrication or abnormal combustion temperatures. Scorch marks on the skirt indicate that blow-by has occurred, perhaps caused by worn bores or piston rings. Burnt areas on the piston crown are usually an indication of pre-ignition, pinking or detonation. In extreme cases, the piston crown may be melted by operating under these conditions. Corrosion pit marks in the piston crown indicate that coolant has seeped into the combustion chamber and/or the crankcase. The faults causing these symptoms must be corrected before the engine is brought back into service, or the same damage will recur.

17 Check the pistons, connecting rods, wrist pins and bearing caps for cracks. Lay the connecting rods on a flat surface, and look along the length to see if it appears bent or twisted. If you have doubts about their condition, get them measured at an engineering workshop. Inspect the small-end bushing bearing for signs of wear or cracking.

18 Using a micrometer, measure the diameter of all four pistons at a point 3/8-inch from the bottom of the skirt, at right angles to the wrist pin axis (see illustration). Compare the measurements with those listed in the Specifications. If the piston diameter is out of the tolerance band listed for its particular size, then it must be replaced. Note: If the cylinder block was re-bored during a previous overhaul, oversize pistons may have been installed. Record the measurements and use them to check the piston clearances when the cylinder bores are measured, later in this Chapter.

19 Hold a new piston ring in the appropriate groove and measure the ring-to-groove clearance using a feeler gauge (see illustration). Note that the rings are of different widths, so use the correct ring for the

groove. Compare the measurements with those listed; if the clearances are outside of the tolerance band, then the piston must be replaced. Confirm this by checking the width of the piston ring with a micrometer.

20 Using internal/external vernier calipers, measure the connecting rod small-end internal diameter and the wrist pin external diameter. Subtract the wrist pin diameter from the small-end diameter to obtain the clearance. If this measurement is outside its specification, then the piston and connecting rod bushing will have to be resized and a new wrist pin installed. An automotive machine shop will have the equipment needed to undertake a job of this nature.

21 The orientation of the piston with respect to the connecting rod must be correct when the two components are reassembled. The piston crown is marked with an arrow (which can easily be obscured by carbon deposits); this must point towards the timing belt end of the engine when the piston is installed in the bore. The connecting rod and its corresponding bearing cap both have recesses machined into them close to their mating surfaces - these recesses must both face in the same direction as the arrow on the piston crown (i.e. towards the timing belt end of the engine) when correctly installed (see illustration). Reassemble the two components to satisfy this requirement. Note: On certain engines, the connecting rod big-ends are provided with offset dowels which locate in holes in the bearing caps.

22 Lubricate the wrist pin and small-end bushing with clean engine oil. Slide the pin into the piston, engaging the connecting rod small-end. Attach two new circlips to the piston at either end of the wrist pin, such that their open ends are facing 180-degrees away from the removal slot in the piston. Repeat this operation for the remaining pistons.

6.4 Measuring crankshaft endplay using a dial indicator

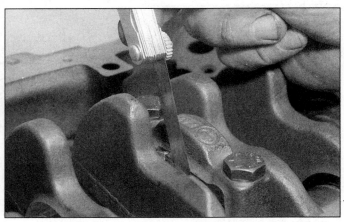

6.5 Measuring crankshaft endplay using feeler gauges

6   Crankshaft - removal and inspection

Removal

Refer to illustrations 6.4, 6.5, 6.6 and 6.8

1   **Note:** *If no work is to be done on the pistons and connecting rods, then removal the cylinder head and pistons will not be necessary. Instead, the pistons need only be pushed far enough up the bores so that they are positioned clear of the crankpins. The use of an engine stand is strongly recommended.*

2   With reference to Chapter 2A or B as applicable, carry out the following:

   a) *Remove the crankshaft timing belt sprocket.*
   b) *Remove the clutch components and flywheel.*
   c) *Remove the oil pan, baffle plate, oil pump and pickup.*
   d) *Remove the front and rear crankshaft oil seals and their housings.*

3   Remove the pistons and connecting rods, as described in Section 5 (refer to the Note above).

4   Carry out a check of the crankshaft endplay, as follows. **Note:** *This can only be accomplished when the crankshaft is still installed in the cylinder block/crankcase, but is free to move.* Set up a dial indicator so that the probe is in line with the crankshaft axis and is in contact with a fixed point on end of the crankshaft. Push the crankshaft along its axis to the end of its travel, and then zero the gauge. Push the crankshaft fully the other way, and record the endplay indicated on the dial **(see illustration)**. Compare the result with the figure given in the Specifications and establish whether new thrustwashers are required.

5   If a dial gauge is not available, feeler gauges can be used. First push the crankshaft fully towards the flywheel end of the engine, then use a feeler gauge to measure the gap between cylinder No. 2 crankpin web and the main bearing thrustwasher **(see illustration)**. Compare the results with the Specifications.

6   Observe the manufacturer's identification marks on the main bearing caps. The number relates to the position in the crankcase, as counted from the timing belt end of the engine **(see illustration)**.

7   Loosen the main bearing cap bolts one quarter of a turn at a time, until they can be removed by hand. Using a soft-faced mallet, strike the caps lightly to free them from the crankcase. Recover the lower main bearing shells, taping them to the cap for safekeeping. Mark them with indelible ink to aid identification, but do not score or scratch them in any way.

8   Carefully lift the crankshaft out, taking care not to dislodge the upper main bearing shells. It would be wise to get an assistant's help, as the crankshaft is quite heavy. Set it down on a clean, level surface and chock it with wooden blocks to prevent it from rolling.

9   Extract the upper main bearing shells from the crankcase, and tape them to their respective bearing caps. Remove the two thrustwasher bearings from either side of No. 3 crank web.

10   With the shell bearings removed, observe the recesses machined into the bearing caps and crankcase - these provide location for the lugs which protrude from the shell bearings and so prevent them from being installed incorrectly.

Inspection

Refer to illustration 6.13

11   Wash the crankshaft in a suitable solvent and allow it to dry. Flush the oil holes thoroughly, to ensure that are not blocked - use a pipe

6.6 Manufacturer's identification markings on the main bearing caps (arrow)

6.8 Lifting the crankshaft from the crankcase

6.13 Use a micrometer to measure the diameter of each main bearing journal

7.2 Check the intermediate shaft endplay using a dial indicator

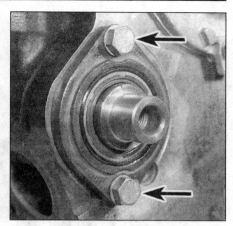

7.3a Loosen the retaining bolts (arrow) . . .

cleaner or a needle brush if necessary. Remove any sharp edges from the edge of the hole which may damage the new bearings when they are installed.

12 Inspect the main bearing and crankpin journals carefully; if uneven wear, cracking, scoring or pitting are evident then the crankshaft should be reground by an automotive machine shop, and installed with undersize bearings.

13 Use a micrometer to measure the diameter of each main bearing journal **(see illustration)**. Taking a number of measurements on the surface of each journal will reveal if it is worn unevenly. Differences in diameter measured at 90-degree intervals indicate that the journal is out of round. Differences in diameter measured along the length of the journal, indicate that the journal is tapered. Again, if wear is detected, the crankshaft must be reground by an automotive machine shop, and undersize bearings will be needed (refer to "Reassembly").

14 Check the oil seal journals at either end of the crankshaft. If they appear excessively scored or damaged, they may cause the new seals to leak when the engine is reassembled. It may be possible to repair the journal; seek the advice of an automotive machine shop or your VW dealer.

15 Measure the crankshaft runout by setting up a dial indicator on the center main bearing and rotating the shaft in V-blocks. The maximum deflection of the gauge will indicate the runout. Take precautions to protect the bearing journals and oil seal mating surfaces from damage during this procedure. A maximum runout figure is not quoted by the manufacturer, but use the figure of 0.002 as a rough guide. If the runout exceeds this figure, crankshaft replacement should be considered - consult your VW dealer or an engine rebuild-

ing specialist for advice.

16 Refer to Section 8 for details of main and rod bearing inspection.

7 Intermediate shaft - removal and installation

Removal

Refer to illustrations 7.2, 7.3a, 7.3b, 7.3c and 7.3d

1 Refer to Chapter 1 and carry out the following:
  a) *Remove the timing belt.*
  b) *Remove the intermediate shaft sprocket.*

2 Before the shaft is removed, the endplay must be checked. Anchor a dial indicator to the cylinder block with its probe in line with the intermediate shaft center axis. Push the shaft into the cylinder block to the end of its travel, zero the dial indicator and then draw the shaft out to the opposite end of its travel. Record the maximum deflection and compare the figure with that listed in Specifications - replace the shaft if endplay exceeds this limit **(see illustration)**.

3 Loosen the retaining bolts and withdraw the intermediate shaft flange. Recover the O-ring seal, then press out the oil seal **(see illustrations)**.

4 Withdraw the intermediate shaft from the cylinder block and inspect the drive gear at the end of the shaft; if the teeth show signs of excessive wear, or are damaged in any way, the shaft should be replaced.

5 If the oil seal has been leaking, check the shaft mating surface for signs of scoring or damage.

2C

7.3b . . . and withdraw the intermediate shaft flange

7.3c Press out the oil seal . . .

7.3d . . . then recover the O-ring seal

8.6 To clean the cylinder block threads, run a correct-size tap into each of the holes

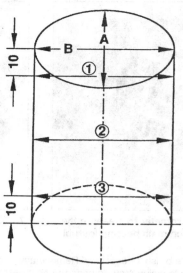

8.12 Bore measurement points

Installation

6   Liberally oil the intermediate shaft bearing surfaces and drive gear, then carefully guide the shaft into the cylinder block and engage the journal at the leading end with its support bearing.

7   Press a new shaft oil seal into its housing in the intermediate shaft flange and install a new O-ring seal to the inner sealing surface of the flange.

8   Lubricate the inner lip of the seal with clean engine oil, and slide the flange and seal over the end of the intermediate shaft. Ensure that the O-ring is correctly seated, then install the flange retaining bolts and tighten them to the specified torque. Check that the intermediate shaft can rotate freely.

9   With reference to Chapter 2A or B, carry out the following:

a) Install the timing belt sprocket to the intermediate shaft and tighten the center bolt to the specified torque.

b) Install the timing belt. On gasoline models, follow the intermediate sprocket alignment instructions carefully to ensure that the distributor drive gear alignment is preserved.

---

8   Cylinder block/crankcase casting - cleaning and inspection

Cleaning

Refer to illustration 8.6

1   Remove all external components and electrical switches/sensors from the block. For complete cleaning, the core plugs should ideally be removed. Drill a small hole in the plugs, then insert a self-tapping screw into the hole. Extract the plugs by pulling on the screw with a pair of grips, or by using a slide hammer.

2   Scrape all traces of gasket and sealant from the cylinder block/crankcase, taking care not to damage the sealing surfaces.

3   Remove all oil gallery plugs (where installed). The plugs are usually very tight - they may have to be drilled out, and the holes re-tapped. Use new plugs when the engine is reassembled.

4   If the casting is extremely dirty, it should be steam-cleaned. After this, clean all oil holes and galleries one more time. Flush all internal passages with warm water until the water runs clear. Dry thoroughly, and apply a light film of oil to all mating surfaces and cylinder bores, to prevent rusting. If you have access to compressed air, use it to speed up the drying process, and to blow out all the oil holes and galleries. **Warning:** *Wear eye protection when using compressed air!*

5   If the castings are not very dirty, you can do an adequate cleaning job with hot, soapy water and a stiff brush. Take plenty of time, and do a thorough job. Regardless of the cleaning method used, be sure to clean all oil holes and galleries very thoroughly, and to dry all compo-

nents well. Protect the cylinder bores as described above, to prevent rusting.

6   All threaded holes must be clean, to ensure accurate torque readings during reassembly. To clean the threads, run the correct-size tap into each of the holes to remove rust, corrosion, thread sealant or sludge, and to restore damaged threads **(see illustration)**. If possible, use compressed air to clear the holes of debris produced by this operation. **Note:** *Take extra care to exclude all cleaning liquid from blind tapped holes, as the casting may be cracked by hydraulic action if a bolt is threaded into a hole containing liquid.*

7   Apply suitable sealant to the new oil gallery plugs, and insert them into the holes in the block. Tighten them securely.

8   If the engine is not going to be reassembled immediately, cover it with a large plastic bag to keep it clean; protect all mating surfaces and the cylinder bores as described above, to prevent rusting.

Inspection

Refer to illustrations 8.12 and 8.17

9   Visually check the casting for cracks and corrosion. Look for stripped threads in the threaded holes. If there has been any history of internal water leakage, it may be worthwhile having an engine overhaul specialist check the cylinder block/crankcase with professional equipment. If defects are found, have them replaced or if possible, repaired.

10   Check the cylinder bores for scuffing or scoring. Any evidence of this kind of damage should be cross-checked with an inspection of the pistons: see Section 5 of this Chapter. If the damage is in its early stages, it may be possible to repair the block by reboring it. Seek the advice of an engineering workshop before you progress.

11   To allow an accurate assessment of the wear in the cylinder bores to be made, their diameter must be measured at a number of points, as follows. Insert a bore gauge into cylinder bore No. 1 and take three measurements in line with the crankshaft axis; one at the top of the bore, roughly 3/8-inch below the bottom of the wear ridge, one halfway down the bore and one at a point roughly 3/8-inch the bottom of the bore. **Note:** *Stand the cylinder block squarely on a workbench during this procedure, inaccurate results may be obtained if the measurements are taken when the engine mounted on a stand.*

12   Rotate the bore gauge through 90-degrees, so that it is at right angles to the crankshaft axis and repeat the measurements detailed in Step 11 **(see illustration)**. Record all six measurements and compare them with the data listed in the Specifications. If the difference in diameter between any two cylinders exceeds the wear limit, or if any one cylinder exceeds its maximum bore diameter, then *all four* cylinders will have to be rebored and oversize pistons will have to be installed.

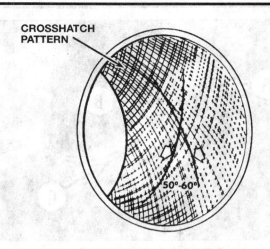

8.17 Cylinder bore honing pattern

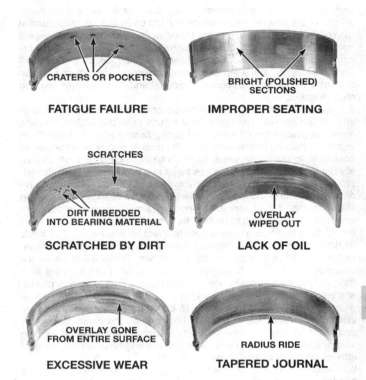

9.1 Typical bearing failures

2C

Note that the imbalances produced by not reboring all the cylinders together would render the engine unusable.

13   Use the piston diameter measurements recorded earlier (see Section 5) to calculate the piston-to-bore clearances. Figures are not available from the manufacturer, so seek the advice of your VW dealer or engine reconditioning specialist.

14   Place the cylinder block on a level work surface, crankcase downwards. Use a straight edge and a set of feeler gauges to measure the distortion of the cylinder head mating surface in both planes. A maximum figure is not quoted by the manufacturer, but use the figure of 0.002-inch as a rough guide. If the measurement exceeds this figure, repair may be possible by machining - consult your dealer for advice.

15   Before the engine can be reassembled, the cylinder bores must be honed. This process involves using an abrasive tool to produce a fine, cross-hatch pattern on the inner surface of the bore. This has the effect of seating the piston rings, resulting in a good seal between the piston and cylinder. There are two types of honing tool available to the home mechanic, both are driven by a rotary power tool, such as a drill. The 'bottle brush' hone is a stiff, cylindrical brush with abrasive stones bonded to its bristles. The more conventional surfacing hone has abrasive stones mounted on spring-loaded legs. For the inexperienced home mechanic, satisfactory results will be achieved more easily using the Bottle Brush hone. **Note:** *If you are unwilling to tackle cylinder bore honing, an engineering workshop will be able to carry out the job for you at a reasonable cost.*

16   Carry out the honing as follows; you will need one of the honing tools described above, a power drill/air wrench, a supply of clean rags, some honing oil and a pair of safety glasses.

17   Install the honing tool in the drill chuck. Lubricate the cylinder bores with honing oil and insert the honing tool into the first bore, compressing the stones to allow it to fit. Turn on the drill and as the tool rotates, move it up and down in the bore at a rate that produces a fine cross-hatch pattern on the surface. The lines of the pattern should ideally cross at about 50 to 60-degrees **(see illustration)**, although some piston ring manufacturer's may quote a different angle; check the literature supplied with the new rings. **Warning:** *Wear safety glasses to protect your eyes from debris flying off the honing tool.*

18   Use plenty of oil during the honing process. Do not remove any more material than is necessary to produce the required finish. When removing the hone tool from the bore, do not pull it out while it is still rotating; maintain the up/down movement until the chuck has stopped, then withdraw the tool while rotating the chuck by hand, in the normal direction of rotation.

19   Wipe out the oil and debris with a rag and proceed to the next bore. When all four bores have been honed, thoroughly clean the whole cylinder block in hot soapy water to remove all traces of honing oil and debris. The block can be considered clean when a clean rag, moistened with new engine oil does not pick up any gray residue when

wiped along the bore.

20   Apply a light coating of engine oil to the mating surfaces and cylinder bores to prevent rust forming. Store the block in a plastic bag until reassembly.

9   Main and connecting rod bearings - inspection and selection

Inspection

Refer to illustration 9.1

1   Even though the main and connecting rod bearings should be replaced during the engine overhaul, the old bearings should be retained for close examination, as they may reveal valuable information about the condition of the engine **(see illustration)**.

2   Bearing failure can occur due to lack of lubrication, the presence of dirt or other foreign particles, overloading the engine, or corrosion. Regardless of the cause of bearing failure, the cause must be corrected before the engine is reassembled, to prevent it from happening again.

3   When examining the bearing shells, remove them from the cylinder block/crankcase, the main bearing caps, the connecting rods and the connecting rod big-end bearing caps. Lay them out on a clean surface in the same general position as their location in the engine. This will enable you to match any bearing problems with the corresponding crankshaft journal. *Do not* touch any shell's internal bearing surface with your fingers while checking it, or the delicate surface may be scratched.

4   Dirt and other foreign matter gets into the engine in a variety of ways. It may be left in the engine during assembly, or it may pass through filters or the crankcase ventilation system. It may get into the oil, and from there into the bearings. Metal chips from machining operations and normal engine wear are often present. Abrasives are sometimes left in engine components after reconditioning, especially when

parts are not thoroughly cleaned using the proper cleaning methods. Whatever the source, these foreign objects often end up embedded in the soft bearing material, and are easily recognized. Large particles will not embed in the bearing, but will score or gouge the bearing and journal. The best prevention for this cause of bearing failure is to clean all parts thoroughly, and keep everything spotlessly-clean during engine assembly. Frequent and regular engine oil and filter changes are also recommended.

5   Lack of lubrication (or lubrication breakdown) has a number of interrelated causes. Excessive heat (which thins the oil), overloading (which squeezes the oil from the bearing face) and oil leakage (from excessive bearing clearances, worn oil pump or high engine speeds) all contribute to lubrication breakdown. Blocked oil passages, which usually are the result of misaligned oil holes in a bearing shell, will also oil-starve a bearing, and destroy it. When lack of lubrication is the cause of bearing failure, the bearing material is wiped or extruded from the steel backing of the bearing. Temperatures may increase to the point where the steel backing turns blue from overheating.

6   Driving habits can have a definite effect on bearing life. Full-throttle, low-speed operation (laboring the engine) puts very high loads on bearings, tending to squeeze out the oil film. These loads cause the bearings to flex, which produces fine cracks in the bearing face (fatigue failure). Eventually, the bearing material will loosen in pieces, and tear away from the steel backing.

7   Short-distance driving leads to corrosion of bearings, because insufficient engine heat is produced to drive off the condensed water and corrosive gases. These products collect in the engine oil, forming acid and sludge. As the oil is carried to the engine bearings, the acid attacks and corrodes the bearing material.

8   Incorrect bearing installation during engine assembly will lead to bearing failure as well. Tight-fitting bearings leave insufficient bearing running clearance, and will result in oil starvation. Dirt or foreign particles trapped behind a bearing shell result in high spots on the bearing, which lead to failure.

9   *Do not* touch any shell's internal bearing surface with your fingers during reassembly; there is a risk of scratching the delicate surface, or of depositing particles of dirt on it.

10   As mentioned at the beginning of this Section, the bearing shells should be replaced as a matter of course during engine overhaul; to do otherwise is false economy.

Selection

11   Main and big-end bearings for the engines described in this Chapter are available in standard sizes and a range of undersizes to suit reground crankshafts - refer to Specifications for details.

12   The running clearances will need to be checked when the crankshaft is with its new bearings. This procedure is described in the Section 11.

10   Engine overhaul - reassembly sequence

1   Before reassembly begins, ensure that all new parts have been obtained, and that all necessary tools are available. Read through the entire procedure to familiarize yourself with the work involved, and to ensure that all items necessary for reassembly of the engine are at hand. In addition to all normal tools and materials, thread-locking compound will be needed. A suitable tube of liquid sealant will also be required for the joint faces that are without gaskets. It is recommended that the manufacturer's own products are used, which are specially formulated for this purpose; the relevant product names are quoted in the text of each Section where they are required.

2   In order to save time and avoid problems, engine reassembly should ideally be carried out in the following order:

a) *Crankshaft.*
b) *Piston/connecting rod assemblies.*
c) *Oil pump (see Chapter 2A or B)*
d) *Oil pan (see Chapter 2A or B)*
e) *Flywheel (see Chapter 2A or B)*

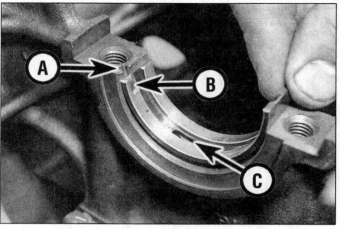

11.3 Bearing shells correctly installed

A   *Recess in bearing saddle*   C   *Oil hole*
B   *Lug on bearing shell*

f) *Cylinder head and gasket (see Chapter 2A or B)*
g) *Timing belt tensioner, sprockets and timing belt (see Chapter 2A or B)*
h) *Engine external components and ancillaries.*
I) *Drivebelts, pulleys and tensioners.*

3   At this stage, all engine components should be absolutely clean and dry, with all faults repaired. The components should be laid out (or in individual containers) on a completely clean work surface.

11   Crankshaft - installation and main bearing oil clearance check

Refer to illustration 11.3

1   Crankshaft installation is the first stage of engine reassembly following overhaul. At this point, it is assumed that the crankshaft, cylinder block/crankcase and bearings have been cleaned, inspected and reconditioned or replaced.

2   Place the cylinder block on a clean, level worksurface, with the crankcase facing upwards. Unbolt the bearing caps and carefully release them from the crankcase; lay them out in order to ensure correct reassembly. If they are still in place, remove the bearing shells from the caps and the crankcase and wipe out the inner surfaces with a clean rag - they must be kept spotlessly clean.

3   Clean the rear surface of the new bearing shells with a rag and lay them on the bearing saddles. Ensure that the orientation lugs on the shells engage with the recesses in the saddles, and that the oil holes are correctly aligned **(see illustration)**. Do not hammer or otherwise force the bearing shells into place. It is critically important that the surfaces of the bearings are kept free from damage and contamination.

4   Give the newly-installed bearing shells and the crankshaft journals a final clean with a rag. Check that the oil holes in the crankshaft are free from dirt, as any left here will become embedded in the new bearings when the engine is first started.

5   Carefully lay the crankshaft in the crankcase, taking care not to dislodge the bearing shells.

Oil clearance check

Refer to illustrations 11.7, 11.8 and 11.11

6   When the crankshaft and bearings are installed, a clearance must exist between them to allow lubricant to circulate. This clearance is impossible to check using feeler gauges, so Plastigage is used. This is a thin strip of soft plastic that is crushed between the bearing shells and journals when the bearing caps are tightened up. The change in its width then indicates the size of the clearance gap.

11.7 Lay a piece of Plastigage on each journal, in line with the crankshaft axis

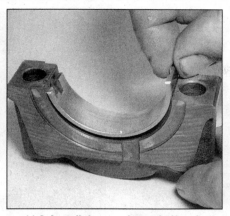

11.8 Install the new lower half main bearing shells on the main bearing caps

11.11 Measure the width of the crushed Plastigage using the scale provided

7 Cut off five pieces of Plastigage, just shorter than the length of the crankshaft journal. Lay a piece on each journal, in line with its axis **(see illustration)**.

8 Wipe off the rear surfaces of the new lower half main bearing shells and install them to the main bearing caps, again ensuring that the locating lugs engage correctly **(see illustration)**.

9 Wipe the front surfaces of the bearing shells and give them a light coating of silicone spray - this will prevent the Plastigage from sticking to the shell. Install the caps in their correct locations on the bearing saddles, using the manufacturer's markings as a guide. Ensure that they are correctly orientated - the caps should be installed such that the recesses for the bearing shell locating lugs are on the same side as those in the bearing saddle.

10 Working from the center bearing cap, tighten the bolts one half turn at a time until they are all correctly torqued *to their first stage only*. Do not allow the crankshaft to rotate at all while the Plastigage is in place. Progressively unbolt the bearing caps and remove them, taking care not to dislodge the Plastigage.

11 The width of the crushed Plastigage can now be measured, using the scale provided **(see illustration)**. Use the correct scale, as both imperial and metric are printed. This measurement indicates the running clearance - compare it with that listed in Specifications. If the clearance is outside the tolerance, it may be due to dirt or debris trapped under the bearing surface; try cleaning them again and repeat the clearance check. If the results are still unacceptable, re-check the journal diameters and the bearing sizes. Note that if the Plastigage is thicker at one end, the journals may be tapered and as such, will require regrinding.

12 When you are satisfied that the clearances are correct, carefully remove the remains of the Plastigage from the journals and bearings faces. Use a soft, plastic or wooden scraper as anything metallic is likely to damage the surfaces.

Crankshaft - final installation

Refer to illustrations 11.16, 11.17, 11.18

13 Lift the crankshaft out of the crankcase. Wipe off the surfaces of the bearings in the crankcase and the bearing caps. Install the thrust bearings either side of the No. 3 bearing saddle, between cylinders No. 2 and 3. Use a small quantity of grease to hold them in place; ensure that they are seated correctly in the machined recesses, with the oil grooves facing outwards

14 Liberally coat the bearing shells in the crankcase with clean engine oil **(see illustration)**.

15 Lower the crankshaft into position so that No. 2 and 3 cylinder crankpins are at TDC; No. 1 and 4 cylinder crankpins will then be at BDC, ready for installing the No. 1 piston.

16 Lubricate the lower bearing shells in the main bearing caps with clean engine oil, then attach the thrustwashers to either side of bearing cap No. 3, noting that the lugs protruding from the washers engage the recesses in the side of the bearing cap **(see illustration)**. Make sure that the locating lugs on the shells are still engaged with the corresponding recesses in the caps.

17 Install the main bearing caps in the correct order and orientation - No. 1 bearing cap must be at the timing belt end of the engine and the bearing shell locating recesses in the bearing saddles and caps must be adjacent to each other **(see illustration)**. Insert the bearing cap bolts and hand tighten them only.

18 Working from the center bearing cap outwards, tighten the retain-

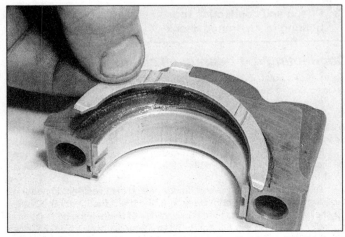

11.16 Installing the thrustwashers to the No. 3 bearing cap

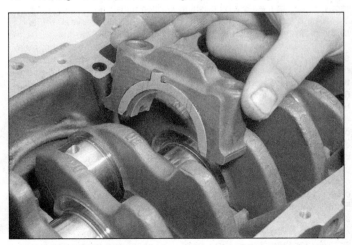

11.17 Installing the main bearing cap and thrustwashers

11.18 Tighten the bearing cap bolts to the specified torque

12.5 Checking a piston ring end gap using a feeler gauge

ing bolts to their specified torques (**see illustration**).

19   Install the crankshaft rear oil seal housing, together with a new oil seal; refer to Part A or B (as applicable) of this Chapter for details.

20   Check that the crankshaft rotates freely by turning it by manually. If resistance is felt, re-check the running clearances, as described above.

21   Carry out a check of the crankshaft endplay as described at the beginning of Section 6. If the thrust surfaces of the crankshaft have been checked and new thrust bearings have been installed, then the endplay should be within specification.

12   Pistons and piston rings - assembly

Refer to illustrations 12.5 and 12.7

1   At this point it is assumed that the pistons have been correctly assembled to their respective connecting rods and that the piston ring-to-groove clearances have been checked. If not, refer to the end of Section 5.

2   Before the rings can be installed on the pistons, the end gaps must be checked with the rings installed into the cylinder bores.

3   Lay out the piston assemblies and the new ring sets on a clean work surface so that the components are kept together in their groups during and after end gap checking. Place the crankcase on the work surface on its side, allowing access to the top and bottom of the bores.

4   Take the No. 1 piston top ring and insert it into the top of the bore. Using the No. 1 piston as a ram, push the ring close to the bottom of the bore, at the lowest point of the piston travel. Ensure that it is perfectly square in the bore by pushing firmly against the piston crown.

5   Use a set of feeler gauges to measure the gap between the ends of the piston ring; the correct blade will just pass through the gap with a minimal amount of resistance (**see illustration**). Compare this measurement with that listed in Specifications. Check that you have the correct ring before deciding that a gap is incorrect. Repeat the operation for all twelve rings.

6   If new rings are being installed, it is unlikely that the end gaps will be too small. If a measurement is found to be undersize, it must be corrected or there is the risk that the ends of the ring may contact each other during operation, possibly resulting in engine damage. This is achieved by gradually filing down the ends of the ring, using a file clamped in a vise. Install the ring over the file such that both its ends contact opposite faces of the file. Move the ring along the file (from the outside-in), removing small amounts of material at a time. Take great care as the rings are brittle and form sharp edges if they fracture. Remember to keep the rings and piston assemblies in the correct order.

7   When all the piston ring end gaps have been verified, they can be installed on the pistons. Work from the lowest ring groove (oil control ring) upwards. Note that the oil control ring comprises two side rails

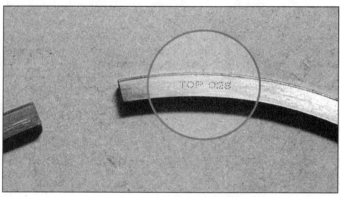

12.7 Piston ring "TOP" marking

separated by a expander ring. Note also that the two compression rings are different in cross section and so must be installed in the correct groove and the right way up, using a piston ring installation tool. Both of the compression rings have marks stamped on one side to indicate the top facing surface. Ensure that these marks face up when the rings are installed (**see illustration**).

8   Distribute the end gaps around the piston, spaced at 120-degree intervals to the each other. **Note:** *If the piston ring manufacturer supplies specific fitting instructions with the rings, follow these exclusively.*

13   Piston and connecting rods - installation and rod bearing oil clearance check

Connecting rod bearing oil clearance check

Note: *At this point, it is assumed that the crankshaft has been installed, as described in Section 11.*

1   As with the main bearings (Section 10), an oil clearance must exist between the connecting rod crankpin and its bearing shells to allow oil to circulate. There are two methods of checking the size of the running clearance, as described in the following Steps.

2   Place the cylinder block on a clean, level work surface, with the crankcase facing upwards. Position the crankshaft such that crankpins No. 1 and 4 are at BDC.

3   The first method is the least accurate and involves bolting bearing caps to the connecting rods, away from the crankshaft, with the bearing shells in place. **Note:** *Correct orientation of the bearing caps is critical; refer to the notes in Section 5. The internal diameter formed by the assembled rod is then measured using internal vernier calipers. The diameter of the respective crankpin is then subtracted from this mea-*

surement and the result is the oil clearance.

4   The second method of carrying out this check involves the use of Plastigage, in the same manner as the main bearing running clearance check (see Section 11) and is much more accurate than the previous method. Clean all four crankpins with a clean rag. With crankpins No. 1 and 4 at BDC initially, place a strand of Plastigage on each crankpin journal.

5   Install the upper bearing shells to the connecting rods, ensuring that the locating lugs and recesses engage correctly. Temporarily install the piston/connecting rod assemblies to the crankshaft; install the bearing caps, using the manufacturer's markings to ensure that they are installed correctly - refer to *"Final installation"* for details.

6   Tighten the bearing cap nuts/bolts as described below. Take care not to disturb the Plastigage or rotate the connecting rod during the tightening process.

7   Disassemble the assemblies without rotating the connecting rods. Use the scale printed on the Plastigage envelope to determine the big-end bearing running clearance and compare it with the figures listed in Specifications.

8   If the clearance is significantly different from that expected, the bearing shells may be the wrong size (or excessively worn, if the original shells are being re-used). Make sure that no dirt or oil was trapped between the bearing shells and the caps or connecting rods when the clearance was measured. Re-check the diameters of the crankpins. Note that if the Plastigage was wider at one end than at the other, the crankpins may be tapered. When the problem is identified, install new bearing shells or have the crankpins reground to a listed undersize, as appropriate.

9   Upon completion, carefully scrape away all traces of the Plastigage material from the crankshaft and bearing shells. Use a plastic or wooden scraper, which will be soft enough to prevent scoring of the bearing surfaces.

Piston and connecting rod assemblies - final installation

Refer to illustration 13.17

10   Note that the following procedure assumes that the crankshaft main bearing caps are in place (see Section 10).

11   Ensure that the bearing shells are correctly installed, as described at the beginning of this Section. If new shells are being installed, ensure that all traces of the protective grease are cleaned off using a solvent. Wipe dry the shells and connecting rods with a lint-free cloth.

12   Lubricate the cylinder bores, the pistons, and piston rings with clean engine oil. Lay out each piston/connecting rod assembly in order on a worksurface. On engines where the rod bolts are captive in the connecting rods, install short sections of rubber hose or tape over the bolt threads, to protect the cylinder bores during reassembly.

13   Start with piston/connecting rod assembly No. 1. Make sure that the piston rings are still spaced as described in Section 12, then clamp them in position with a piston ring compressor.

14   Insert the piston/connecting rod assembly into the top of cylinder No. 1. Lower the rod in first, guiding it to protect the big-end bolts and the cylinder bores **(see illustration)**.

15   Ensure that the orientation of the piston in its cylinder is correct - the piston crown, connecting rods and bearing caps have markings, which must point towards the timing belt end of the engine when the piston is installed in the bore - refer to Section 5 for details.

16   Using a block of wood or hammer handle against the piston crown, tap the assembly into the cylinder until the piston crown is flush with the top of the cylinder.

17   Ensure that the bearing shell is still correctly installed. Liberally lubricate the crankpin and both bearing shells with clean engine oil. Taking care not to mark the cylinder bores, tap the piston/connecting rod assembly down the bore and onto the crankpin. Install the bearing cap, tightening its retaining nuts/bolts finger-tight at first **(see illustration)**. Note that the orientation of the bearing cap with respect to the connecting rod must be correct when the two components are reassembled. The connecting rod and its corresponding bearing cap both have recesses machined into them, close to their mating surfaces

13.17 Installing a connecting rod bearing cap

A Dowel	B Locating hole

- these recesses must both face in the same direction as the arrow on the piston crown (i.e. towards the timing belt end of the engine) when correctly installed - refer to the illustrations in Section 5 for details.

Note: *On certain engines, the connecting rod caps are provided with offset dowels which locate in holes in the bearing caps.*

18   Working progressively around each bearing cap, tighten the retaining nuts half a turn at a time to the specified torque.

19   Install the remaining three piston/connecting rod assemblies in the same way.

20   Rotate the crankshaft by hand. Check that it turns freely; some stiffness is to be expected if new components have been installed, but there should be no indication of binding or tight spots.

Diesel engines

21   If new pistons are to be installed, or if a new short block is to be installed, the projection of the piston crowns above the cylinder head at TDC must be measured, to determine the type of head gasket that should be installed.

22   Turn the cylinder block over (so that the crankcase is facing downwards) and rest it on a stand or wooden blocks. Anchor a dial indicator to the cylinder block, and zero it on the head gasket mating surface. Rest the gauge probe on No. 1 piston crown and turn the crankshaft slowly by hand so that the piston reaches and then passes through TDC. Measure and record the maximum deflection at TDC.

23   Repeat the measurement at piston No. 4, then turn the crankshaft through 180-degrees and take measurements at pistons Nos. 2 and 3.

24   If the measurements differ from piston to piston, take the highest figure and use this to determine the head gasket type that must be used - refer to the Specifications for details.

25   Note that if the original pistons have been installed, then a new head gasket of the same type as the original item must be installed; refer to Chapter 2B for details of how to identify different head gasket types.

14 Engine - initial start-up after overhaul and reassembly

Warning: *Have a fire extinguisher on hand when starting the engine for the first time.*

1   Install the remainder of the engine components in the order listed in Section 10 of this Chapter, referring to Part A or B where necessary. Install the engine (and transaxle) to the vehicle as described in Section 2 of this Chapter. Double-check the engine oil and coolant levels and make a final check that everything has been reconnected. Make sure that there are no tools or rags left in the engine compartment.

Gasoline models

2   Remove the spark plugs, referring to Chapter 1 for details.

3   The ignition must be disabled so the engine can be turned over using the starter motor, without starting - disable the fuel pump by unplugging the fuel pump power relay from the relay board; refer to the relevant Part of Chapter 4 for further information.

4   Turn the engine using the starter motor until the oil pressure warning lamp goes out. If the lamp fails to go off after several seconds of cranking, check the engine oil level and oil filter security. Assuming these are correct, check the security of the oil pressure switch cabling - do not progress any further until you are satisfied that oil is being pumped around the engine at sufficient pressure.

5   Install the spark plugs, and reconnect the fuel pump relay.

Diesel models

6   Disconnect the wire from the fuel cut-off valve at the fuel injection pump - refer to Chapter 4C for details.

7   Turn the engine using the starter motor until the oil pressure warning lamp goes out.

8   If the lamp fails to go off after several seconds of cranking, check the engine oil level and oil filter security. Assuming these are correct, check the security of the oil pressure switch cabling - do not progress any further until you are satisfied that oil is being pumped around the engine at sufficient pressure.

9   Reconnect the fuel cut-off valve wire.

All models

10   Start the engine, but be aware that as fuel system components have been disturbed, the cranking time may be a little longer than usual.

11   While the engine is idling, check for fuel, water and oil leaks. Don't be alarmed if there are some odd smells and the occasional plume of smoke as components heat up and burn off oil deposits.

12   Assuming all is well, keep the engine idling until hot water is felt circulating through the top hose.

13   On gasoline models, check the ignition timing, idle speed and idle mixture setting as described in Chapter 1A, then switch the engine off.

14   On diesel models, check the fuel injection pump timing and engine idle speed, as described in Chapter 4C and Chapter 1B.

15   After a few minutes, recheck the oil and coolant levels, and top-up as necessary.

16   There is no need to re-tighten the cylinder head bolts once the engine has been run following reassembly.

17   If new pistons, rings or crankshaft bearings have been installed, the engine must be treated as new, and run-in for the first 600 miles. *Do not* operate the engine at full-throttle, or allow it to labor at low engine speeds in any gear. It is recommended that the engine oil and filter are changed at the end of this period.

Chapter 3
Cooling, heating and air conditioning systems

Contents

3

Specifications

General
Expansion tank cap opening pressure .... 19 to 23 psi

Thermostat
Opening temperatures
  Starts to open .... 185°F
  Fully open .... 221°F

Electric cooling fan(s)
Cooling fan(s) cut in
  Stage 1 speed
    Switches on .... 198 to 207°F
    Switches off .... 183 to 196°F
  Stage 2 speed
    Switches on .... 210 to 221°F
    Switches off .... 196 to 208°F
  Stage 3 speed (where installed - see Section 5))
    Switches on .... 230 to 239°F
    Switches off .... 221 to 230°F

Torque specifications
	Ft-lbs
Alternator mounting bracket nuts	22
Water pump	
Retaining bolts	7
Pulley bolts	18
Water pump/thermostat housing retaining bolts/studs	
Stage 1	15
Stage 2	Angle-tighten a further 90-degrees
Cooling fan retaining nuts	7
Cooling fan thermostatic switch	26
Radiator mounting bolts	7
Thermostat cover bolts	7

1 General information

The cooling system is of pressurized type, comprising of a pump, an aluminum crossflow radiator, an electric cooling fan, and a thermostat. The system functions as follows. Cold coolant from the radiator passes through the hose to the water pump where it is pumped around the cylinder block and head passages. After cooling the cylinder bores, combustion surfaces and valve seats, the coolant reaches the underside of the thermostat, which is initially closed. The coolant passes through the heater and is returned through the cylinder block to the water pump.

When the engine is cold the coolant circulates only through the cylinder block, cylinder head, expansion tank and heater. When the coolant reaches a predetermined temperature, the thermostat opens and the coolant passes through to the radiator. As the coolant circulates through the radiator it is cooled by the inrush of air when the car is in forward motion. Airflow is supplemented by the action of the electric cooling fan(s) when necessary. Upon reaching the radiator, the coolant is now cooled and the cycle is repeated.

The electric cooling fan(s) mounted on the rear of the radiator are controlled by a thermostatic switch. At a preset coolant temperature, the switch actuates the fan(s).

Refer to Section 11 for information on the air conditioning system.

Precautions

Refer to illustration 2.3

Warning 1: *Do not attempt to remove the expansion tank filler cap or disturb any part of the cooling system while the engine is hot, as there is a high risk of scalding. If the expansion tank filler cap must be removed before the engine and radiator have fully cooled (even though this is not recommended) the pressure in the cooling system must first be relieved. Cover the cap with a thick layer of cloth, to avoid scalding, and slowly unscrew the filler cap until a hissing sound can be heard. When hissing has stopped, indicating that the pressure has reduced, slowly unscrew the filler cap until it can be removed; if more hissing sounds are heard, wait until they have stopped before unscrewing the cap completely. At all times keep well away from the filler cap opening.*

Do not allow antifreeze to come into contact with skin or painted surfaces of the vehicle. Rinse off spills immediately with plenty of water. Never leave antifreeze lying around in an open container or in a puddle in the driveway or garage floor. Children and pets are attracted to its sweet smell. Antifreeze can be fatal if ingested.

If the engine is hot, the electric cooling fan may start rotating even if the engine is not running, so be careful to keep hands, hair and loose clothing well clear when working in the engine compartment.

Refer to Section 11 for precautions to be observed when working on models with air conditioning.

Warning 2: *On models equipped with airbags, always disable the airbag system before working in the vicinity of the impact sensors, steering column or instrument panel to avoid the possibility of accidental deployment of the airbag, which could cause personal injury (see Chapter 12).*

2 Cooling system hoses - removal and replacement

Refer to illustration 2.3

Note: *Refer to the Warnings given in Section 1 of this Chapter before proceeding.*

1   If the checks described in Chapter 1 reveal a faulty hose, it must be replaced as follows.

2   First drain the cooling system (see Chapter 1). If the coolant is not due for replacement, it may be re-used if it is collected in a clean container.

3   To disconnect a hose, release its retaining clamps, then move them along the hose, clear of the relevant inlet/outlet fitting **(see illustration)**. Carefully work the hose free. While the hoses can be removed with relative ease when new or hot, do not attempt to disconnect any

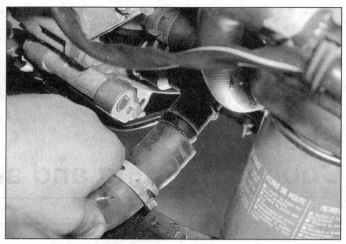

2.3 Disconnect the thermostat housing hose (1.9 liter diesel shown)

part of the system while it is still hot.

4   Note that the radiator inlet and outlet fittings are fragile; do not use excessive force when attempting to remove the hoses. If a hose proves to be difficult to remove, try to release it by rotating the hose ends before attempting to free it.

5   When fitting a hose, first slide the clamps onto the hose, then work the hose into position. If clamp type clamps were originally installed, it is a good idea to replace them with screw type clamps when installing the hose. If the hose is stiff, use a little soapy water as a lubricant, or soften the hose by soaking it in hot water. **Note:** *If all else fails, cut the hose with a sharp knife, then slit it so that it can be peeled off in two pieces. Although this may prove more expensive if the hose is otherwise undamaged, it is preferable to buying a new radiator.*

6   Work the hose into position, checking that it is correctly routed, then slide each clip along the hose until it passes over the flared end of the relevant inlet/outlet fitting, before securing it in position with the retaining clip.

7   Refill the cooling system (see Chapter 1).

8   Check thoroughly for leaks as soon as possible after disturbing any part of the cooling system.

3 Radiator - removal, inspection and installation

Note: *If leakage is the reason for wanting to remove the radiator, bear in mind that minor leaks can often be cured using a radiator sealant with the radiator in place.*

Removal

Refer to illustrations 3.5a, 3.5b, 3.8a, 3.8b, 3.8c and 3.8d

1   Disconnect the negative battery cable. **Caution:** *If the stereo in your vehicle is equipped with an anti-theft system, make sure you have the correct activation code before disconnecting the battery.*

2   Drain the cooling system (see Chapter 1).

3   Remove the front bumper as described in Chapter 11.

4   Remove both headlights as described in Chapter 12.

5   Release the retaining clamps and disconnect the coolant hoses from the radiator **(see illustrations)**.

6   Disconnect the wiring connector from the cooling fan switch on the left-hand end of the radiator.

7   On models equipped with air conditioning, in order to gain the clearance required to remove the radiator carry out the following. Unscrew the retaining nuts and release the air conditioning system fluid reservoir/drier assembly from its mounting bracket. Release the refrigerant lines from all the relevant retaining clamps then undo the retaining bolts and move the condenser forwards as far as possible, taking great care not to place any excess strain on the refrigerant lines.

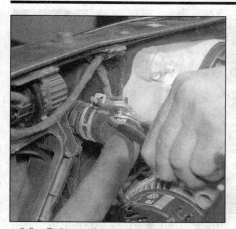

3.5a Release the retaining clamps and disconnect the top . . .

3.5b . . . and bottom hoses from the radiator

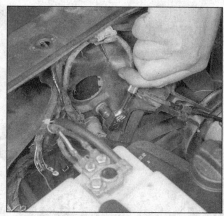

3.8a Removing a radiator retaining bolt

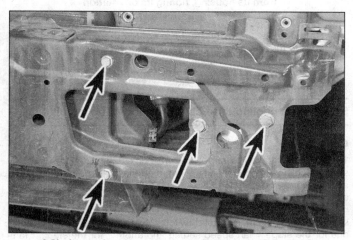

3.8b It may be necessary to undo the bolts (left-hand bolts) (arrows) . . .

Note: *Do not disconnect the refrigerant lines from the condenser (refer to the Warnings given in Section 11).*

8   On all models, loosen and remove the four retaining bolts from the rear of the radiator then maneuver the radiator out from the front of the vehicle. On some models, it will be necessary to unbolt the crossmember from the front of the vehicle to gain the necessary clearance required to withdraw the radiator **(see illustrations)**.

Inspection

9   If the radiator has been removed due to suspected blockage, reverse flush it as described in Chapter 1. Clean dirt and debris from the radiator fins, using an air line (in which case, wear eye protection) or a soft brush. Be careful, as the fins are sharp and easily damaged.

10   If necessary, a radiator specialist can perform a "flow test" on the radiator, to establish whether an internal blockage exists.

11   A leaking radiator must be referred to a specialist for permanent repair. Do not attempt to weld or solder a leaking radiator, as damage may result.

12   In an emergency, minor leaks from the radiator can be cured using a suitable radiator sealant in accordance with the manufacturers instructions with the radiator *in place*.

13   If the radiator is to be sent for repair or replaced, remove the cooling fan switch.

Installation

14   Maneuver the radiator into position and install its retaining bolts, tightening them to the specified torque setting. Where necessary, install the crossmember and securely tighten all its mounting bolts.

15   On models with air conditioning, seat the condenser in position and securely tighten its retaining bolts. Install the reservoir/drier retaining nuts and ensure that all refrigerant lines are retained by all the relevant clamps.

16   Reconnect the hoses to the radiator fittings and securely tighten their clamps.

17   Reconnect the wiring connector to the cooling fan switch.

3.8c . . . and remove the crossmember to remove the radiator

3.8d Removing the radiator

3

4.6 Undo the retaining bolts and remove the thermostat cover

4.7 Recover the sealing ring, then withdraw the thermostat from its housing, noting its orientation

18   Install the headlights and front bumper as described in Chapters 11 and 12.
19   On completion, reconnect the battery and refill the cooling system (see Chapter 1).

---

4   Thermostat - removal, testing and installation

Removal

Refer to illustrations 4.6 and 4.7
1   Disconnect the negative battery cable. **Caution:** *If the stereo in your vehicle is equipped with an anti-theft system, make sure you have the correct activation code before disconnecting the battery.*
2   Drain the cooling system (see Chapter 1).
3   On these models the thermostat is in the base of the water pump housing which is on the front, right-hand end of the engine.
4   On models with power steering, remove the pump drivebelt (see Chapter 1). Loosen and remove the bolts securing the power steering pump mounting bracket to the engine and position the pump assembly clear of the engine. It is not necessary to disconnect the hydraulic hose/pipe from the pump.
5   Release the retaining clamp and disconnect the coolant hose from the thermostat cover.
6   Loosen and remove the two retaining bolts and remove the thermostat cover from the water pump housing **(see illustration)**.
7   Recover the sealing ring and withdraw the thermostat. Discard the sealing ring; a new one should be used on installation **(see illustration)**.

Testing

8   A rough test of the thermostat may be made by suspending it with a piece of string in a container full of water. Heat the water to bring it to the boil - the thermostat must open by the time the water boils. If not, replace it.
9   If a thermometer is available, the precise opening temperature of the thermostat may be determined, and compared with the figures given in the Specifications. The opening temperature is also marked on the thermostat.
10   A thermostat which fails to close as the water cools must also be replaced.

Installation

11   Installation is the reverse of the removal sequence noting the following points:
  a) *Ensure that the thermostat is correctly located in the housing and install the new sealing ring.*
  b) *Tighten the cover bolts to the specified torque setting.*

  c) *On models with power steering, tighten the mounting bracket bolts to the specified torque (see Chapter 10) and install the drivebelt as described in Chapter 1.*
  d) *On completion refill the cooling system as described in Chapter 1.*

---

5   Electric cooling fan - check, removal and installation

Check

1   The cooling fan is supplied with current through the ignition switch, cooling fan control unit (mounted on the left-hand front suspension tower), the relay(s) and fuses/fusible link (see Chapter 12). The circuit is completed by the cooling fan thermostatic switch, which is mounted in the left-hand end of the radiator. The cooling fan has two speed settings; the thermostatic switch actually contains two switches, one for the stage 1 fan speed setting and another for the stage 2 fan speed setting. **Note 1:** *On some models equipped with air conditioning, there is also a second switch (fitted into one of the coolant outlet housings/hoses on the cylinder head). This switch controls the cooling fan stage 3 speed setting.* Testing of the cooling fan circuit is as follows noting that the following check should be carried out on both the stage 1 speed circuit and speed 2 circuit (see wiring diagrams at the end of Chapter 12). **Note 2:** *On models with a twin cooling fan arrangement, if only one fan is working, the drivebelt linking the fans has broken.*
2   If a fan does not appear to work, first check the fuses/fusible links. If they are good, run the engine until normal operating temperature is reached, then allow it to idle. If the fan does not cut in within a few minutes, switch off the ignition and disconnect the wiring plug from the cooling fan switch. Bridge the relevant two contacts in the wiring plug using a length of spare wire, and switch on the ignition. If the fan now operates, the switch is probably faulty and should be replaced.
3   If the switch appears to work, the motor can be checked by disconnecting the motor wiring connector and connecting a 12 volt supply directly to the motor terminals. If the motor is faulty, it must be replaced.
4   If the fan still fails to operate, check that the cooling fan circuit wiring (Chapter 12). Check each wire for continuity and ensure all connections are clean and free of corrosion.
5   If no fault can be found with the fuses/fusible links, wiring, fan switch, or fan motor then it is likely that the cooling fan control unit is faulty. Testing of the unit should be entrusted to a VW dealer; if the unit is faulty it must be replaced.

Removal

Refer to illustrations 5.6, 5.7, 5.8a, 5.8b and 5.8c
6   Remove the radiator (refer to Section 3). Disconnect the wiring

5.6 Disconnect the radiator cooling fan wiring connector . . .

5.7 . . . then remove the fan retaining ring

connector from the rear of the cooling fan motor **(see illustration)**.

7   Press out the pins from the center of the fan retaining ring fasteners and unclip the ring from the shroud **(see illustration)**.

8   Loosen and remove the motor retaining nuts and remove the cooling fan assembly from the front of the vehicle **(see illustrations)**. On models with twin cooling fans, as the motor is removed free it from the drivebelt linking the fans and remove the belt; if necessary undo the retaining nuts and remove the second fan. If the unit is faulty, it must be replaced.

Installation

9   Installation is a reversal of removal, tightening the cooling fan retaining nuts to the specified torque. On models with twin fans, prior to installation inspect the fan drivebelt for signs of damage or deterioration and replace if necessary.

10   Install the radiator (see Section 3) then start the engine and run it until it reaches normal operating temperature. Continue to run the engine and check that the cooling fan cuts in and functions correctly.

6   Cooling system electrical switches - check, removal and installation

Electric cooling fan thermostatic switch

Check

1   Testing of the switch is described in Section 5, as part of the electric cooling fan test procedure.

Removal

2   The switch is located in the left-hand side of the radiator. The engine and radiator should be cold before removing the switch.

3   Disconnect the negative battery cable. **Caution:** *If the stereo in your vehicle is equipped with an anti-theft system, make sure you have the correct activation code before disconnecting the battery.*

4   Either drain the cooling system to below the level of the switch (as described in Chapter 1), or have ready a suitable plug which can be used to plug the switch aperture in the radiator while the switch is removed. If a plug is used, take great care not to damage the radiator, and do not use anything which will allow foreign matter to enter the radiator.

5   Disconnect the wiring plug from the switch.

6   Carefully unscrew the switch from the radiator.

Installation

7   Installation is a reversal of removal, applying a smear of suitable sealant to the threads of the switch and tightening it to the specified torque setting. On completion, refill the cooling system as described in Chapter 1.

8   Start the engine and run it until it reaches normal operating temperature, then continue to run the engine and check that the cooling fan cuts in and functions correctly.

Coolant temperature gauge sensor

Note: *The sender unit is an integral part of the fuel system/preheating system (as applicable) temperature sender unit.*

3

5.8a Loosen and remove the fan retaining nuts from the rear of the shroud

5.8b On models with twin cooling fans, release the drivebelt from the fan pulley . . .

5.8c . . . and remove the fan from the vehicle

7.5a Loosen the retaining bolts . . .

7.5b . . . and remove the pulley from the water pump

Check

9   The coolant temperature gauge, mounted in the instrument panel, is fed with a stabilized voltage supply from the instrument panel feed (through the ignition switch and a fuse), and its ground is controlled by the sensor.

10   The sensor unit is clipped into the coolant outlet elbow on the front of the cylinder head. The sensor contains a thermistor, which consists of an electronic component whose electrical resistance decreases at a predetermined rate as its temperature rises. When the coolant is cold, the sensor resistance is high, current flow through the gauge is reduced, and the gauge needle points towards the "cold" end of the scale. If the sensor is faulty, it must be replaced.

11   If the gauge develops a fault, first check the other instruments; if they do not work at all, check the instrument panel electrical feed. If the readings are erratic, there may be a fault in the instrument panel assembly. If the fault lies in the temperature gauge alone, check it as follows.

12   If the gauge needle remains at the "cold" end of the scale, disconnect the wiring connector from the sensor unit, and ground the temperature gauge wire (see "Wiring diagrams" for details) to the cylinder head. If the needle then deflects when the ignition is switched on, the sensor unit is proved faulty, and should be replaced. If the needle still does not move, remove the instrument panel (Chapter 12) and check the continuity of the wiring between the sensor unit and the gauge, and the feed to the gauge unit. If continuity is shown, and the fault still exists, then the gauge is faulty and should be replaced.

13   If the gauge needle remains at the "hot" end of the scale, disconnect the sensor wire. If the needle then returns to the "cold" end of the scale when the ignition is switched on, the sensor unit is proved faulty and should be replaced. If the needle still does not move, check the remainder of the circuit as described previously.

Removal

14   Either partially drain the cooling system to just below the level of the sensor (as described in Chapter 1), or have ready a suitable plug which can be used to plug the sensor aperture while it is removed. If a plug is used, take great care not to damage the sensor unit aperture, and do not use anything which will allow foreign matter to enter the cooling system.

15   Disconnect the negative battery cable. **Caution:** *If the stereo in your vehicle is equipped with an anti-theft system, make sure you have the correct activation code before disconnecting the battery.*

16   Disconnect the wiring from the sensor.

17   Depress the sensor unit and slide out its retaining clip. Withdraw the sensor from the coolant elbow and recover its sealing ring.

Installation

18   Install a new sealing ring to the sensor unit. Push the sensor fully into the coolant elbow and secure it in position with the retaining clip.

19   Reconnect the wiring connector then refill the cooling system as described in Chapter 1.

Fuel injection/preheating system coolant temperature sensor

20   The sensor is combined with the coolant temperature gauge sensor (see above). Testing of the sensor should be entrusted to a VW dealer.

---

7   Water pump - removal and installation

Removal

Refer to illustrations 7.5a, 7.5b, 7.8a, 7.8b, 7.8c and 7.9

Note: *New water pump/thermostat housing assembly retaining studs/bolts will be required on installation.*

1   Drain the cooling system as described in Chapter 1. Disconnect the negative battery cable. **Caution:** *If the stereo in your vehicle is equipped with an anti-theft system, make sure you have the correct activation code before disconnecting the battery.*

2   Remove the alternator as described in Chapter 5.

3   On models equipped with power steering, remove the power steering pump as described in Chapter 10.

4   On models equipped with air conditioning, unbolt the compressor from its mounting bracket and position it clear of the engine. **Note:** *Do not disconnect the refrigerant lines from the compressor (see Warnings in Section 11).*

5   Loosen and remove the retaining bolts and remove the pulley from the water pump **(see illustrations)**.

6   Loosen and remove the nuts securing the alternator mounting bracket assembly to the side of the cylinder block and remove the bracket.

7   Release the retaining clamps and disconnect the coolant hoses from the back of the water pump housing and the thermostat housing.

8   Unscrew the retaining studs/bolts (as applicable) securing the water pump/thermostat housing to the block and remove the housing assembly from the engine. **Note:** *On some engines it will be necessary to unscrew the bolt(s) that secure the timing belt cover to the housing assembly (see Chapter 2).* Recover the sealing ring which is installed between the housing and block and discard it; a new one should be used on installation **(see illustrations)**.

9   With the assembly on a bench, unscrew the retaining bolts and remove the pump from the housing. Discard the gasket, a new one must be used on installation **(see illustration)**.

Installation

10   Ensure that the pump and housing mating surfaces are clean and dry and position a new gasket on the housing.

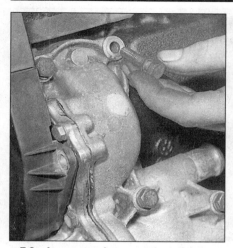

7.8a Loosen and remove the retaining bolts/studs . . .

7.8b . . . then remove water pump/ thermostat housing from the block . . .

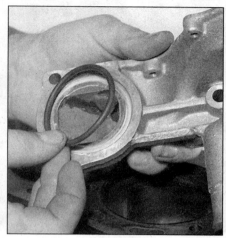

7.8c . . . and recover its sealing ring

11   Install the water pump to the housing and evenly tighten its retaining bolts to the specified torque setting.

12   Install the new sealing ring to the housing assembly recess and install the housing to the cylinder block. Install the retaining studs/bolts (as applicable) and tighten them to the specified Stage 1 torque setting and then through the specified Stage 2 angle.

13   Connect the coolant hoses to the housing and securely tighten their retaining clamps.

14   Install the alternator mounting bracket to the engine and tighten its retaining nuts to the specified torque setting.

15   Install the pulley to the water pump and tighten its retaining bolts to the specified torque setting (this can be done once the drivebelt is reinstalled and tensioned).

16   Where necessary, install the power steering pump as described in Chapter 10 and the air conditioning compressor.

17   Install the alternator as described in Chapter 5.

18   On completion refill the cooling system as described in Chapter 1.

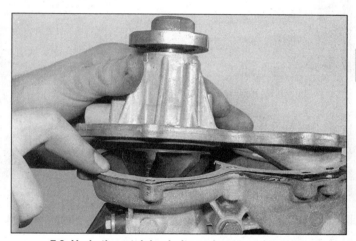

7.9 Undo the retaining bolts and remove the pump and gasket from the housing

8   Heating and ventilation system - general information

1   The heating/ventilation system consists of a four-speed blower motor (housed in the passenger compartment), face-level vents in the center and at each end of the dash, and air ducts to the front and rear footwells.

2   The control unit is located in the dash, and the controls operate flap valves to deflect and mix the air flowing through the various parts of the heating/ventilation system. The flap valves are contained in the air distribution housing, which acts as a central distribution unit, passing air to the various ducts and vents.

3   Cold air enters the system through the grille at the rear of the engine compartment. On some models (depending on specification) a pollen filter is installed to the ventilation inlet to filter out dust, soot, pollen and spores from the air entering the vehicle.

4   The airflow, which can be boosted by the blower, then flows through the various ducts, according to the settings of the controls. Stale air is expelled through ducts behind the rear bumper. If warm air is required, the cold air is passed through the heater core, which is heated by the engine coolant.

5   If necessary, the outside air supply can be closed off, allowing the air inside the vehicle to be recirculated. This can be useful to prevent unpleasant odors entering from outside the vehicle, but should only be used briefly, as the recirculated air inside the vehicle will soon deteriorate.

6   Certain models may be equipped with heated front seats. The heat is produced by electrically-heated mats in the seat and backrest

cushions (see Chapter 12). The temperature is regulated automatically by a thermostat, and cannot be adjusted.

9   Heater/ventilation components - removal and installation

Warning: *On models equipped with airbags, always disable the airbag system before working in the vicinity of the impact sensors, steering column or instrument panel to avoid the possibility of accidental deployment of the airbag, which could cause personal injury (see Chapter 12).*

Models without air conditioning

Heater/ventilation control unit

Refer to illustrations 9.4a and 9.4b

1   Disconnect the negative battery cable. **Caution:** *If the stereo in your vehicle is equipped with an anti-theft system, make sure you have the correct activation code before disconnecting the battery.*

2   Remove the cigarette lighter/dash switch panel as described in Section 11 of Chapter 12.

3   Withdraw the control unit from the dash and disconnect its wiring connectors.

9.4a Release the outer cable retaining clamp . . .

9.4b . . . and detach the control cable from the heater control unit

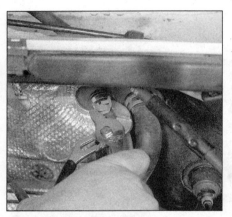

9.15a Release the retaining clamps . . .

9.15b . . . and detach the coolant hoses from the heater core fittings

9.17 Unscrew the nut and detach the ground block and strap (arrow) from the dash

4   Unclip the control cables and release each cable from the control unit, noting each cable's correct installed location and routing; to avoid confusion on installation, label each cable as it is disconnected. The outer cables are released by simply lifting the retaining clamps **(see illustrations)**.

5   Installation is reversal of removal. Ensure that the control cables are correctly routed and reconnected to the control panel, as noted before removal. Clip the outer cables in position and check the operation of each knob/lever before installing the center dash panel.

Heater/air conditioning control cables

6   Remove the heater/ventilation control unit from the dash as described above in paragraphs 1 to 4, detaching the relevant cable from the control unit.

7   Loosen and remove the passenger side dash shelf retaining screws. Move the shelf downwards, to release its upper retaining clamps and remove it from the dash.

8   Release the retaining clamps and remove the insulating sheet from underneath the air distribution/blower motor housing.

9   Follow the run of the cable behind the dash, taking note of its routing, and disconnect the cable from the lever on the air distribution/blower motor housing. Note that the method of fastening is the same as that used at the control unit.

10   Install the new cable, ensuring that it is correctly routed and free from kinks and obstructions.

11   Connect the cable to the control unit and air distribution/blower motor housing making sure the outer cable is clipped securely in position.

12   Check the operation of the control knob then install the control unit as described previously in this Section. Finally install the insulating sheeting and dash shelf.

Heater core

Refer to illustrations 9.15a, 9.15b, 9.17, 9.23a, 9.23b, 9.24a, 9.24b, 9.25a, 9.25b, 9.26a, 9.26b, 9.27a, 9.27b, 9.29, 9.31a and 9.31b

13   Unscrew the expansion tank cap (referring to the Warning note in Section 1) to release any pressure present in the cooling system then securely install the cap.

14   Clamp both heater hoses as close to the firewall as possible to minimize coolant loss. Alternatively, drain the cooling system as described in Chapter 1.

15   Release the retaining clamps and disconnect both hoses from the heater core fittings which are located in the center of the engine compartment firewall **(see illustrations)**.

16   Remove the dash panel as described in Chapter 12.

17   Loosen the retaining nuts and free the ground terminal block and ground strap from both the left- and right-hand ends of the dash mounting frame **(see illustration)**.

18   Release the plastic retaining clamps securing the fusebox/relay plate assembly to the base of the dash mounting frame. Unhook the fusebox/relay plate assembly pivots and position it clear of the dash frame.

19   Unscrew the center screw from the passenger side sill trim panel retaining clip then remove the retaining clip. Press down on the top of the front trim panel, to release its lower edge from the sill, then pull the

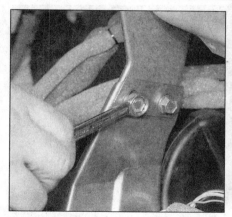

9.23a Loosen the bolts securing the dash frame to its lower center bracket . . .

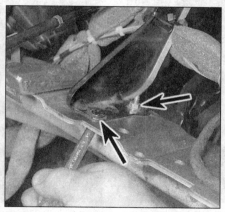

9.23b . . . and the nuts (arrows) securing it to the upper center bracket

9.24a Undo the bolts (arrows) securing the mounting frame to the door pillars . . .

9.24b . . . and remove the frame from the vehicle

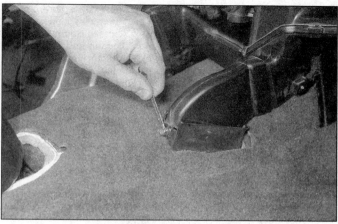

9.25a Undo the retaining screw . . .

3

panel upwards and remove it from the vehicle.
20 Undo the passenger side footwell panel retaining screw and remove the panel from the vehicle.
21 Using a pencil or felt tip pen, mark the outline of the outer dash mounting frame retaining bolts relative to the door pillars, to use as a guide on installation.
22 Release the wiring harness from any retaining clamps securing it to the dash frame.
23 Loosen and remove the nuts and bolts securing the center mounting brackets to the frame and the two bolts securing the frame

to the pedal bracket assembly **(see illustrations)**.
24 With the aid of an assistant, undo the four bolts securing the frame to the door pillars then ease the frame rearwards and remove it from the vehicle **(see illustrations)**.
25 Loosen the retaining screws and remove the rear footwell duct joining pieces from the base of the air distribution housing **(see illustrations)**.
26 Remove the retaining screw and fastener and remove the front footwell duct assembly from the base of the air distribution housing **(see illustrations)**.

9.25b . . . and remove the rear footwell duct pieces from the distribution housing

9.26a Loosen and remove the retaining screw and fastener (arrow) . . .

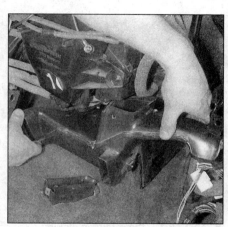

9.26b . . . and remove the front footwell duct assembly

9.27a Release the wiring connector from the side of the air distribution housing . . .

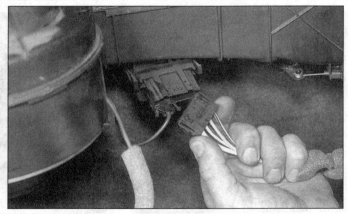

9.27b . . . and disconnect the wiring from the blower motor resistor

27   Release the retaining clip and free the wiring connector from the side of the air distribution housing. Disconnect the wiring connector and those connected to the blower motor and resistor **(see illustrations)**.

28   From within the engine compartment, loosen and remove the three nuts securing the air distribution/blower motor housing in position, there are two nuts securing the air distribution housing to the firewall and a single nut securing the blower motor housing in position. **Note:** *The nuts are hidden behind the engine compartment firewall sound insulation material; if the insulation material is examined closely it will be seen that access holes are already provided.*

29   From inside the vehicle maneuver the air distribution/blower motor housing assembly out of position **(see illustration)**. **Note:** *Keep the core fittings uppermost as the core is removed to prevent coolant spillage. Mop up any spilled coolant immediately and wipe the affected area with a damp cloth to prevent staining.*

30   Recover the seal which is installed between the core fitting and firewall; the seal should be replaced if it shows signs of damage or deterioration.

31   Release the retaining clamps and withdraw the core from the air distribution housing **(see illustrations)**.

32   Installation is a reversal of removal, bearing in mind the following points:

a) *Ensure that the core is clipped securely into the air distribution housing.*

b) *Prior to installation, make sure the foam seals on the top of the housings are in good condition and install the seal to the core fittings.*

c) *When tightening the air distribution/blower motor housing retaining nuts, have an assistant hold the assembly fully upwards to ensure an airtight seal between the housings and firewall.*

9.29 Removing the air distribution/blower motor housing assembly

d) *Maneuver the dash mounting frame into position and install all its retaining bolts. Align the end retaining bolts with the marks made prior to removal then go around and tighten all the retaining bolts securely.*

e) *On completion, refill the cooling system as described in Chapter 1.*

Heater blower motor

Refer to illustrations 9.34 and 9.37

33   Disconnect the negative battery cable. **Caution:** *If the stereo in your vehicle is equipped with an anti-theft system, make sure you have the correct activation code before disconnecting the battery.*

9.31a Release the retaining clamps . . .

9.31b . . . and withdraw the heater core from the housing

9.34 Removing the passenger side dash shelf

9.37 Removing the heater blower motor

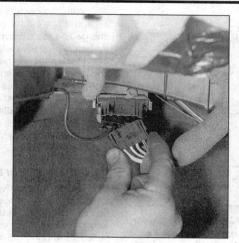

9.40a Disconnect the wiring connectors . . .

9.40b . . . then release the clamps and withdraw the blower motor resistor

34   Loosen and remove the passenger side dash shelf retaining screws. Move the shelf downwards, to release its upper retaining clamps and remove it from the dash **(see illustration).**
35   Release the retaining clamps and remove the insulating sheet from underneath the air distribution/blower motor housing.
36   Disconnect the motor wiring connectors from the resistor and the ground block on the dash frame.
37   Release the retaining clip then rotate the motor assembly and lower it out from the base of the housing **(see illustration).**
38   Installation is a reversal of the removal procedure making sure the motor is correctly clipped into the housing.

Heater blower motor resistor

Refer to illustrations 9.40a and 9.40b
39   Carry out the operations described in paragraphs 33 to 35.
40   Disconnect the wiring connectors from the resistor then release the retaining clamps and withdraw the resistor from the housing **(see illustrations).**
41   Installation is the reverse of removal.

Models with air conditioning

Note: *The following information is only applicable to manually controlled air conditioning systems. At the time of writing no information was available on models with the automatic "Climatronic" system.*

Heater control unit

42   Refer to the information given in paragraphs 1 to 5.

Heater core

43   On models equipped with air conditioning it is not possible to remove the heater core without opening the refrigerant circuit (See Section 11). Therefore this task must be entrusted to a VW dealer.

Heater blower motor

44   Carry out the operations described in paragraphs 33 to 35.
45   Disconnect the wiring connector from the motor then undo the retaining screws and lower the motor assembly out of position.
46   Installation is the reverse of removal.

Heater blower motor resistor

47   On models equipped with a passenger side airbag, remove the airbag unit as described in Chapter 12.
48   On models not equipped with a passenger side airbag, remove the glovebox as described in Chapter 11, Section 28.
49   Disconnect the wiring connector then undo the retaining screw and remove the resistor from the housing.
50   Installation is the reverse of removal.

10 Heater/air conditioning vents and housings - removal and installation

Warning: *On models equipped with airbags, always disable the airbag system before working in the vicinity of the impact sensors, steering column or instrument panel to avoid the possibility of accidental deployment of the airbag, which could cause personal injury (see Chapter 12).*

Vents

Refer to illustration 10.1
1   All vents can be carefully levered out of position with a small flat-bladed screwdriver, taking great care not to mark the vent housing **(see illustration).**
2   On installation, carefully maneuver the vent back into position ensuring it is correctly engaged with the locating pegs.

Driver's side dash vent housing

3   Remove the lighting switch as described in Chapter 12.
4   Remove the vent as described in paragraph 1.
5   Undo the retaining screws and remove the vent housing from the dash.
6   Installation is the reverse of removal.

10.1 Removing a dash vent

Passenger's side dash vent housing

7   Remove the vent as described in paragraph 1.
8   Undo the retaining screws and withdraw the vent housing from the dash.
9   Installation is the reverse of removal.

Central dash vent housing

10   Remove the radio/cassette unit as described in Chapter 12.
11   Loosen and remove the vent housing retaining screws and withdraw the housing from the dash.
12   Installation is the reverse of removal.

---

11   Air conditioning and heating system - check and maintenance

---

Warning: *The air conditioning system is under high pressure. Do not loosen any hose fittings or remove any components until after the system has been discharged by a dealer service department or service station. Always wear eye protection when disconnecting air conditioning system fittings.*

1   The following maintenance checks should be performed on a regular basis to ensure the air conditioner continues to operate at peak efficiency.

   a) *Check the compressor drivebelt. If it's worn or deteriorated, replace it* (see Chapter 1).
   b) *Check the drivebelt tension and, if necessary, adjust it* (see Chapter 1).
   c) *Check the system hoses. Look for cracks, bubbles, hard spots and deterioration. Inspect the hoses and all fittings for oil bubbles and seepage. If there's any evidence of wear, damage or leaks, replace the hose(s).*
   d) *Inspect the condenser fins for leaves, bugs and other debris. Use a "fin comb" or compressed air to clean the condenser.*
   e) *Make sure the system has the correct refrigerant charge.*
   f) *Check the evaporator housing drain tube for blockage.*

2   It's a good idea to operate the system for about 10 minutes at least once a month, particularly during the winter. Long term non-use can cause hardening, and subsequent failure, of the seals.
3   Because of the complexity of the air conditioning system and the special equipment necessary to service it, in-depth troubleshooting and repairs are not included in this manual (refer to the *Haynes Automotive Heating and Air Conditioning Repair Manual*). However, simple checks and component replacement procedures are provided in this Chapter.
4   The most common cause of poor cooling is simply a low system refrigerant charge. If a noticeable drop in cool air output occurs, the following quick check will help you determine if the refrigerant level is low.

Checking the refrigerant charge

5   Warm the engine up to normal operating temperature.
6   Place the air conditioning temperature selector at the coldest setting and the blower at the highest setting. Open the doors (to make sure the air conditioning system doesn't cycle off as soon as it cools the passenger compartment).

7   With the compressor engaged - the clutch will make an audible click and the center of the clutch will rotate. If the compressor discharge line feels warm and the compressor inlet pipe feels cool, the system is properly charged.
8   Place a thermometer in the dashboard vent nearest the evaporator and monitor the system. If the ambient (outside) air temperature is very high, say 110-degrees F, the duct air temperature may be as high as 60-degrees F, but generally the air conditioning is 30 to 40-degrees F cooler than the ambient air, down to approximately 40-degrees F.
Note: *Humidity of the ambient air also affects the cooling capacity of the system. Higher ambient humidity lowers the effectiveness of the air conditioning system.*

Adding refrigerant

9   Buy an automotive charging kit at an auto parts store. A charging kit includes a 14-ounce can of refrigerant, a tap valve and a short section of hose that can be attached between the tap valve and the system low side service valve. Because one can of refrigerant may not be sufficient to bring the system charge up to the proper level, it's a good idea to buy a couple of additional cans. Make sure that one of the cans contains red refrigerant dye. If the system is leaking, the red dye will leak out with the refrigerant and help you pinpoint the location of the leak. **Caution:** *There are two types of refrigerant, R-12, used on vehicles up to 1992, and the more environmentally-friendly R-134a used in all the models covered by this manual. These two refrigerants (and their appropriate refrigerant oils) are not compatible and must never be mixed or components will be damaged. Use only R-134a refrigerant in the models covered by this manual.* **Warning:** *Never add more than two cans of refrigerant to the system.*
10   Hook up the charging kit by following the manufacturer's instructions. **Warning:** *DO NOT hook the charging kit hose to the system high side! The fittings on the charging kit are designed to fit **only** on the low side of the system.*
11   Back off the valve handle on the charging kit and screw the kit onto the refrigerant can, making sure first that the O-ring or rubber seal inside the threaded portion of the kit is in place. **Warning:** *Wear protective eyewear when dealing with pressurized refrigerant cans.*
12   Remove the dust cap from the low-side charging connection and attach the quick-connect fitting on the kit hose.
13   Warm up the engine and turn on the air conditioner. Keep the charging kit hose away from the fan and other moving parts. **Note:** *The charging process requires the compressor to be running. Your compressor may cycle off if the pressure is low due to a low charge. If the clutch cycles off, you can pull the low-pressure cycling switch plug and attach a jumper wire. This will keep the compressor ON.*
14   Hold the can upright, turn the valve handle on the kit until the stem pierces the can, then back the handle out to release the refrigerant. You should be able to hear the rush of gas. Add refrigerant to the low side of the system until both the receiver-drier surface and the evaporator inlet pipe feel about the same temperature. Allow stabilization time between each addition.
15   When the can is empty, turn the valve handle to the closed position and release the connection from the low-side port. Replace the dust cap.
16   Remove the charging kit from the can and store the kit for future use with the piercing valve in the UP position, to prevent inadvertently piercing the can on the next use.

Chapter 4 Part A
Fuel system - single-point fuel injection

Contents

Specifications

System type...	Bosch Mono-Motronic
Fuel pump type..	Electric, immersed in fuel tank
Regulated fuel pressure...	11.6 to 17.4 psi
Engine idle speed (non-adjustable, electronically controlled)	700 to 1000 rpm
Idle CO content (non-adjustable, electronically controlled).....................	0.2 to 1.2 percent
Injector electrical resistance	1.2 to 1.6 ohms at 60 to 80-degrees F

Torque specifications

	Ft-lbs (unless otherwise indicated)
Fuel tank strap bolts	18
Injector cap/inlet air temperature sensor housing screw	53 in-lbs
Intake manifold heater retaining screws..	84 in-lbs
Intake manifold retaining bolts	84 in-lbs
Oxygen sensor..	37
Throttle body air box retaining screw	84 in-lbs
Throttle body retaining screws	84 in-lbs
Throttle valve positioning module screws	53 in-lbs

1   General information and precautions

General information

The Bosch Mono-Motronic system installed on the 1.8L gasoline engine is a self-contained engine management system, which controls both the fuel injection and ignition. This Chapter deals with the fuel injection system components only - refer to Chapter 5B for details of the ignition system components.

The fuel injection system comprises a fuel tank, an electric fuel pump, a fuel filter, fuel supply and return lines, a throttle body with an integral electronic fuel injector, and an Electronic Control Unit (ECU) together with its associated sensors, actuators and wiring.

The fuel pump delivers a constant supply of fuel through a cartridge filter to the throttle body, at a slightly higher pressure than required - the fuel pressure regulator (integral with the throttle body) maintains a constant fuel pressure at the fuel injector and returns excess fuel to the tank via the return line. This constant flow system also helps to reduce fuel temperature and prevents vaporization.

The fuel injector is opened and closed by an Electronic Control Unit (ECU), which calculates the injection timing and duration according to engine speed, throttle position and rate of opening, inlet air temperature, coolant temperature, road speed and exhaust gas oxygen content information, received from sensors mounted on the engine.

Inlet air is drawn into the engine through the air cleaner, which contains a renewable paper filter element. The inlet air temperature is regulated by a vacuum operated valve mounted in the air cleaner, which blends air at ambient temperature with hot air, drawn from over the exhaust manifold.

Idle speed control is achieved partly by an electronic throttle positioning module, mounted on the side of the throttle body and partly by the ignition system, which gives fine control of the idle speed by altering the ignition timing. As a result, manual adjustment of the engine idle speed is not necessary.

To improve cold starting and idling, an electric heating element is mounted on the underside of the intake manifold; this prevents fuel vapor condensation when the engine is cold. Power is supplied to the heater by a relay, which is in turn controlled by the ECU.

The exhaust gas oxygen content is constantly monitored by the ECU via the Oxygen sensor, which is mounted in the exhaust pipe. The ECU then uses this information to modify the injection timing and duration to maintain the optimum air/fuel ratio - a result of this is that manual adjustment of the idle exhaust CO content is not necessary. In addition, certain models are equipped with an exhaust catalyst - see Chapter 6 for details.

In addition, the ECU controls the operation of the activated charcoal filter evaporative loss system - refer to Chapter 6 for further details.

It should be noted that fault diagnosis of the Bosch Mono-Motronic system is only possible with dedicated electronic test equipment. Problems with the systems operation should therefore be referred to a VW dealer for assessment. Once the fault has been identified, the removal/installation sequences detailed in the following Sections will then allow the appropriate component(s) to be replaced as required.

Precautions

Warning: *Gasoline is extremely flammable, so take extra precautions when you work on any part of the fuel system. Don't smoke or allow open flames or bare light bulbs near the work area, and don't work in a garage where a natural gas-type appliance (such as a water heater or clothes dryer) with a pilot light is present. Since gasoline is carcinogenic, wear latex gloves when there's a possibility of being exposed to fuel, and, if you spill any fuel on your skin, rinse it off immediately with soap and water. Mop up any spills immediately and do not store fuel-soaked rags where they could ignite. The fuel system is under constant pressure, so, if any fuel lines are to be disconnected, the fuel pressure in the system must be relieved first. When you perform any kind of work on the fuel system, wear safety glasses and have a Class B type fire extinguisher on hand.*

Many of the operations described in this Chapter involve the disconnection of fuel lines, which may cause an amount of fuel spillage. Before commencing work, refer to the above **Warning** and the information in *Safety First!* at the beginning of this manual.

Residual fuel pressure always remain in the fuel system, long after the engine has been switched off. This pressure must be relieved in a controlled manner before work can commence on any component in the fuel system - refer to Section 9 for details.

When working with fuel system components, pay particular attention to cleanliness - dirt entering the fuel system may cause blockages which will lead to poor running.

In the interests of personal safety and equipment protection, many of the procedures in this Chapter suggest that the negative cable be removed from the battery terminal. This firstly eliminates the possibility of accidental short circuits being caused as the vehicle is being worked upon, and secondly prevents damage to electronic components (e.g. sensors, actuators, ECU's) which are particularly sensitive to the power surges caused by disconnection or reconnection of the wiring harness while they are still "live".

It should be noted, however, that many of the engine management systems described in this Chapter (and Chapter 5B) have a "learning" capability, that allows the system to adapt to the engine's running characteristics as it wears with use. This "learned" information is lost when the battery is disconnected and the system will then take a short period of time to "re-learn" the engine's characteristics - this may be manifested (temporarily) as rough idling, reduced throttle response and possibly a slight increase in fuel consumption, until the system re-adapts. The re-adaptation time will depend on how often the vehicle is used and the driving conditions encountered.

Caution: *If the stereo in your vehicle is equipped with an anti-theft system, make sure you have the correct activation code before disconnecting the battery.*

2   Air cleaner and inlet system - removal and installation

Removal

Refer to illustration 2.2

1   Loosen the worm drive clamps and disconnect the air ducting from the air cleaner assembly.

2   Lift off the plastic cap, remove the retaining screw **(see illustration)** and lift off the throttle body air box, recovering the seal.

3   Disconnect the vacuum hoses from the intake air temperature regulator vacuum switch, noting their order of installation.

4   Unhook the rubber loops from the lugs on the chassis member.

5   Pull the air cleaner towards the engine and withdraw the air intake hose from the port on the inner fender.

6   Lift the air cleaner out of the engine bay.

7   Pry open the retaining clamps and lift the top cover from the air cleaner. Remove the air cleaner filter element (see Chapter 1 for more details).

Installation

8   Install the air cleaner by following the removal procedure in reverse.

3   Intake air temperature regulator - removal and installation

Removal

Refer to illustration 3.3

1   Disconnect the vacuum hoses from the temperature regulator, noting their order of connection.

2   Remove the throttle body air box/air cleaner, as described in Section 2.

2.2 Remove the throttle body air box retaining screws (arrows)

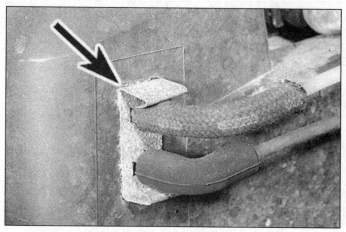

3.3 Temperature regulator metal retaining plate (arrow)

3 Pry off the metal retaining plate **(see illustration)** and remove the temperature regulator from the throttle body air box/air cleaner. Recover the gasket.

Installation

4 Install the regulator by following the removal procedure in reverse.

4 Accelerator cable - removal, installation and adjustment

Removal

Refer to illustration 4.2

1 Remove the throttle body air box/air cleaner as described in Section 2.
2 At the throttle body, disconnect the accelerator cable inner from the throttle valve spindle plate **(see illustration)**.
3 Remove the adjustment clip and extract the cable outer from the mounting bracket **(see illustration 4.2)**.
4 Refer to Chapter 11 and remove the dash trim panels from underneath the steering column.
5 Working under the dash, depress the accelerator pedal slightly, then unclip the accelerator cable end from the pedal extension lever.
6 At the point where the cable passes through the firewall, unscrew the cap from the two-piece grommet so that the cable can move freely.
7 Release the cable from its securing clamps and guide it out through the firewall grommet.

Installation

8 Install the accelerator cable by following the removal procedure in reverse.

Adjustment

Vehicles with manual transaxle

9 At the throttle body, fix the position of the cable outer in its mounting bracket by inserting the metal clip in one of the locating slots, such that when the accelerator is depressed fully, the throttle valve is held wide open to its end stop.

Vehicles with automatic transaxle

10 On vehicles with automatic transaxle, place a block of wood approximately 9/16-inch thick between the underside of the accelerator pedal and the stop on the floorpan, then hold the accelerator pedal down onto the block of wood.
11 At the throttle body, fix the position of the cable outer in its mounting bracket by inserting the metal clip in one of the locating

slots, such that when the accelerator is depressed fully (onto the block of wood), the throttle valve is held wide open to its end stop.
12 Remove the block of wood and release the accelerator pedal. Refer to Chapter 7B and using a continuity tester, check that the kick-down switch contacts close as the accelerator pedal is depressed past the full throttle position, just before it contacts the stop on the floorpan.

5 Bosch Mono Motronic engine management system components - removal and installation

Warning: *Observe the precautions in Section 1 before working on any component in the fuel system.*

Throttle body

Removal

Refer to illustration 5.3

1 Refer to Section 2 and remove the air cleaner/throttle body air box.
2 Refer to Section 9 and depressurize the fuel system, then disconnect the battery negative cable and position it away from the terminal.
Caution: *If the stereo in your vehicle is equipped with an anti-theft system, make sure you have the correct activation code before disconnecting the battery.*

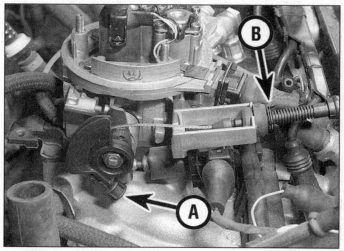

4.2 Throttle body accelerator cable mounting arrangement

A Throttle valve spindle plate   *B Adjustment clip*

4A

5.3 Typical throttle body fuel supply and return ports

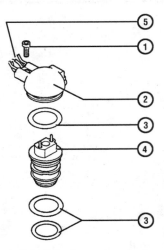

5.11 Injector components

1	Screw	3	O-ring seals
2	Injector cap/intake air temperature sensor housing	4	Injector
		5	Wiring harness connection

3   Disconnect the fuel supply and return hoses from the ports on the side of the throttle body. Note the arrows that denote the direction of fuel flow, and mark the hoses accordingly **(see illustration)**.
4   Unplug the wiring harness from the throttle body at the connectors, labeling them to aid correct installation later.
5   Refer to Section 4 and disconnect the accelerator cable from the throttle body.
6   Remove the through-bolts and lift the throttle body away from the intake manifold, recovering the gasket.

Installation

7   Installation is a reversal of removal; replace all gaskets where appropriate. On completion, check and if necessary adjust the accelerator cable. If the lower section of the throttle body has been replaced (with integral throttle potentiometer) in vehicles with electronic automatic transaxle control, the new potentiometer must be matched to the transaxle ECU; refer to a VW dealer for advice as this operation requires access to dedicated test equipment.

Fuel injector

Removal

Refer to illustration 5.11
8   Refer to Section 2 and remove the air cleaner/throttle body air box.
9   Refer to Section 9 and depressurize the fuel system, then disconnect the battery negative cable and position it away from the terminal. **Caution:** *If the stereo in your vehicle is equipped with an anti-theft system, make sure you have the correct activation code before disconnecting the battery.*
10   Unplug the wiring harness from the injector at the connector(s), labeling them to aid correct installation later.
11   Remove the screw and lift off the injector retaining cap/intake air temperature sensor housing **(see illustration)**.
12   Lift the injector out of the throttle body, recovering the O-ring seals.
13   Check the injector electrical resistance using a multimeter and compare the result with the Specifications.

Installation

14   Install the injector by following the removal procedure in reverse, renewing all O-ring seals. Tighten the retaining screw to the specified torque.

Intake air temperature sensor

15   The intake air temperature sensor is an integral part of the injector retaining cap. Removal is as described in the previous sub-Section. **(refer to illustration 5.11)**.

Fuel pressure regulator

Removal

Refer to illustration 5.20
16   If the operation of the fuel pressure regulator is in question, disassemble the unit as described below, then check the cleanliness and integrity of the internal components.
17   Remove the air cleaner/throttle body air box, with reference to Section 2.
18   Refer to Section 9 and depressurize the fuel system, then disconnect the battery negative cable and position it away from the terminal. **Caution:** *If the stereo in your vehicle is equipped with an anti-theft system, make sure you have the correct activation code before disconnecting the battery.*
19   With reference to the relevant sub-Section, remove the screw and lift off the inlet air temperature/injector cap.
20   Loosen and withdraw the retaining screws and lift off the fuel pressure regulator retaining frame **(see illustration)**.
21   Lift out the upper cover, spring and membrane.
22   Clean all the components thoroughly, then inspect the membrane for cracks or splits - replace it if necessary.

Installation

23   Reassemble the pressure regulator by following the removal procedure in reverse.

Throttle valve positioning module

Removal

24   Disconnect the battery negative cable and position it away from the terminal. **Caution:** *If the stereo in your vehicle is equipped with an anti-theft system, make sure you have the correct activation code before disconnecting the battery.* Remove the air cleaner/throttle body air box, with reference to Section 2.
25   Refer to Section 4 and disconnect the accelerator cable from the throttle body.
26   Unplug the connector from the side of the throttle valve positioning module.
27   Remove the retaining screws and lift the module together with the accelerator cable outer mounting bracket away from the throttle body.

Installation

28   Installation is a reversal of removal. Note that if a new module has

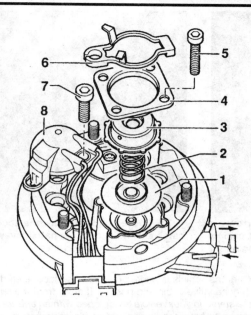

5.20 Fuel pressure regulator components

1	Membrane	5	Screws
2	Spring	6	Cable guide
3	Upper cover	7	Injector retaining screw
4	Retaining frame	8	Injector cap

been installed, the adjustment of the idle switch will need to be checked - refer to a VW dealer for advice as this operation requires access to dedicated test equipment.

Throttle valve potentiometer

29   Refer to the relevant sub-Section and remove the throttle body. The throttle valve potentiometer is an integral part of the lower section of the throttle body and cannot be replaced separately.
30   Where a new lower throttle body (with throttle potentiometer) has been installed in vehicles with electronic automatic transaxle control, the potentiometer must be matched to the transaxle ECU; refer to a VW dealer for advice as this operation requires access to dedicated test equipment.

Idle switch

31   Refer to the relevant sub-Section and remove the throttle valve positioning module. The idle switch is an integral part of the module and cannot be replaced separately.
32   Where a new throttle valve positioning module has been installed, the adjustment of the idle switch will need to be checked - refer to a VW dealer for advice as this operation requires access to dedicated test equipment.

Oxygen sensor

Removal

Refer to illustration 5.33

33   The oxygen sensor is threaded into the exhaust manifold **(see illustration)**. Refer to Chapter 6 for details.
34   Disconnect the battery negative cable and position it away from the terminal. **Caution:** *If the stereo in your vehicle is equipped with an anti-theft system, make sure you have the correct activation code before disconnecting the battery.* Unplug the wiring harness from the oxygen sensor at the connector, located adjacent to the right hand rear engine mounting.
35   **Note:** *If the correct size wrench is not available, a slotted socket will be required to remove the sensor.* Working under the vehicle,

5.33 Typical oxygen sensor

loosen and withdraw the sensor, taking care to avoid damaging the sensor probe as it is removed.

Installation

36   Apply a little anti-seize compound to the sensor threads - avoid contaminating the probe tip.
37   Install the sensor to its housing, tightening it to the specified torque. Restore the harness connection.

Coolant temperature sensor

Removal

Refer to illustration 5.39

38   Disconnect the battery negative cable and position it away from the terminal. **Caution:** *If the stereo in your vehicle is equipped with an anti-theft system, make sure you have the correct activation code before disconnecting the battery.* Refer to Chapter 3 and drain approximately one quarter of the coolant from the engine.
39   The sensor is mounted at the top coolant outlet elbow, at the front of the cylinder head **(see illustration)**.
40   Unscrew/unclamp the sensor from its housing and recover the sealing washer(s) and O-ring - be prepared for an amount of coolant loss.

Installation

41   Install the sensor by reversing the removal procedure, using new sealing washers and rings where appropriate. Refer to Chapter 1 and top-up the cooling system.

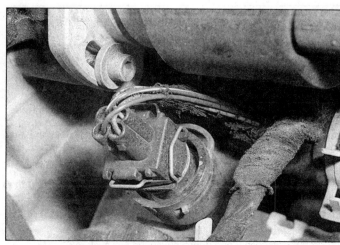

5.39 Typical coolant temperature sensor

4A

6.4 Fuel filter hose clamps (arrows)

6.6 Note the direction of flow arrow marked on the side of the filter unit casing - this must point towards the engine when installed

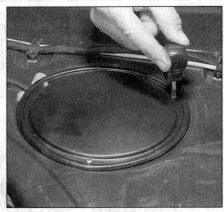

7.5 Loosen and withdraw the access hatch screws and lift the hatch away from the floorpan

6   Fuel filter - removal and installation

Refer to illustrations 6.4 and 6.6
Warning: *Observe the precautions in Section 1 before working on any component in the fuel system.*

Removal

1   The fuel filter is mounted in the fuel supply line, in front of the fuel tank. Access is from the underside of the vehicle.
2   Refer to Section 9 and depressurize the fuel system.
3   Park the vehicle on a level surface, apply the parking brake, select first gear (manual transaxle) or 'PARK' (automatic transaxle) and chock the front wheels. Raise the rear of the vehicle, support it securely on axle stands and remove the wheels; refer to *"Jacking and vehicle support"* for guidance.
4   Loosen the hose clamps and disconnect the fuel lines from either side of the filter unit **(see illustration)**. If the clamps are of the crimp type, snip them off with cutters and replace them with equivalent size worm drive clamps upon reconnection.
5   Release the filter retaining clamp, remove the cover bracket (where applicable) and lower the filter unit away from its mounting bracket.

Installation

6   Installation is a reversal of removal. Note the direction of flow arrow marked on the side of the filter unit casing - this must point towards the engine when installed **(see illustration)**.

7   Fuel pump and gauge sender unit - removal and installation

Warning: *Observe the precautions in Section 1 before working on any component in the fuel system.*

General information

1   The fuel pump and gauge sender unit are combined in one assembly, which is mounted on the top of the fuel tank. Access is via a hatch provided in the load space floor. The unit protrudes into the fuel tank and its removal involves exposing the contents of the tank to the atmosphere.
Warning: *Avoid direct skin contact with fuel - wear protective clothing and gloves when handling fuel system components. Ensure that the work area is well ventilated to prevent the build up of fuel vapor.*

Removal

Refer to illustrations 7.5 and 7.6
2   Refer to Section 9 and depressurize the fuel system.

3   Ensure that the vehicle is parked on a level surface, then disconnect the battery negative cable and position it away from the terminal. **Caution:** *If the stereo in your vehicle is equipped with an anti-theft system, make sure you have the correct activation code before disconnecting the battery.*
4   Refer to Chapter 11 and remove the trim from the load space floor.
5   Loosen and withdraw the access hatch screws and lift the hatch away from the floorpan **(see illustration)**.
6   Unplug the wiring harness connector from the pump/sender unit **(see illustration)**.
7   Pad the area around the supply and return fuel hoses with rags to absorb any spilled fuel, then loosen the hose clamps and remove them from the ports at the sender unit. Observe the supply and return arrows markings on the ports - label the fuel hoses accordingly to ensure correct installation later.
8   Unscrew the plastic securing ring and lift it out. **Note:** *Use a pair of water pump pliers to grip and rotate the plastic securing ring.* Turn the pump/sender unit to the left to release it from its bayonet fitting and lift it out, holding it above the level of the fuel in the tank until the excess fuel has drained out. Recover the rubber seal.
9   Remove the pump/sender unit from the vehicle and lay it on an absorbent card or rag. Inspect the float at the end of the sender unit swinging arm for punctures and fuel ingress - replace the unit if it appears damaged.
10   The fuel pick-up incorporated in the assembly is spring loaded to ensure that it always draws fuel from the lowest part of the tank. Check that the pick-up is free to move under spring tension with respect to the sender unit body.
11   Inspect the rubber seal from the fuel tank aperture for signs of fatigue - replace it if necessary.
12   Inspect the sender unit wiper and track; clean off any dirt and debris that may have accumulated and look for breaks in the track.

Installation

Refer to illustrations 7.13a, 7.13b and 7.13c
13   Install the sender unit by following the removal procedure in reverse, noting the following points:
  a) *The arrow markings on the sender unit body and the fuel tank must be aligned.*
  b) *Smear the tank aperture rubber seal with clean fuel before fitting it in position.*
  c) *When correctly installed, the pump/sender unit float arm must point towards the front left hand side of the vehicle, by an angle that depends on vehicle model.*
  d) *Reconnect the fuel hoses to the correct ports - observe the direction of flow arrow markings* **(see illustrations)**.

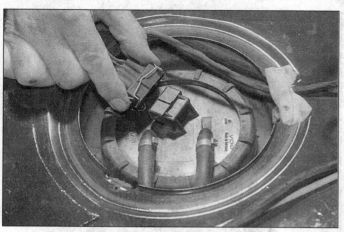

7.6 Unplug the wiring harness connector from the pump/sender unit

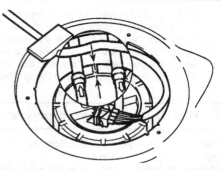

7.13a The arrow markings on the sender unit body and the fuel tank must be aligned

7.13c Reconnect the fuel hoses to the correct ports - observe the direction of flow arrow markings

4A

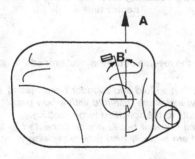

7.13b When correctly installed, the pump/sender unit float arm must point towards the front left hand side of the vehicle, by an angle that depends on the vehicle model

A   To front of vehicle   B   Angle = 5-degrees

8   Fuel tank - removal and installation

Warning: *Observe the precautions in Section 1 before working on any component in the fuel system.*

Removal

Refer to illustration 8.6

1   Before the tank can be removed, it must be drained of as much fuel as possible. As no drain plug is provided, it is preferable to carry out this operation with the tank almost empty.

2   Disconnect the battery negative cable and position it away from the terminal. **Caution:** *If the stereo in your vehicle is equipped with an anti-theft system, make sure you have the correct activation code before disconnecting the battery.* Using a hand pump or siphon, remove any remaining fuel from the bottom of the tank.

3   Refer to Section 7 and carry out the following:

   a) *Disconnect the wiring harness from the top of the pump sender unit at the multi-way connector.*

   b) *Disconnect the fuel supply and return hoses from the pump/sender unit.*

4   Position a floor jack under the center of the tank. Insert a block of wood between the jack head and the tank to prevent damage to the tank surface. Raise the jack until it just takes the weight of the tank.

5   Working inside the rear right hand wheel arch, loosen and withdraw the screws that secure the tank filler neck inside of the wheel arch. Open the fuel filler flap and peel the rubber sealing flange away from the bodywork.

6   Remove the retaining screws from the tank securing straps **(see illustration)**, keeping one hand on the tank to steady it, as it is released from its mountings.

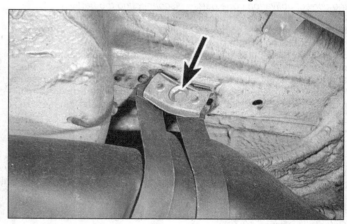

8.6 Remove the retaining screws from the tank securing straps (arrow)

7   Lower the jack and tank away from the underside of the vehicle; disconnect the charcoal canister vent pipe from the port on the filler neck as it is exposed. Locate the ground strap and disconnect it from the terminal at the filler neck.

8   If the tank is contaminated with sediment or water, remove the fuel pump/sender unit (see Section 7) and swill the tank out with clean fuel. The tank is injection molded from a synthetic material and if damaged, it should be replaced. However, in certain cases it may be possible to have small leaks or minor damage repaired. Seek the advice of a suitable specialist before attempting to repair the fuel tank.

Installation

9   Installation is the reverse of the removal procedure noting the following points:

   a) *When lifting the tank back into position make sure the mounting rubbers are correctly positioned and take great care to ensure that none of the hoses become trapped between the tank and vehicle body.*

b) *Ensure that all pipes and hoses are correctly routed and securely held in position with their retaining clips.*
c) *Reconnect the ground strap to its terminal on the filler neck.*
d) *Tighten the tank retaining strap bolts to the specified torque.*
e) *On completion, refill the tank with fuel and exhaustively check for signs of leakage prior to taking the vehicle out on the road.*

9   Fuel injection system - depressurization

Warning 1: *Observe the precautions in Section 1 before working on any component in the fuel system.*
Warning 2: *The following procedure will merely relieve the pressure in the fuel system - remember that fuel will still be present in the system components and take precautions accordingly before disconnecting any of them.*

1   The fuel system referred to in this Section is defined as the tank-mounted fuel pump, the fuel filter, the fuel injector, the throttle body-mounted fuel pressure regulator and the metal pipes and flexible hoses of the fuel lines between these components. All these contain fuel which will be under pressure while the engine is running and/or while the ignition is switched on. The pressure will remain for some time after the ignition has been switched off and must be relieved before any of these components are disturbed for servicing work. Ideally, the engine should be allowed to cool completely before work commences.
2   Refer to Chapter 12 and locate the fuel pump relay. Remove the relay from its housing, then crank the engine for a few seconds. The engine may fire and run for a while, but continue cranking until it stops. The fuel injector should have opened enough times during cranking to considerably reduce the line fuel pressure.
3   Disconnect the battery negative terminal. **Caution:** *If the stereo in your vehicle is equipped with an anti-theft system, make sure you have the correct activation code before disconnecting the battery.*
4   Place a suitable container beneath the relevant connection/union to be disconnected, and have a large rag ready to soak up any escaping fuel not being caught by the container.
5   Slowly loosen the connection or union nut (as applicable) to avoid a sudden release of pressure and position the rag around the connection to catch any fuel spray which may be expelled. Once the pressure has been released, disconnect the fuel line and insert plugs to minimize fuel loss and prevent the entry of dirt into the fuel system.

10   Intake manifold - removal and installation

Warning: *Observe the precautions in Section 1 before working on any component in the fuel system.*

Removal
Refer to illustration 10.7
1   Disconnect the battery negative cable and position it away from the terminal, then refer to Chapter 3 and drain the coolant from the engine.
2   With reference to Section 5, remove the throttle body from the intake manifold. Recover and discard the gasket and where applicable, remove the intermediate flange.
3   Loosen the clamps and remove the coolant hoses from the intake manifold.
4   Refer to Chapter 9 and disconnect the brake booster vacuum hose from the port on the intake manifold.
5   Disconnect the harness wiring from the intake manifold heater at the connector.
6   Progressively loosen and remove the intake manifold nuts and screws, then loosen the manifold from the cylinder head.
7   Make a final check to ensure that nothing remains connected to the manifold, then maneuver it out of the engine bay and recover the gasket. If required, remove the retaining screws and lift out the manifold heater unit **(see illustration)**.

10.7 Remove the retaining screws and lift out the manifold heater unit

Installation
8   Installation is the reverse of the removal procedure, noting the following points:
a) *Ensure that the manifold and cylinder head mating surfaces are clean and dry. Install the manifold with a new gasket and tighten its retaining nuts to the specified torque setting.*
b) *Ensure that all relevant hoses are reconnected to their original positions and are securely held (where necessary) by their retaining clamps.*
c) *Install the throttle body as described in Section 5.*
d) *On completion, refill the cooling system as described in Chapter 1.*

11   Fuel injection system - testing and adjustment

1   If a fault appears in the fuel injection system first ensure that all the system wiring connectors are securely connected and free of corrosion. Then ensure that the fault is not due to poor maintenance; i.e., check that the air cleaner filter element is clean, the spark plugs are in good condition and correctly gapped, the cylinder compression pressures are correct, the ignition timing is correct and the engine breather hoses are clear and undamaged, referring to Chapter 1, Chapter 2 Part A and Chapter 5 Part B for further information.
2   If these checks fail to reveal the cause of the problem the vehicle should be taken to a suitably equipped VW dealer for testing. A diagnostic connector is incorporated in the engine management system wiring harness, into which a dedicated electronic test equipment can be plugged. The test equipment is capable of "interrogating" the engine management system ECU electronically and accessing its internal fault log. In this manner, faults can be pinpointed quickly and simply, even if their occurrence is intermittent. Testing all the system components individually in an attempt to locate the fault by elimination is a time consuming operation that is unlikely to be fruitful (particularly if the fault occurs dynamically) and carries high risk of damage to the ECU's internal components.
3   Experienced home mechanics equipped with an accurate tachometer and a carefully-calibrated exhaust gas analyzer may be able to check the exhaust gas CO content and the engine idle speed; if these are found to be out of specification, then the vehicle must be taken to a suitably-equipped VW dealer for assessment. Neither the air/fuel mixture (exhaust gas CO content) nor the engine idle speed are manually adjustable; incorrect test results indicate a fault within the fuel injection system.

Chapter 4 Part B
Fuel system - multi-point fuel injection

Contents

Specifications

System type	Bosch Motronic
Fuel pump type	Electric, immersed in fuel tank
Regulated fuel pressure	36 psi
Engine idle speed (non-adjustable, electronically controlled)	
With air conditioning off	800 to 880 rpm
With air conditioning on	830 to 910 rpm
Idle CO content (non-adjustable, electronically controlled)	0.3 to 1.2-percent
Injector electrical resistance	14 to 21.5 ohms

Torque specifications

	Ft-lbs (unless otherwise indicated)
Fuel rail mounting bolts	84 in-lbs
Intake air temperature sensor	84 in-lbs
Lower intake manifold to cylinder head	18
Oxygen sensor	37
Throttle body through-bolts (M6 bolts)	84 in-lbs
Throttle body through-bolts (M8 bolts)	15
Upper intake manifold-to-lower intake manifold bolts	15

1 General information and precautions

General information

The Bosch Motronic system installed on the 2.0L gasoline engine is a self-contained engine management system, which controls both the fuel injection and ignition. This Chapter deals with the fuel injection system components only - refer to Chapter 5B for details of the ignition system components.

The fuel injection system comprises a fuel tank, an electric fuel pump, a fuel filter, fuel supply and return lines, a throttle body, a fuel rail, a fuel pressure regulator, four electronic fuel injectors, and an Electronic Control Unit (ECU) together with its associated sensors, actuators and wiring.

The fuel pump delivers a constant supply of fuel through a car-

tridge filter to the fuel rail, at a slightly higher pressure than required - the fuel pressure regulator maintains a constant fuel pressure to the fuel injectors and returns excess fuel to the tank via the return line. This constant flow system also helps to reduce fuel temperature and prevents vapor lock.

The fuel injectors are opened and closed by an Electronic Control Unit (ECU), which calculates the injection timing and duration according to engine speed, crankshaft position, throttle position and rate of opening intake air volume, intake air temperature, coolant temperature, road speed and exhaust gas oxygen content information, received from sensors mounted on and around the engine. Refer to the relevant Section for specific details of the components utilized in each system.

Intake air is drawn into the engine through the air cleaner, which contains a replaceable paper filter element. The intake air temperature is regulated by a vacuum operated valve mounted in the air cleaner, which blends air at ambient temperature with hot air, drawn from over the exhaust manifold.

Idle speed control is achieved partly by an electronic throttle valve positioning module, mounted on the side of the throttle body and partly by the ignition system, which gives fine control of the idle speed by altering the ignition timing. As a result, manual adjustment of the engine idle speed is not necessary or possible.

The exhaust gas oxygen content is constantly monitored by the ECU via the oxygen sensor, which is mounted in the exhaust pipe. The ECU then uses this information to modify the injection timing and duration to maintain the optimum air/fuel ratio - a result of this is that manual adjustment of the idle exhaust CO content is not necessary or possible. In addition, certain models are equipped with an exhaust catalyst - see Chapter 6 for details. If this is the case, the ECU controls the operation of the activated charcoal filter evaporative loss system - refer to Chapter 6 for further details.

It should be noted that fault diagnosis of all the engine management systems described in this Chapter is only possible with dedicated electronic test equipment. Problems with the systems operation should therefore be referred to a VW dealer for assessment. Once the fault has been identified, the removal/installation sequences detailed in the following Sections will then allow the appropriate component(s) to be replaced as required.

Precautions

Warning: *Gasoline is extremely flammable, so take extra precautions when you work on any part of the fuel system. Don't smoke or allow open flames or bare light bulbs near the work area, and don't work in a garage where a natural gas-type appliance (such as a water heater or clothes dryer) with a pilot light is present. Since gasoline is carcinogenic, wear latex gloves when there's a possibility of being exposed to fuel, and, if you spill any fuel on your skin, rinse it off immediately with soap and water. Mop up any spills immediately and do not store fuel-soaked rags where they could ignite. The fuel system is under constant pressure, so, if any fuel lines are to be disconnected, the fuel pressure in the system must be relieved first. When you perform any kind of work on the fuel system, wear safety glasses and have a Class B type fire extinguisher on hand.*

Many of the operations described in this Chapter involve the disconnection of fuel lines, which may cause an amount of fuel spillage. Before commencing work, refer to the above Warning and the information in "Safety First!" at the beginning of this manual.

Residual fuel pressure always remain in the fuel system, long after the engine has been switched off. This pressure must be relieved in a controlled manner before work can commence on any component in the fuel system - refer to Section 8 for details.

When working with fuel system components, pay particular attention to cleanliness - dirt entering the fuel system may cause blockages which will lead to poor running.

In the interests of personal safety and equipment protection, many of the procedures in this Chapter suggest that the negative cable be removed from the battery terminal. **Caution:** *If the stereo in your vehicle is equipped with an anti-theft system, make sure you have the correct activation code before disconnecting the battery.* This firstly eliminates the possibility of accidental short circuits being caused as the vehicle is being worked upon, and secondly prevents damage to electronic components (sensors, actuators, ECU) which are particularly sensitive to the power surges caused by disconnection or reconnection of the wiring harness while they are still "live".

It should be noted, however, that the engine management system has a "learning" capability, that allows the system to adapt to the engine's running characteristics as it wears with normal use. This "learned" information is lost when the battery is disconnected and on reconnection, the system will then take a short period of time to "re-learn" the engine's characteristics - this may be manifested (temporarily) as rough idling, reduced throttle response and possibly a slight increase in fuel consumption, until the system re-adapts. The re-adaptation time will depend on how often the vehicle is used and the driving conditions encountered.

2   Air cleaner and intake system - removal and installation

Removal

1   Loosen the clamps and disconnect the air ducting from the air cleaner assembly.

2   Where applicable, loosen the clip and disconnect the warm air duct from the base of the air cleaner.

3   Refer to Section 5 and remove the airflow meter from the air cleaner. **Caution:** *The airflow meter is a delicate component - handle it carefully.*

4   Disconnect the vacuum hoses from the intake air temperature regulator vacuum switch, noting their positions.

5   Unhook the rubber loops from the lugs on the chassis member.

6   Pull the air cleaner towards the engine and withdraw the air intake hose from the port on the inner fender.

7   Lift the air cleaner out of the engine bay and recover the rubber mountings.

8   Pry open the retaining clamps and lift the top cover from the air cleaner. Remove the air cleaner filter element (see Chapter 1 for more details).

Installation

9   Install the air cleaner by following the removal procedure in reverse.

3   Intake air temperature regulation system - general information and component replacement

General information

1   The intake air regulation system consists of a temperature controlled vacuum switch, mounted in the air cleaner housing, a vacuum operated flap valve and several lengths of interconnecting vacuum hose. The switch senses the temperature of the intake air and opens when a preset lower limit is reached. It then directs the manifold vacuum to the flap valve which opens, allowing warm air drawn from around the exhaust manifold to blend with the intake air.

Component replacement

Temperature switch

2   With reference to Section 2, release the clips and remove the top cover from the air cleaner.

3   Disconnect the vacuum hoses from the temperature switch, noting their order of connection to ensure correct installation.

4   Pry the metal retaining clip off the temperature switch ports, then press the switch body through into the top of the air cleaner. Recover the gasket.

5   Installation is a reversal of removal.

Flap valve

6   The flap valve is integrated with the lower section of the air cleaner and cannot be replaced separately.

4   Accelerator cable - removal, installation and adjustment

Refer to illustrations 4.2a, 4.2b, 4.3a and 4.3b

Warning: *Observe the precautions in Section 1 before working on any component in the fuel system.*

Removal

1   Remove the throttle body air box/air cleaner as described in Section 2.

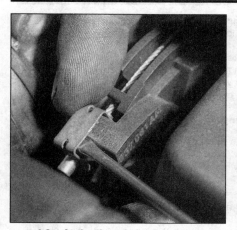

4.2a At the throttle body, pry off the clip . . .

4.2b . . . and disconnect the accelerator cable from the throttle valve

4.3a Remove the metal clip . . .

2   At the throttle body, pry off the clip and disconnect the accelerator cable inner from the throttle valve spindle **(see illustrations)**.
3   Remove the metal clip and extract the cable casing from the mounting bracket **(see illustrations)**.
4   Refer to Chapter 11 and remove the dash trim panels from underneath the steering column.
5   Depress the accelerator pedal slightly, then unclip the accelerator cable end from the pedal extension lever.
6   At the point where the cable passes through the firewall, unscrew the cap from the two-piece grommet so that the cable can move freely.
7   Release the cable from its securing clips and guide it out through the firewall grommet.

Installation
8   Install the accelerator cable by following the removal procedure in reverse.

Adjustment
Vehicles with manual transaxle
9   At the throttle body, fix the position of the cable casing in its mounting bracket by inserting the metal clip in one of the locating slots, such that when the accelerator is depressed fully, the throttle valve is held wide open to its end stop.

Vehicles with automatic transaxle
10   On vehicles with an automatic transaxle, place a block of wood approximately 9/16-inch thick between the underside of the accelerator pedal and the stop on the floorpan, then hold the accelerator pedal down onto the block of wood.
11   At the throttle body, fix the position of the cable casing in its mounting bracket by inserting the metal clip in one of the locating slots, such that when the accelerator is depressed fully (onto the block of wood), the throttle valve is held wide open to its end stop.
12   Remove the block of wood and release the accelerator pedal. Refer to Chapter 7B and using a continuity tester, check that the kickdown switch contacts close as the accelerator pedal is depressed past the full throttle position, just before it contacts the stop on the floorpan.

5   Bosch Motronic engine management system components - removal and installation

Note: *Observe the precautions in Section 1 before working on any component in the fuel system.*

Airflow meter
Removal
1   Disconnect the battery negative cable and position it away from

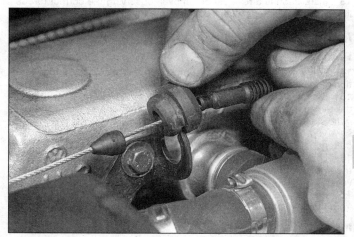

4.3b . . . and extract the cable casing from the mounting bracket

the terminal. **Caution:** *If the stereo in your vehicle is equipped with an anti-theft system, make sure you have the correct activation code before disconnecting the battery.*
2   With reference to Section 2 , loosen the clamps and disconnect the air ducting from the airflow meter, at the rear of the air cleaner housing.
3   Unplug the harness connector from the airflow meter.
4   Remove the retaining screws and extract the meter from the air cleaner housing. Recover the O-ring seal. **Caution:** *Handle the airflow meter carefully - its internal components are easily damaged.*

Installation
5   Installation is a reversal of removal. Replace the O-ring seal if it appears damaged.

Throttle valve potentiometer
Removal
6   Disconnect the battery negative cable and position it away from the terminal. **Caution:** *If the stereo in your vehicle is equipped with an anti-theft system, make sure you have the correct activation code before disconnecting the battery.*
7   Unplug the harness connector from the potentiometer.
8   Remove the retaining screws and lift the potentiometer away from the throttle body. Recover the O-ring seal.

Installation
9   Installation is a reversal of removal, noting the following:
a) *Replace the O-ring seal if it appears damaged.*

4B

b) *Ensure that the potentiometer drive engages correctly with the throttle spindle extension.*

c) *On vehicles with an automatic transaxle, the potentiometer must be matched to the automatic transaxle Electronic Control Unit (ECU) - this operation requires access to dedicated electronic test equipment (refer to a VW dealer for advice).*

Intake air temperature sensor

Removal

10   The sensor is threaded into the right hand side of the upper section of the intake manifold.

11   Disconnect the battery negative cable and position it away from the terminal. Unplug the harness connector from the sensor. **Caution:** *If the stereo in your vehicle is equipped with an anti-theft system, make sure you have the correct activation code before disconnecting the battery.*

12   Unscrew the sensor from the manifold.

Installation

13   Installation is a reversal of removal, observing the correct torque listed in this Chapter's Specifications.

Idling stabilization valve

Removal

14   The valve is mounted on a bracket on the rear of the upper section of the intake manifold.

15   Disconnect the battery negative cable and position it away from the terminal. Unplug the harness connector from the sensor. **Caution:** *If the stereo in your vehicle is equipped with an anti-theft system, make sure you have the correct activation code before disconnecting the battery.*

16   Loosen the clamps and disconnect the intake air duct hose and the intake muffler unit hose from the ports on the idle stabilization valve.

17   Loosen the mounting bracket retaining clip and remove the valve from the manifold.

Installation

18   Installation is a reversal of removal.

Road speed sensor

19   The road speed sensor is mounted on the transaxle - refer to Chapter 7A or B as applicable.

Coolant temperature sensor

Removal

Warning: *Wait until the engine is completely cool before beginning this procedure.*

20   The coolant temperature sensor is mounted in the coolant outlet elbow on the front of the cylinder head (see Chapter 3).

21   Disconnect the battery negative cable and position it away from the terminal, then unplug the harness connector from the sensor. **Caution:** *If the stereo in your vehicle is equipped with an anti-theft system, make sure you have the correct activation code before disconnecting the battery.*

22   Refer to Chapter 3 and drain approximately one quarter of the coolant from the engine.

23   Extract the retaining clip and lift the sensor from the coolant elbow - be prepared for an amount of coolant loss. Recover the O-ring.

Installation

24   Install the sensor by reversing the removal procedure, using a new O-ring. Refer to Chapter 1 and top-up the cooling system.

Engine speed sensor

Removal

25   The engine speed sensor is mounted on the front cylinder block,

adjacent to the mating surface of the block and transaxle bellhousing. If necessary, drain the engine oil and remove the oil filter and cooler to improve access.

26   Disconnect the battery negative cable and position it away from the terminal, then unplug the harness connector from the sensor. **Caution:** *If the stereo in your vehicle is equipped with an anti-theft system, make sure you have the correct activation code before disconnecting the battery.*

27   Remove the retaining screw and withdraw the sensor from the cylinder block.

Installation

28   Install the sensor by reversing the removal procedure.

Throttle body

29   Refer to Section 4 and detach the accelerator cable from the throttle valve lever.

30   Loosen the clamps and detach the intake air ducting from the throttle body.

31   Disconnect the battery negative cable and position it away from the terminal, then unplug the harness connector from the throttle potentiometer. **Caution:** *If the stereo in your vehicle is equipped with an anti-theft system, make sure you have the correct activation code before disconnecting the battery.*

32   Disconnect the vacuum hose from the port on the throttle body, then release the wiring harness from the guide clip.

33   Loosen and withdraw the through-bolts, then lift the throttle body away from the intake manifold. Recover and discard the gasket.

34   If required, refer to the relevant sub-Section and remove the throttle potentiometer.

Installation

35   Installation is a reversal of removal, noting the following:

a) *Use a new throttle body-to-intake manifold gasket.*

b) *Observe the correct tightening torque when installing the throttle body through-bolts.*

c) *Ensure that all vacuum hoses and electrical connectors are installed securely.*

d) *With reference to Section 4, check and if necessary adjust the accelerator cable.*

Fuel injectors and fuel rail

Warning: *Observe the precautions in Section 1 before working on any component in the fuel system.*

Removal

Refer to illustration 5.42

36   Disconnect the battery negative cable and position it away from the terminal. **Caution:** *If the stereo in your vehicle is equipped with an anti-theft system, make sure you have the correct activation code before disconnecting the battery.*

37   Refer to the relevant sub-Section in this Chapter and remove the throttle body, then refer to Section 10 and remove the upper section of the intake manifold.

38   Unplug the injector harness connectors, labeling them to aid correct installation later.

39   Refer to Section 8 and depressurize the fuel system.

40   Disconnect the vacuum hose from the port on the top of the fuel pressure regulator.

41   Loosen the clamps and disconnect the fuel supply and return hoses from the end of the fuel rail. *Carefully* note the positions of the hoses - the supply hose is marked with a white arrow and the return hose is marked with a blue arrow.

42   Unscrew the fuel rail retaining bolts, then carefully lift the rail away from the intake manifold, together with the injectors. Recover the injector lower O-ring seals as they emerge from the manifold **(see illustration)**.

43   The injectors can be removed individually from the fuel rail by extracting the relevant metal clip and easing the injector out of the rail. Recover the injector upper O-ring seals.

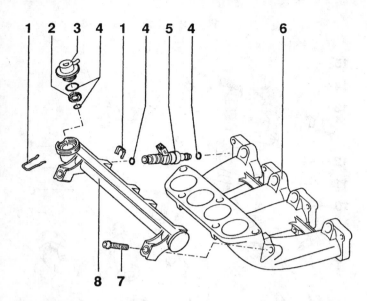

5.42 Fuel rail, fuel pressure regulator, injectors and associated components (Bosch Motronic system)

1	Clip	5	Injector
2	Strainer	6	Lower intake manifold
3	Fuel pressure regulator	7	Fuel rail bolts
4	O-ring seals	8	Fuel rail

44   If required, remove the fuel pressure regulator, referring to the relevant sub-Section for guidance.

45   Check the electrical resistance of the injector using a multimeter and compare it with the Specifications. **Note:** *If a faulty injector is suspected, before condemning the injector, it is worth trying the effect of one of the proprietary injector-cleaning treatments.*

Installation

46   Install the injectors and fuel rail by following the removal procedure in reverse, noting the following points:

a)   *Replace the injector O-ring seals if they appear worn or damaged (it's a good idea to do this regardless of the condition of the O-rings).*

b)   *Ensure that the injector retaining clamps are securely seated. Be sure to tighten the fuel rail mounting bolts to the torque listed in this Chapter's Specifications.*

c)   *Check that the fuel supply and return hoses are reconnected correctly - refer to the color coding described in "Removal".*

d)   *Use a new gasket when installing the upper section of the intake manifold.*

e)   *Check that all vacuum and electrical connections are remade correctly and securely.*

f)   *On completion, check exhaustively for fuel leaks before bringing the vehicle back into service.*

Fuel pressure regulator

Warning: *Observe the precautions in Section 1 before working on any component in the fuel system.*

Removal

47   Disconnect the battery negative cable and position it away from the terminal. **Caution:** *If the stereo in your vehicle is equipped with an anti-theft system, make sure you have the correct activation code before disconnecting the battery.*

48   Remove the retaining screws and lift the deflector plate away from the regulator housing.

49   Refer to Section 10 and depressurize the fuel system.

50   Disconnect the vacuum hose from the port on the top of the fuel pressure regulator.

51   Loosen the clip and disconnect the fuel supply hose from the end of the fuel rail **(refer to illustration 5.42)**. This will allow the majority of fuel in the regulator to drain out. Be prepared for an amount of fuel loss - position a small container and some old rags underneath the fuel regulator housing. **Note:** *The supply hose is marked with a white arrow.*

52   Extract the retaining clip from the side of the regulator housing and lift out the regulator body, recovering the O-ring seals and the strainer plate.

53   Examine the strainer plate for contamination and clean it if necessary, using clean fuel.

Installation

54   Install the fuel pressure regulator by following the removal procedure in reverse, noting the following points:

a)   *Replace the O-ring seals if they appear worn or damaged.*

b)   *Ensure that the regulator retaining clip is securely seated.*

c)   *Install the regulator vacuum hose securely.*

Oxygen sensor

Removal

55   The oxygen sensor is threaded into the exhaust pipe, at the front of the catalytic converter. Refer to Chapter 6 for details

56   Disconnect the battery negative cable and position it away from the terminal, then unplug the wiring harness from the oxygen sensor at the connector, located adjacent to the right hand rear engine mounting. **Caution:** *If the stereo in your vehicle is equipped with an anti-theft system, make sure you have the correct activation code before disconnecting the battery.*

57   Working under the vehicle, unscrew the sensor, taking care to avoid damaging the sensor probe as it is removed.

Installation

58   Apply a little anti-seize compound to the sensor threads - avoid contaminating the probe tip.

59   Install the sensor to its housing, tightening it to the specified torque. Restore the harness connection.

4B

6   Fuel filter - replacement

Warning: *Observe the precautions in Section 1 before working on any component in the fuel system.*

   Refer to the information given in Chapter 4A, Section 6.

7   Fuel pump and gauge sender unit - removal and installation

Warning: *Observe the precautions in Section 1 before working on any component in the fuel system.*

   Refer to the information given in Chapter 4A, Section 7.

8   Fuel injection system - depressurization

Warning: *Observe the precautions in Section 1 before working on any component in the fuel system.*

   Refer to the information given in Chapter 4A, Section 9.

9   Fuel tank - removal and installation

Warning: *Observe the precautions in Section 1 before working on any component in the fuel system.*

   Refer to the information given in Chapter 4A, Section 8.

10 Intake manifold - removal and installation

Refer to illustration 10.7
Warning: *Observe the precautions in Section 1 before working on any component in the fuel system.*

Removal

1   Refer to Section 10 and depressurize the fuel system, then disconnect the battery negative cable and position it away from the terminal. **Caution:** *If the stereo in your vehicle is equipped with an anti-theft system, make sure you have the correct activation code before disconnecting the battery.*
2   Refer to Section 2 and disconnect the air intake ducting from the throttle body.
3   Refer to Section 5 and remove the throttle body from the upper section of the intake manifold.
4   Remove the retaining screws and lift the deflector plate away from the fuel pressure regulator.
5   Unplug the wiring harness from the intake air temperature sensor (see Section 5).
6   Disconnect the brake booster vacuum hose from the port on the side of the intake manifold.
7   Loosen and withdraw the through-bolts, then separate the upper and lower sections of the intake manifold **(see illustration)**. **Note:** *Remove the retaining screw from the support bracket between the upper section of the intake manifold and the cylinder head.* Check that nothing remains connected to the upper section, then lift it out of the engine bay. Recover the gasket.
8   Refer to Section 5 and remove the fuel rail and fuel injectors. **Note:** *The fuel rail may be moved to one side, leaving the fuel lines connected to it, but take care to avoid straining them.*
9   Progressively loosen and remove the intake manifold-to-cylinder head bolts **(see illustration 5.42)**. Lift the manifold away from the head and recover the gasket.

Installation

10   Install the intake manifold by following the removal procedure in reverse, noting the following points:

   a) *Use new manifold gaskets and make sure all mating surfaces are clean.*
   b) *Tighten the manifold-to-cylinder head and upper-to-lower manifold bolts to the specified torque.*
   c) *Check that all vacuum and electrical connections are remade correctly and securely.*
   d) *On completion, check exhaustively for fuel leaks before bringing the vehicle back into service.*

11 Fuel injection system - testing and adjustment

1   If a fault appears in the fuel injection system first ensure that all the system wiring connectors are securely connected and free of corrosion. Then ensure that the fault is not due to poor maintenance; i.e., check that the air cleaner filter element is clean, the spark plugs are in good condition and correctly gapped, the cylinder compression pressures are correct, the ignition timing is correct and the engine breather hoses are clear and undamaged, referring to Chapter 1, Chapter 2A and Chapter 5B for further information.
2   If these checks fail to reveal the cause of the problem the vehicle should be taken to a suitably equipped VW dealer for testing. A diagnostic connector is incorporated in the engine management system

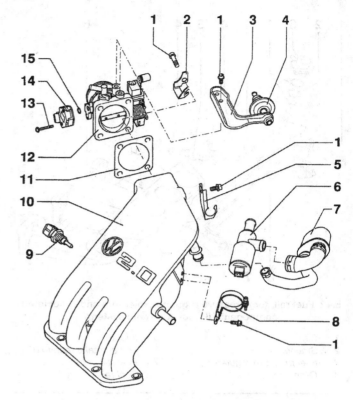

10.7 Upper intake manifold and throttle body details

1	Bolt	9	Intake Air Temperature
2	Bracket		(IAT) sensor
3	Bracket	10	Upper intake manifold
4	Deceleration damper	11	Gasket
5	Accelerator cable bracket	12	Throttle body
6	Idle Air Control (IAC)	13	Screw
	valve	14	Throttle Position Sensor
7	Noise damper		(TPS)
8	Support ring	15	O-ring

wiring harness, into which a dedicated electronic test equipment can be plugged. The test equipment is capable of "interrogating" the engine management system ECU electronically and accessing its internal fault log. In this manner, faults can be pinpointed quickly and simply, even if their occurrence is intermittent. Testing all the system components individually in an attempt to locate the fault by elimination is a time consuming operation that is unlikely to be fruitful (particularly if the fault occurs dynamically) and carries high risk of damage to the ECU's internal components.
3   Experienced home mechanics equipped with an accurate tachometer and a carefully-calibrated exhaust gas analyzer may be able to check the exhaust gas CO content and the engine idle speed; if these are found to be out of specification, then the vehicle must be taken to a VW dealer or other qualified repair shop for assessment. Neither the air/fuel mixture (exhaust gas CO content) nor the engine idle speed are manually adjustable; incorrect test results indicate a fault within the fuel injection system.

Chapter 4 Part C
Fuel system - diesel-engine models

Contents

4C

Specifications

General
Firing order	1-3-4-2
Maximum engine speed	5200 ± 100 rpm
Engine idle speed	900 ± 30 rpm
Engine fast idle speed	1050 ± 50 rpm

Fuel injection pump
Injection pump timing, dial indicator reading
Test	0.73 ± 0.87 mm (0.028 ± 0.034 inch)
Setting	0.80 ± 0.02 mm (0.031 ± 0.008 inch

Torque specifications
Ft-lb (unless otherwise indicated)
Fuel tank retaining strap bolts	18
Injection pump fuel supply and return banjo bolts	18
Injection pump fuel union lock nuts	15
Injection pump head fuel unions	18
Injection pump timing plug	132 in-lbs
Injection pump to front support bracket bolts	18
Injection pump to rear support bracket bolts	18
Injector fuel line fittings	18
Injectors	52

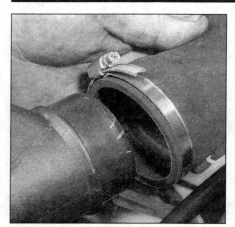

2.1 Loosen the worm drive clamps and disconnect the air ducting from the air cleaner assembly

2.2 Unhook the rubber loops from the lugs on the chassis member

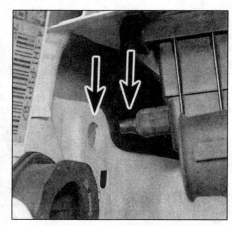

2.6 Engage the mounting lug with the recess (arrows) in the inner fender

1 General information and precautions

General information

The fuel system comprises a fuel tank, a fuel injection pump, an engine-bay mounted fuel filter with an integral water separator, fuel supply and return lines and four fuel injectors.

The injection pump is driven at half crankshaft speed by the camshaft timing belt. Fuel is drawn from the fuel tank, through the filter by the injection pump, which then distributes the fuel under very high pressure to the injectors via separate delivery lines.

The injectors are spring loaded mechanical valves, which open when the pressure of the fuel supplied to them exceeds a specific limit. Fuel is then sprayed from the injector nozzle into the cylinder via a swirl chamber (indirect injection). The engine is equipped with two-stage injectors which open in steps as the supplied fuel pressure rises, an effect which improves the engines combustion characteristics.

The basic injection timing is set by the position of the injection pump on its mounting bracket. When the engine is running, the injection timing is advanced and retarded mechanically by the injection pump itself and is influenced primarily by the accelerator position and engine speed.

The engine is stopped by means of a solenoid operated fuel cut-off valve which interrupts the flow of fuel to the injection pump when de-activated.

The engine idle speed can be raised manually by means of a cold start accelerator cable, controlled via a knob on the dash. Pulling the cable out to its first detent should increase the engine idle speed approximately 60 rpm. Pulling the cable out fully should raise the engine speed to approximately 1050 rpm.

Precautions

Many of the operations described in this Chapter involve the disconnection of fuel lines, which may cause an amount of fuel spillage. Before commencing work, refer to the **Warnings and Cautions** given below and the information in *Safety First!* at the beginning of this manual.

Warning 1: *When working on any part of the fuel system, avoid direct contact skin contact with diesel fuel - wear protective clothing and gloves when handling fuel system components. Ensure that the work area is well ventilated to prevent the build up of diesel fuel vapor.*

Warning 2: *Fuel injectors operate at extremely high pressures and the jet of fuel produced at the nozzle is capable of piercing skin, with potentially fatal results. When working with pressurized injectors, take great to avoid exposing any part of the body to the fuel spray. It is recommended that any pressure testing of the fuel system components should be carried out by a diesel fuel systems specialist.*

Caution: *Under no circumstances should diesel fuel be allowed to come into contact with coolant hoses - wipe off accidental spillage immediately. Hoses that have been contaminated with fuel for an extended period should be replaced. Diesel fuel systems are particularly sensitive to contamination from dirt, air and water. Pay particular attention to cleanliness when working on any part of the fuel system, to prevent the entry of dirt. Thoroughly clean the area around fuel unions before disconnecting them. Store dismantled components in sealed containers to prevent contamination and the formation of condensation. Only use lint-free cloths and clean fuel for component cleansing. Avoid using compressed air when cleaning components in place.*

2 Air cleaner assembly - removal and installation

Removal

Refer to illustrations 2.1 and 2.2

1   Loosen the worm drive clamps and disconnect the air ducting from the air cleaner assembly **(see illustration)**.

2   Unhook the rubber loops from the lugs on the chassis member **(see illustration)**.

3   Pull the air cleaner towards the engine and withdraw the air intake hose from the port on the inner fender.

4   Lift the air cleaner out of the engine bay.

5   Pry open the retaining clips and lift the top cover from the air cleaner. Remove the air cleaner filter element (see Chapter 1 for more details).

Installation

Refer to illustration 2.6

6   Install the air cleaner by following the removal procedure in reverse. Engage the mounting lug with the recess in the inner fender **(see illustration)**.

3 Accelerator cable - removal, installation and adjustment

Removal

Refer to illustrations 3.2, 3.4 and 3.5

1   Refer to Chapter 11 and remove the dash trim panels from underneath the steering column.

2   Depress the accelerator pedal slightly, then unclamp the accelerator cable end from the pedal extension lever **(see illustration)**.

3   At the point where the cable passes through the firewall, unscrew

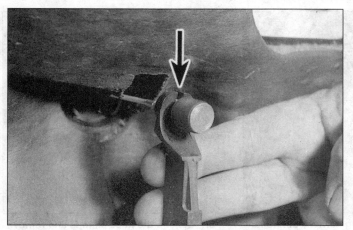

3.2 Depress the accelerator pedal slightly, then unclamp the accelerator cable end (arrow) from the pedal extension lever

3.4 Working in the engine bay, remove the clamp and detach the end of the accelerator cable inner from the fuel injection pump lever

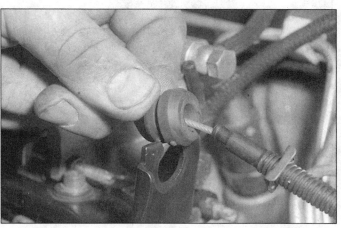

3.5 Slide back the rubber grommet and extract the accelerator cable outer from the mounting bracket

the cap from the two-piece grommet so that the cable can move freely.

4   Working in the engine bay, remove the clamp and detach the end of the accelerator cable inner from the fuel injection pump lever **(see illustration)**.

5   Slide back the rubber grommet and extract the accelerator cable outer from the mounting bracket **(see illustration)**.

6   Release the cable from its securing clamps and guide it out through the firewall grommet.

Installation

7   Install the accelerator cable by following the removal procedure in reverse.

Adjustment

8   At the fuel injection pump, fix the position of the cable outer in its mounting bracket by inserting the metal clamp in one of the locating slots, such that when the accelerator is depressed fully, the throttle lever is held wide open to its end stop.

4   Cold Start Accelerator (CSA) cable - removal, installation and adjustment

Removal

1   Working in the engine bay at the fuel injection pump, loosen the locking screw and disconnect the CSA cable inner from the injection pump lever.

2   Pry off the retaining clip and withdraw the cable outer from the mounting bracket on the side of the injection pump. Recover the washer.

3   Release the cable from the clamps that secure it in position in the engine bay.

4   With reference to Chapter 11, remove the trim panels from underneath the steering column, to gain access to the inside of the dash.

5   Pull the cold start knob out to expose its rear surface, then pry off the clamp and remove the knob from the cable inner.

6   Loosen and remove the retaining nut to release the cable outer from the dash.

7   Pull the cable through into the cabin, guiding it through the firewall grommet.

Installation

8   Install the CSA cable by reversing the removal procedure.

Adjustment

9   Push the cold start knob into the "fully off" position.

10   Thread the CSA cable inner through the drilling in the lever on the injection pump. Hold the injection pump cold start lever in the closed position, then pull the cable inner taught to take up the slack and tighten the locking screw.

11   Operate the cold start knob from the cabin and check that it is possible to move the injection pump lever through its full range of travel.

12   Push the cold start knob in to its "fully off" position, then start the engine and check the idle speed, as described in Chapter 1.

13   Pull the cold start knob fully out and check that the idle speed rises to approximately 1050 rpm. Adjust the cable if necessary.

5   Fuel tank sender unit - removal and installation

Refer to illustrations 5.5, 5.6, 5.7, 5.8a, 5.8b, 5.8c, 5.8d and 5.12

1   The fuel tank sender unit is mounted on the top of the fuel tank and is accessible via a hatch provided in the load space floor. It provides a variable voltage signal that drives the dash mounted fuel gauge and also serves as a connection point for the fuel supply and return hoses.

2   The unit protrudes into the fuel tank and its removal involves exposing the contents of the tank to the atmosphere. **Warning:** *Avoid direct contact skin contact with diesel fuel - wear protective clothing and gloves when handling fuel system components. Ensure that the work area is well ventilated to prevent the build up of diesel fuel vapor.*

4C

5.5 Loosen and withdraw the access hatch screws and lift the hatch away from the floorpan

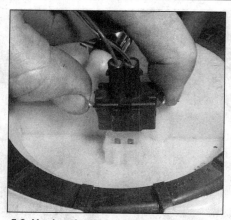

5.6 Unplug the wiring harness connector from the sender unit

5.7 Loosen the hose clamps and remove the fuel lines from the ports at the sender unit

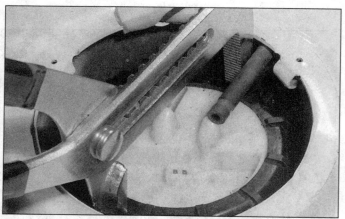

5.8a Use a pair of water pump pliers to grip and rotate the plastic securing ring

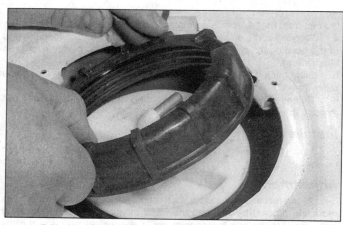

5.8b Unscrew the plastic securing ring and lift it out

Removal

3   Ensure that the vehicle is parked on a level surface, then disconnect the battery negative cable and position it away from the terminal. **Caution:** *If the stereo in your vehicle is equipped with an anti-theft system, make sure you have the correct activation code before disconnecting the battery.*

4   Refer to Chapter 11 and remove the trim from the load space floor.

5   Loosen and withdraw the access hatch screws and lift the hatch

away from the floorpan **(see illustration)**.

6   Unplug the wiring harness connector from the sender unit **(see illustration)**.

7   Pad the area around the supply and return fuel hoses with rags to absorb any spilt fuel, then loosen the hose clamps and remove them from the ports at the sender unit **(see illustration)**. Observe the supply and return arrows markings on the ports - label the fuel hoses accordingly to ensure correct installation later.

8   Unscrew the plastic securing ring and lift it out. **Note:** *Use a pair of water pump pliers to grip and rotate the plastic securing ring* **(see**

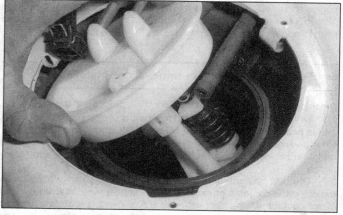

5.8c Lift out the sender unit . . .

5.8d . . . and recover the rubber seal

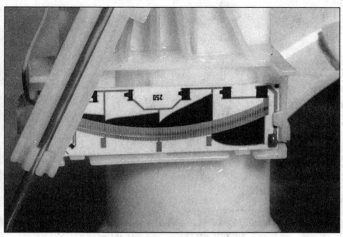

5.12 Look for breaks in the sender unit wiper track

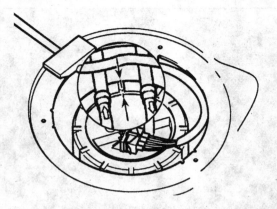

5.13 The arrow markings on the sender unit body and the fuel tank must be aligned

illustration). Where applicable, turn the sender unit to left to release it from its bayonet fitting and lift it out, holding it above the level of the fuel in the tank until the excess fuel has drained out. Recover the rubber seal **(see illustrations)**.

9   Remove the sender unit from the vehicle and lay it on an absorbent card or rag. Inspect the float at the end of the swinging arm for punctures and fuel entry - replace the sender unit if it appears damaged.

10   The fuel pick-up incorporated in the sender unit is spring loaded to ensure that it always draws fuel from the lowest part of the tank. Check that the pick-up is free to move under spring tension with respect to the sender unit body.

11   Recover the rubber seal from the fuel tank aperture and inspect for signs of fatigue - replace it if necessary **(see illustration)**.

12   Inspect the sender unit wiper and track; clean off any dirt and debris that may have accumulated and look for breaks in the track **(see illustration)**. An electrical specification for the sender unit is not quoted by the manufacturer, however the integrity of the wiper and track may be verified by connecting a multimeter, set to the resistance measurement function, across the sender unit connector terminals. The resistance should vary as the float arm is moved up and down, and an open circuit reading indicates that the sender is faulty and should be replaced.

Installation

13   Install the sender unit by following the removal procedure in reverse, noting the following points:
   a) *The arrow markings on the sender unit body and the fuel tank must be aligned* **(see illustration)**.
   b) *Smear the tank aperture rubber seal with clean fuel before fitting it in position.*

6   Fuel tank - removal and installation

Warning: *Observe the precautions in Section 1 before working on any component in the fuel system.*

Removal

1   Before the tank can be removed, it must be drained of as much fuel as possible. As no drain plug is provided, it is preferable to carry out this operation with the tank almost empty.

2   Disconnect the battery negative cable and position it away from the terminal. **Caution:** *If the stereo in your vehicle is equipped with an anti-theft system, make sure you have the correct activation code before disconnecting the battery.* Using a hand pump or siphon, remove any remaining fuel from the bottom of the tank.

3   Refer to Section 5 and carry out the following:

   a) *Disconnect the wiring harness from the top of the sender unit at the multi-way connector.*
   b) *Disconnect the fuel supply and return hoses from the sender unit.*

4   Position a floor jack under the center of the tank. Insert a block of wood between the jack head and the tank to prevent damage to the tank surface. Raise the jack until it just takes the weight of the tank (see *"Jacking and vehicle support"*).

5   Working inside the rear right hand wheelarch, loosen and withdraw the screws that secure the tank filler neck inside of the wheelarch. Open the fuel filler flap and peel the rubber sealing flange away from the bodywork.

6   Remove the retaining screws from the tank securing straps, keeping one hand on the tank to steady it, as it is released from its mountings.

7   Lower the jack and tank away from the underside of the vehicle; disconnect the breather hose(s) from the port on the filler neck as they are exposed. Locate the ground strap and disconnect it from the terminal at the filler neck.

8   If the tank is contaminated with sediment or water, remove the sender unit (see Section 5) and clean the tank out with clean diesel fuel. The tank is injection molded from a synthetic material and if damaged, it should be replaced. However, in certain cases it may be possible to have small leaks or minor damage repaired. Seek the advice of a suitable specialist before attempting to repair the fuel tank.

Installation

9   Installation is the reverse of the removal procedure noting the following points:

   a) *When lifting the tank back into position make sure the mounting rubbers are correctly positioned and take great care to ensure that none of the hoses become trapped between the tank and vehicle body.*
   b) *Ensure that all lines and hoses are correctly routed and securely held in position with their retaining clips.*
   c) *Reconnect the ground strap to its terminal on the filler neck.*
   d) *Tighten the tank retaining strap bolts to the specified torque.*
   e) *On completion, refill the tank with fuel and exhaustively check for signs of leakage prior to taking the vehicle out on the road.*

7   Fuel injection pump - removal and installation

Removal

Refer to illustrations 7.4, 7.5a, 7.5b, 7.6a, 7.6b, 7.7, 7.8, 7.9 and 7.13

1   Disconnect the battery negative cable and position it away from the terminal. **Caution:** *If the stereo in your vehicle is equipped with an anti-theft system, make sure you have the correct activation code before disconnecting the battery.*

4C

7.4 Attach a two-legged puller to the injection pump sprocket

7.5a Lift off the pump sprocket . . .

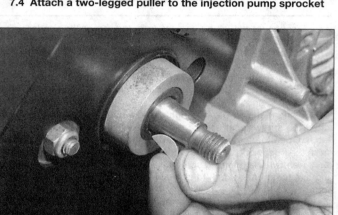

7.5b . . . and recover the Woodruff key

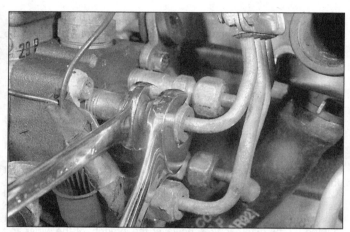

7.6a Loosen the rigid fuel line fittings at the rear of the injection pump

2   With reference to Chapter 2B, carry out the following:

a) *Remove the air cleaner and the associated ducting.*
b) *Remove the valve cover and timing belt outer cover.*
c) *Set the engine to TDC on cylinder No. 1.*
d) *Remove the timing belt from the camshaft and fuel injection pump sprockets.*

3   Loosen the nut that secures the timing belt sprocket to the injection pump shaft. The sprocket must be braced while the nut is loosened - a home made tool can easily be fabricated for this purpose; refer to Section 5 of Chapter 2B for further details.

4   Attach a two-legged puller to the injection pump sprocket, then gradually tighten the puller until the sprocket is under firm tension **(see illustration). Caution:** *To prevent damage to the injection pump shaft, insert a piece of scrap metal between the end of the shaft and the puller center bolt.*

5   Tap sharply on the puller center bolt with a hammer - this will free the sprocket from the tapered shaft. Detach the puller, then fully loosen and remove the sprocket nut, lift off the sprocket and recover the Woodruff key **(see illustrations)**.

6   Using a pair of wrenches, loosen the rigid fuel line fittings at the rear of the injection pump and at each end of the injectors, then lift the fuel line assembly away from the engine **(see illustrations). Caution 1:** *Be prepared for an amount of fuel leakage during this operation, position a small container under the union to be loosened and pad the area with old rags, to catch any spilt diesel.* **Caution 2:** *Take great care to avoid stressing the rigid fuel lines as they are removed.*

7   Cover the open lines and ports to prevent the entry of dirt and excess fuel leakage. **Note:** *Cut the fingertips from an old pair of rubber*

7.6b Lift the fuel line assembly away from the engine

gloves and secure them over the fuel ports with elastic bands **(see illustration)**.

8   Loosen the fuel supply and return banjo bolts at the injection pump ports, again taking precautions to minimize fuel spillage. Cover the open lines and ports to prevent the entry of dirt and excess fuel leakage. **Note:** *Install a short length of hose over the banjo bolt so that the drillings are covered, then thread the bolt back into its injection pump port* **(see illustration)**.

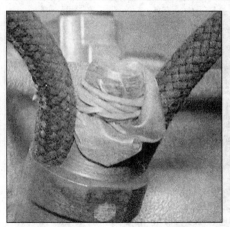

7.7 Cut the fingertips from an old pair of rubber gloves and secure them over the fuel ports with elastic bands

7.8 Install a short length of hose over the banjo bolt (arrow) so that the drillings are covered, then thread the bolt back into its injection pump port

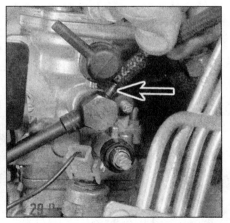

7.9 Disconnect the injector bleed hose from the port on the fuel return fitting (arrow)

9 Disconnect the injector bleed hose from the port on the fuel return union (see illustration).
10 Refer to Section 12 and disconnect the cabling from the stop control valve.
11 With reference to Sections 3 and 4, disconnect the cold start accelerator cable and accelerator cable from the injection pump.
12 Use a scriber or a pen to mark the relationship between the injection pump body and the front mounting bracket. This will allow an approximate injection timing setting to be achieved when the pump is refitted.
13 Loosen and withdraw the bolt that secures the injection pump to the rear mounting bracket (see illustration). Caution: Do not loosen the pump distributor head bolts, as this could cause serious internal damage to the injection pump.
14 Loosen and withdraw the three nuts/bolts that secure the injection pump to the front mounting bracket. Note that where the mounting bolts are used, the two outer bolts are held captive with metal brackets. Support the pump body as the last mounting is removed.
15 Check that nothing remains connected to the injection pump, then lift it away from the engine.

Installation

Refer to illustration 7.23
16 Install the injection pump to the engine, then insert the injection pump-to-rear support bracket bolt and tighten it to the specified torque.

17 Insert the injection pump-to-front support bracket bolts and tighten them to the specified torque. Note: *The mounting holes are elongated to allow adjustment - if a new pump is being installed, then mount it such that the bolts are initially at the center of the holes to allow the maximum range of pump timing adjustment. Alternatively, if the existing pump is being refitted, use the markings made during removal for alignment.*
18 Reconnect the fuel injector delivery lines to the injectors and injection pump head, then tighten the unions to the correct torque using a pair of wrenches.
19 Reconnect the fuel supply and return lines to the fuel injection pump and tighten the banjo bolts to the specified torque, use new sealing washers. Note: *The inside diameter of the banjo bolt for the fuel return line is smaller than that of the fuel supply line and is marked "OUT".*
20 Push the injector bleed hose onto the port on the return hose fitting.
21 Install the timing belt sprocket to the injection pump shaft, ensuring that the Woodruff key is correctly seated. Install the washer and retaining nut, hand tightening the nut only at this stage.
22 Lock the injection pump sprocket in position by inserting a bar or bolt through its alignment hole and into the drilling in the pump front mounting bracket. Ensure that there is minimal play in the sprocket, once it has been locked in position.
23 With reference to Chapter 2B, install the timing belt, then check and adjust the injection pump to camshaft timing. On completion, ten-

7.13 Withdraw the injection pump rear mounting bolt

7.23 Tightening the fuel injection pump sprocket, using a home-made locking tool

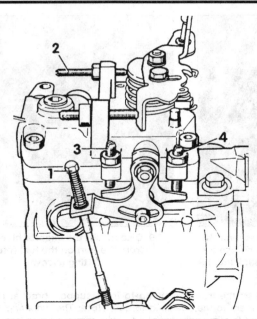

8.3 Fuel injection pump adjustment points

1   Idling speed adjustment   3   Minimum idling speed
    screw   stop screw
2   Maximum engine speed   4   Maximum idling speed
    adjustment screw   stop screw

sion the timing belt and tighten the fuel injection pump sprocket to the specified torque **(see illustration)**. Install the timing belt outer cover and valve cover, using a new gasket where necessary.
24   The remainder of the installation procedure is a direct reversal of the removal procedure, noting the following points:
  a) *Reconnect all electrical connections to the pump, using the labels made during removal.*
  b) *Reconnect the accelerator and cold start accelerator cables to the pump and adjust them as necessary.*
  c) *Install the air cleaner and its associated ducting.*
  d) *Reconnect the battery negative cable.*
25   Carry out the following:
  a) *Check and if necessary adjust the injection pump static timing as described in Section 10.*
  b) *Check and if necessary adjust the engine idling speed as described in Chapter 1.*
  c) *Check and if necessary adjust the maximum no-load engine speed, as described in Section 8.*
  d) *Check and if necessary adjust the engine fast idle speed, as described in Section 9.*

8   Maximum engine speed - checking and adjustment

Refer to illustration 8.3
Warning: *Observe the precautions in Section 1 before working on any component in the fuel system. This operation should not be carried out if the condition of the camshaft timing belt is questionable.*
1   Start the engine and with the parking brake applied and the transaxle in neutral, have an assistant depress the accelerator fully.
2   Using a diesel tachometer, check that the maximum engine speed is as quoted in the Specifications. **Warning:** *Do not maintain maximum engine speed for more than 2 or 3 seconds.*
3   If necessary, adjust the maximum engine speed by loosening the lock nut and rotating the adjusting screw **(see illustration)**.
4   On completion, tighten the lock nut.

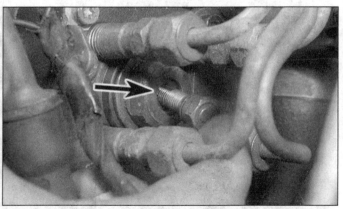

10.3 Unscrew the plug (arrow) from the pump head and recover the seal

9   Engine fast idle speed - checking and adjustment

Warning: *Observe the precautions in Section 1 before working on any component in the fuel system.*
1   With reference to Chapter 1, check and if necessary adjust the engine idling speed.
2   Pull the dash cold start knob fully out and using a diesel tachometer, check that the idle speed rises to that given in the Specifications.
3   If necessary, adjust the setting by loosening the lock nut and rotating the adjusting screw **(see illustration 8.3)**.
4   On completion, tighten the lock nut.

10   Fuel injection pump timing - testing and adjustment

Warning: *Observe the precautions in Section 1 before working on any component in the fuel system.*

Testing

Refer to illustrations 10.3 and 10.4
1   Disconnect the battery negative cable and position it away from the terminal. **Caution:** *If the stereo in your vehicle is equipped with an anti-theft system, make sure you have the correct activation code before disconnecting the battery.*
2   With reference to Chapter 2B, set the engine to TDC on cylinder No. 1 then check the valve timing, adjusting it if necessary. On completion, reset the engine to TDC on cylinder No. 1.
3   At the rear of the injection pump, unscrew the plug from the pump head and recover the seal **(see illustration)**.
4   Using a suitably threaded adapter, screw a dial indicator into the pump head **(see illustration)**. Pre-load the gauge by a reading of approximately 2.5 mm (0.098-inch).
5   Using a socket and wrench on the crankshaft bolt, slowly rotate the crankshaft counterclockwise; the dial indicator will indicate movement - keep turning the crankshaft until the movement just ceases.
6   Zero the dial indicator, with a pre-load of approximately 1.0 mm (0.040-inch).
7   Now rotate the crankshaft clockwise to bring the engine back up to TDC on cylinder No. 1. Observe the reading indicated by the dial indicator and compare it with the Specifications.
8   If the reading is within the test tolerance quoted in the Specifications, remove the dial indicator and install the pump head plug. Use a new seal and tighten the plug to the specified torque.
9   If the reading is out of tolerance, proceed as described in the next sub-Section.

Adjustment

10   Loosen the pump securing bolts at the front and rear brackets

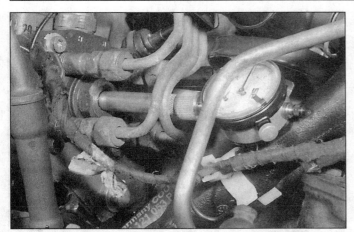

10.4 Using a suitably threaded adapter, screw a dial indicator into the pump head

11.6 Removing an injector from the cylinder head

(see Section 7).

11   Rotate the injection pump body until the "Setting" reading (see Specifications) is indicated on the dial indicator.

12   On completion, tighten the pump securing bolts to the specified torque.

13   Remove the dial indicator and install the pump head plug. Use a new seal and tighten the plug to the specified torque.

11   Injectors - general information, removal and installation

Warning: *Exercise extreme caution when working on the fuel injectors. Never expose the hands or any part of the body to injector spray, as the high working pressure can cause the fuel to penetrate the skin, with possibly fatal results. You are strongly advised to have any work which involves testing the injectors under pressure carried out by a dealer or fuel injection specialist. Refer to the precautions given in Section 1 of this Chapter before proceeding.*

General information

1   Injectors do deteriorate with prolonged use and it is reasonable to expect them to need reconditioning or replacement after 60,000 miles (100,000 km) or so. Accurate testing, overhaul and calibration of the injectors must be left to a specialist. A defective injector which is causing knocking or smoking can be located without disassembly as follows.

2   Run the engine at a fast idle. Loosen each injector fitting in turn,

11.7 Recover the heat shield washer

placing rag around the fitting to catch spilt fuel and being careful not to expose the skin to any spray. When the fitting on the defective injector is loosened, the knocking or smoking will stop.

Removal

Refer to illustrations 11.6 and 11.7

Note: *Take great care not to allow dirt into the injectors or fuel lines during this procedure. Do not drop the injectors or allow the needles at their tips to become damaged. The injectors are precision-made to fine limits and must not be handled roughly.*

3   Disconnect the battery negative lead and cover the alternator with a clean cloth or plastic bag to prevent the possibility of fuel being spilt onto it. **Caution:** *If the stereo in your vehicle is equipped with an anti-theft system, make sure you have the correct activation code before disconnecting the battery.*

4   Carefully clean around the injectors and line fitting nuts and disconnect the return line from the injector.

5   Wipe clean the line fittings then loosen the fitting nut securing the relevant injector lines to each injector and the relevant fitting nuts securing the lines to the rear of the injection pump (lines are removed as one assembly); as each pump fitting nut is loosened, retain the adapter with a suitable open-ended wrench to prevent it being unscrewed from the pump. With the fitting nuts undone remove the injector lines from the engine. Cover the injector and line fittings to prevent the entry of dirt into the system. Cut the fingertips from an old rubber glove and secure them over the open fittings with elastic bands to prevent the entry of dirt **(see illustration 7.7)**.

6   Unscrew the injector, using a deep socket or box wrench, and remove it from the cylinder head **(see illustration)**.

7   Recover the heat shield washer **(see illustration)**.

Installation

Refer to illustrations 11.7 and 11.9

8   Install a new heat shield washer to the cylinder head, noting that it must be installed with its convex side facing downwards (towards the cylinder head) **(see illustration)**.

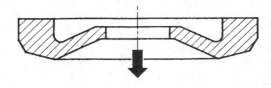

11.8 The heat shield washer must be installed with its convex side facing downwards (arrow faces the cylinder head)

4C

11.9 Screw the injector into position and tighten it to the specified torque

12.2 Fuel cut-off valve connector (arrow)

9   Screw the injector into position and tighten it to the specified torque **(see illustration)**.
10   Install the injector lines and tighten the fitting nuts to the specified torque setting. Position any clamps attached to the lines as noted before removal.
11   Reconnect the return line securely to the injector.
12   Restore the battery connection and check the running of the engine.

12 Fuel cut-off solenoid - removal and installation

Warning: *Observe the precautions in Section 1 before working on any component in the fuel system.*

Removal
Refer to illustration 12.2
1   The fuel cut-off valve is located at the rear of the injection pump.
2   Disconnect the battery negative cable and position it away from the terminal. Unplug the harness from the connector at the top of the valve **(see illustration)**. **Caution:** *If the stereo in your vehicle is equipped with an anti-theft system, make sure you have the correct activation code before disconnecting the battery.*
3   Loosen and withdraw the valve body from the injection pump. Recover the sealing washer, O-ring seal and the plunger.

Installation
4   Installation is a reversal of removal. Use a new sealing washer and O-ring seal.

Chapter 5 Part A
Starting and charging systems

Contents

Specifications

General
System type	12 volt, negative ground
Alternator minimum brush length	13/64-inch

Torque specifications
	Ft-lbs
Alternator mounting bolts	18
Starter motor mounting bolts:	
Lower bolt	33
Upper bolt	44
Stud	44

1 General information and precautions

General information

The engine electrical system consists mainly of the charging and starting systems. Because of their engine-related functions, these components are covered separately from the body electrical devices such as the lights, instruments, etc. (which are covered in Chapter 12). On gasoline engine models refer to Part B of this Chapter for information on the ignition system, and on diesel models refer to Part C for information on the pre-heating system.

The electrical system is of the 12-volt negative ground type.

The battery may of the low maintenance or "maintenance-free" (sealed for life) type and is charged by the alternator, which is belt-driven from the crankshaft pulley.

The starter motor is of the pre-engaged type incorporating an integral solenoid. On starting, the solenoid moves the drive pinion into engagement with the flywheel ring gear before the starter motor is energized. Once the engine has started, a one-way clutch prevents the motor armature being driven by the engine until the pinion disengages from the flywheel.

Precautions

Further details of the various systems are given in the relevant Sections of this Chapter. While some repair procedures are given, the usual course of action is to replace the component concerned. The owner whose interest extends beyond mere component replacement

should obtain a copy of the "Automobile Electrical & Electronic Systems Manual", available from the publishers of this manual.

It is necessary to take extra care when working on the electrical system to avoid damage to semi-conductor devices (diodes and transistors), and to avoid the risk of personal injury. In addition to the precautions given in "Safety first!", observe the following when working on the system:

Always remove rings, watches, etc. before working on the electrical system. Even with the battery disconnected, capacitive discharge could occur if a component's live terminal is grounded through a metal object. This could cause a shock or nasty burn. **Caution:** If the stereo in your vehicle is equipped with an anti-theft system, make sure you have the correct activation code before disconnecting the battery.

Do not reverse the battery connections. Components such as the alternator, electronic control units, or any other components having semi-conductor circuitry could be irreparably damaged.

If the engine is being started using jump leads and a slave battery, connect the batteries positive-to-positive and negative-to-negative (see "Booster battery (jump) starting"). This also applies when connecting a battery charger.

Never disconnect the battery terminals, the alternator, any electrical wiring or any test instruments when the engine is running.

Do not allow the engine to turn the alternator when the alternator is not connected.

Never "test" for alternator output by "flashing" the output lead to ground.

Never use an ohmmeter of the type incorporating a hand-cranked generator for circuit or continuity testing.

Always ensure that the battery negative lead is disconnected when working on the electrical system.

Before using electric-arc welding equipment on the car, disconnect the battery, alternator and components such as the fuel injection/ignition electronic control unit to protect them from the risk of damage.

2 Battery - testing and charging

Standard and low maintenance battery - testing

1   If the vehicle covers a small annual mileage it is worthwhile checking the specific gravity of the electrolyte every three months to determine the state of charge of the battery. Use a hydrometer to make the check and compare the results with the following table.

	Ambient temperature above 25°C (77°F)	Ambient temperature below 25°C (77°F)
Fully charged	1.210 to 1.230	1.270 to 1.290
70% charged	1.170 to 1.190	1.230 to 1.250
Fully discharged	1.050 to 1.070	1.110 to 1.130

Note that the specific gravity readings assume an electrolyte temperature of 15°C (60°F); for every 10°C (48°F) below 15°C (60°F) subtract 0.007. For every 10°C (48°F) above 15°C (60°F) add 0.007.
2   If the battery condition is suspect, first check the specific gravity of electrolyte in each cell. A variation of 0.040 or more between any cells indicates loss of electrolyte or deterioration of the internal plates.
3   If the specific gravity variation is 0.040 or more, the battery should be replaced. If the cell variation is satisfactory but the battery is discharged, it should be charged as described later in this Section.

Maintenance-free battery - testing

4   In cases where a "sealed for life" maintenance-free battery is installed, topping-up and testing of the electrolyte in each cell is not possible. The condition of the battery can therefore only be tested using a battery condition indicator or a voltmeter.
5   Certain models my be equipped with a maintenance-free battery, with a built-in charge condition indicator. The indicator is located in the top of the battery casing, and indicates the condition of the battery from its color. If the indicator shows green, then the battery is in a good state of charge. If the indicator turns darker, eventually to black, then the battery requires charging, as described later in this Section. If the indicator shows clear/yellow, then the electrolyte level in the battery is too low to allow further use, and the battery should be replaced. Do not attempt to charge, load or jump start a battery when the indicator shows clear/yellow.
6   If testing the battery using a voltmeter, connect the voltmeter across the battery. The test is only accurate if the battery has not been subjected to any kind of charge for the previous six hours. If this is not the case, switch on the headlights for 30 seconds, then wait four to five minutes before testing the battery after switching off the headlights. All other electrical circuits must be switched off, so check that the doors and tailgate are fully shut when making the test.
7   If the voltage reading is less than 12.2 volts, then the battery is discharged, while a reading of 12.2 to 12.4 volts indicates a partially discharged condition.
8   If the battery is to be charged, remove it from the vehicle and charge it as described later in this Section.

Standard and low maintenance battery - charging

Note: The following is intended as a guide only. Always refer to the manufacturer's recommendations (often printed on a label attached to the battery) before charging a battery.
9   Charge the battery at a rate equivalent to 10% of the battery capacity (e.g. for a 45 Ah battery charge at 4.5 A) and continue to charge the battery at this rate until no further rise in specific gravity is noted over a four hour period.

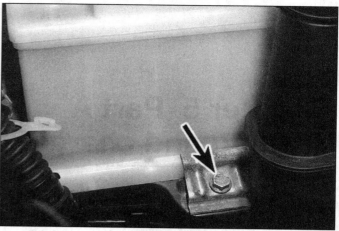

3.4 Battery clamping plate screw (arrow)

10   Alternatively, a trickle charger charging at the rate of 1.5 amps can safely be used overnight.
11   Specially rapid "boost" charges which are claimed to restore the power of the battery in 1 to 2 hours are not recommended, as they can cause serious damage to the battery plates through overheating.
12   While charging the battery, note that the temperature of the electrolyte should never exceed 37.8°C (100°F).

Maintenance-free battery - charging

Note: The following is intended as a guide only. Always refer to the manufacturer's recommendations (often printed on a label attached to the battery) before charging a battery.
13   This battery type takes considerably longer to fully recharge than the standard type, the time taken being dependent on the extent of discharge, but it can take anything up to three days.
14   A constant voltage type charger is required, to be set, when connected, to 13.9 to 14.9 volts with a charger current below 25 amps. Using this method, the battery should be useable within three hours, giving a voltage reading of 12.5 volts, but this is for a partially discharged battery and, as mentioned, full charging can take considerably longer.
15   If the battery is to be charged from a fully discharged state (condition reading less than 12.2 volts), have it recharged by your VW dealer or local automotive electrician, as the charge rate is higher and constant supervision during charging is necessary.

3 Battery - removal and installation

Refer to illustrations 3.4 and 3.7

Removal

1   **Caution:** *If the stereo in your vehicle is equipped with an anti-theft system, make sure you have the correct activation code before disconnecting the battery.*
2   Loosen the clamp screw and disconnect the battery negative cable from the terminal.
3   Unclip the plastic cover and disconnect the battery positive cable in the same manner.
4   At the base of the battery, loosen and withdraw the retaining screw, then lift off the clamping plate **(see illustration)**.
5   Remove the battery from the engine bay.
6   To remove the battery mounting tray, first remove the power steering fluid reservoir mounting screws (where applicable). Lay the reservoir on the inner fender, taking care not to stress the fluid hoses.
7   Loosen and withdraw the four retaining screws **(see illustration)**, then lift off the battery mounting tray, extracting the wiring harness from clips on the underside as they become exposed.

3.7 Battery mounting tray retaining screws (arrows)

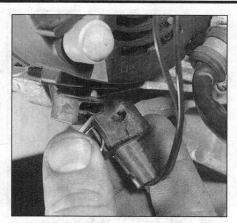

5.3 Unplug the sense cable from the alternator at the connector

5.4a Remove the protective cap . . .

Installation

8   Install the battery by following the removal procedure in reverse. Tighten the battery mounting tray and battery clamping plate screws securely.

4   Alternator/charging system - testing in vehicle

Note: *Refer to the warnings given in "Safety first!" and in Section 1 of this Chapter before starting work.*
1   If the ignition warning light fails to illuminate when the ignition is switched on, first check the alternator wiring connections for security. If satisfactory, check that the warning light bulb has not blown, and that the bulbholder is secure in its location in the instrument panel. If the light still fails to illuminate, check the continuity of the warning light feed wire from the alternator to the bulbholder. If all is satisfactory, the alternator is at fault and should be replaced or taken to an auto-electrician for testing and repair.
2   If the ignition warning light illuminates when the engine is running, stop the engine and check that the drivebelt is correctly tensioned (see Chapter 2A or B) and that the alternator connections are secure. If all is so far satisfactory, check the alternator brushes and slip rings as described in Section 8. If the fault persists, the alternator should be replaced, or taken to an auto-electrician for testing and repair.
3   If the alternator output is suspect even though the warning light functions correctly, the regulated voltage may be checked as follows.
4   Connect a voltmeter across the battery terminals and start the engine.

5   Increase the engine speed until the voltmeter reading remains steady; the reading should be approximately 12 to 13 volts, and no more than 14 volts.
6   Switch on as many electrical accessories (e.g., the headlights, heated rear window and heater blower) as possible, and check that the alternator maintains the regulated voltage at around 13 to 14 volts.
7   If the regulated voltage is not as stated, the fault may be due to worn brushes, weak brush springs, a faulty voltage regulator, a faulty diode, a severed phase winding or worn or damaged slip rings. The brushes and slip rings may be checked (see Section 6), but if the fault persists, the alternator should be replaced or taken to an auto-electrician for testing and repair.

5   Alternator - removal and installation

Refer to illustrations 5.3, 5.4a, 5.4b, 5.4c and 5.5

Removal

1   Disconnect the battery negative cable and position it away from the terminal. **Caution:** *If the stereo in your vehicle is equipped with an anti-theft system, make sure you have the correct activation code before disconnecting the battery.*
2   Refer to Chapter 2A or B as applicable and remove the drivebelt from the alternator pulley.
3   Unplug the sense cable from the alternator at the connector **(see illustration)**.
4   Remove the protective cap, loosen and withdraw the nut and washers, then disconnect the power cable from the alternator at the

5.4b . . . loosen and withdraw the nut and washers, then disconnect the power cable from the alternator at the screw terminal post

5.4c Where applicable, unbolt and remove the cable guide

5.5 Lifting the alternator away from its mounting bracket (diesel engine shown)

6.3a Remove the retaining screws (arrows) . . .

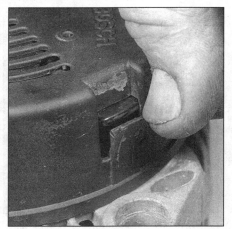

6.3b . . . then pry open the clips . . .

6.3c . . . and lift the plastic cover from the rear of the alternator

6.4a Loosen and withdraw the brush holder/voltage regulator module screws . . .

6.4b . . . then lift the module away from the alternator

screw terminal post. Where applicable, unbolt and remove the cable guide **(see illustrations)**.

5   Loosen and remove first the lower, then the upper mounting bolts, then lift the alternator away from its mounting bracket **(see illustration)**. Where applicable, pivot the tensioner roller out of the way to gain access to the lower mounting bolt.

6   Refer to Section 6 if the removal of the brush holder/voltage regulator module is required.

Installation

7   Installation is a reversal of removal. Refer to Chapter 2A or B as applicable for details of installation and tensioning the drivebelt.

8   On completion, tighten the alternator mounting bolts to the specified torque.

6   Alternator - brush holder/regulator module replacement

Refer to illustrations 6.3a, 6.3b, 6.3c, 6.4a, 6.4b, 6.5 and 6.6

1   Remove the alternator, as described in Section 5.

2   Place the alternator on a clean work surface, with the pulley facing down.

3   Remove the retaining screws, then pry open the clips and lift the plastic cover from the rear of the alternator **(see illustrations)**.

4   Loosen and withdraw the brush holder/voltage regulator module

screws, then lift the module away from the alternator **(see illustrations)**.

5   Measure the free length of the brush contacts - take the measure-

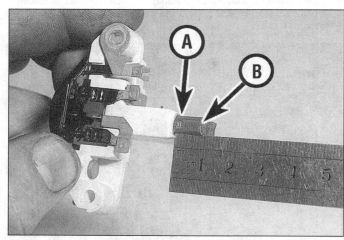

6.5 Measure the free length of the brush contacts - take the measurement from the manufacturers emblem (A) etched on the side of the brush contact, to the shallowest part of the curved end face of the brush (B)

6.6 Inspect the surfaces of the slip rings (arrows), at the end of the alternator shaft

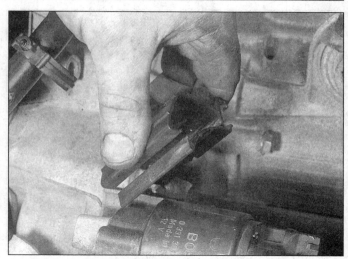

8.3a Remove the cable guide from above the solenoid housing . . .

ment from the manufacturers emblem etched on the side of the brush contact, to the shallowest part of the curved end face of the brush. Check the measurement with the Specifications; replace the module if the brushes are worn below the minimum limit **(see illustration)**.

6   Inspect the surfaces of the slip rings, at the end of the alternator shaft **(see illustration)**. If they appear excessively worn, burnt or pitted, then replacement must be considered; refer to an automobile electrical system specialist for further guidance.

7   Reassemble the alternator by following the dismantling procedure in reverse. On completion, refer to Section 5 and install the alternator.

7   Starting system - testing

Note: *Refer to the precautions given in "Safety first!" and in Section 1 of this Chapter before starting work.*

1   If the starter motor fails to operate when the ignition key is turned to the appropriate position, the following possible causes may be to blame:
   a) *The battery is faulty.*
   b) *The electrical connections between the switch, solenoid, battery and starter motor are somewhere failing to pass the necessary current from the battery through the starter to ground.*
   c) *The solenoid is faulty.*
   d) *The starter motor is mechanically or electrically defective.*

2   To check the battery, switch on the headlights. If they dim after a few seconds, this indicates that the battery is discharged - recharge (see Section 3) or replace the battery. **Caution:** *If the stereo in your vehicle is equipped with an anti-theft system, make sure you have the correct activation code before disconnecting the battery.* If the headlights glow brightly, operate the ignition switch and observe the lights. If they dim, then this indicates that current is reaching the starter motor, therefore the fault must lie in the starter motor. If the lights continue to glow brightly (and no clicking sound can be heard from the starter motor solenoid), this indicates that there is a fault in the circuit or solenoid - see following paragraphs. If the starter motor turns slowly when operated, but the battery is in good condition, then this indicates that either the starter motor is faulty, or there is considerable resistance somewhere in the circuit.

3   If a fault in the circuit is suspected, disconnect the battery leads (including the ground connection to the body), the starter/solenoid wiring and the engine/transmission ground strap. Thoroughly clean the connections, and reconnect the leads and wiring, then use a voltmeter or test lamp to check that full battery voltage is available at the battery positive lead connection to the solenoid, and that the ground is sound.

Smear petroleum jelly around the battery terminals to prevent corrosion - corroded connections are amongst the most frequent causes of electrical system faults.

4   If the battery and all connections are in good condition, check the circuit by disconnecting the wire from the solenoid blade terminal. Connect a voltmeter or test lamp between the wire end and a good ground (such as the battery negative terminal), and check that the wire is live when the ignition switch is turned to the "start" position. If it is, then the circuit is sound - if not the circuit wiring can be checked as described in Chapter 12.

5   The solenoid contacts can be checked by connecting a voltmeter or test lamp between the battery positive feed connection on the starter side of the solenoid, and ground. When the ignition switch is turned to the "start" position, there should be a reading or lighted bulb, as applicable. If there is no reading or lighted bulb, the solenoid is faulty and should be replaced.

6   If the circuit and solenoid are proved sound, the fault must lie in the starter motor. Begin checking the starter motor by removing it (see Section 8), and checking the brushes. If the fault does not lie in the brushes, the motor windings must be faulty. In this event, it may be possible to have the starter motor overhauled by a specialist, but check on the availability and cost of spares before proceeding, as it may prove more economical to obtain a new or exchange motor.

8   Starter motor - removal and installation

The starter is situated on the bellhousing at the front of the engine and shares its mounting bolts with the front engine mounting bracket. **Note:** *Removal of the front engine mounting bracket involves supporting the engine with either a lifting beam or an engine hoist while the bracket is removed - refer to Chapter 2A or B as applicable for greater detail.*

Removal

Refer to illustrations 8.3a, 8.3b, 8.5 and 8.7

1   Disconnect the battery negative cable and position it away from the terminal. **Caution:** *If the stereo in your vehicle is equipped with an anti-theft system, make sure you have the correct activation code before disconnecting the battery.*

2   Refer to Chapter 2A or B as applicable and remove the front engine mounting bracket from the starter motor.

3   Unhook the wiring connector from the cable guide above the solenoid housing, then remove the cable guide. Unplug the solenoid supply cabling at the connector **(see illustrations)**.

5A

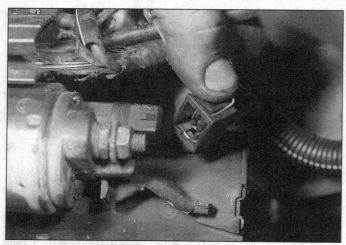

8.3b . . . then unplug the solenoid supply cabling at the connector

8.5 Remove the nut and washer from the power cable terminal post and take off the power cables

4   Where applicable, unbolt the PAS hose guide from the starter motor mountings.
5   At the rear of the solenoid housing, remove the nut and washer from the power cable terminal post and take off the power cables **(see illustration)**.
6   Remove the starter upper mounting bolt, then loosen and remove the nut from the mounting stud underneath the starter motor.
7   Guide the starter and solenoid assembly out of the bellhousing aperture **(see illustration)**.

Installation

8   Install the starter motor by following the removal procedure in reverse. Tighten the mounting bolts to the specified torque. Where applicable, refer to Chapter 2A or B and install the front engine mounting bracket.

9   Starter motor - testing and overhaul

If the starter motor is thought to be defective, it should be removed from the vehicle and taken to an auto-electrician for assessment. In the majority of cases, new starter motor brushes can be installed at a reasonable cost. However, check the cost of repairs first as it may prove more economical to purchase a new or exchange motor.

8.7 Guide the starter and solenoid assembly out of the bellhousing aperture (diesel engine shown)

Chapter 5 Part B
Ignition system - gasoline-engine models

Contents

Specifications

General

Type

2.0L engine .... Bosch Motronic
1.8L engine .... Bosch Mono-Motronic

Ignition coil

Primary resistance .... 0.5 to 0.7 ohms
Secondary resistance .... 3 to 4 k-ohms

Distributor

Type .... Breakerless
Ignition timing .... Controlled by engine management system

Spark plugs

See the Chapter 1 Specifications

Torque specifications

Ft-lbs (unless otherwise indicated)

Distributor clamp plate bolt 18
Distributor clamp bolt .... 84 in-lbs
Knock sensor mounting bolt 15
Spark plugs
2.0L engine 22
1.8L engine 18

1 General information

The Bosch Motronic and Mono-Motronic systems are self-contained engine management systems, which control both the fuel injection and ignition. This Chapter deals with the ignition system components only - refer to Chapter 4A or B for details of the fuel injection system components.

The ignition system comprises four spark plugs, five spark plug wires, the distributor, an electronic ignition coil, and an Electronic Control Unit (ECU) together with its associated sensors, actuators and wiring. The component layout varies from system to system but the basic operation is the same for all models.

The basic operation is as follows: the ECU supplies a voltage to the input stage of the ignition coil which causes the primary windings in the coil to be energized. The supply voltage is periodically interrupted by the ECU and this results in the collapse of primary magnetic field, which then induces a much larger voltage in the coil secondary windings. This voltage is directed, by the distributor via the spark plug wires, to the spark plug in the cylinder currently on its ignition stroke. The spark plug electrodes form a gap small enough for the secondary voltage to arc across, and the resulting spark ignites the fuel/air mixture in the cylinder. The timing of this sequence of events is critical and is regulated solely by the ECU.

The ECU calculates and controls the ignition timing and dwell angle primarily according to engine speed, crankshaft position and intake manifold depression (or inlet air volume flow rate, depending on system type) information, received from sensors mounted on and around the engine. Other parameters that affect ignition timing are throttle position and rate of opening, inlet air temperature, coolant temperature and on certain systems, engine knock. Again, these are monitored via sensors mounted on the engine.

On systems where knock control is employed, knock sensor(s) are mounted on the cylinder block - these have the ability to detect engine pre-ignition (or 'pinging') before it actually becomes audible. If pre-ignition occurs, the ECU retards the ignition timing of the cylinder that is pre-igniting in steps until the pre-ignition ceases. The ECU then advances the ignition timing of that cylinder in steps until it is restored to normal, or until pre-ignition occurs again.

Idle speed control is achieved partly by an electronic throttle valve positioning module, mounted on the side of the throttle body and partly by the ignition system, which gives fine control of the idle speed by altering the ignition timing. As a result, manual adjustment of the engine idle speed is not necessary or possible.

On certain systems, the ECU has the ability to perform multiple ignition cycles during cold starting. During cranking, each spark plug

3.3 Unplug the coil wire from the ignition coil

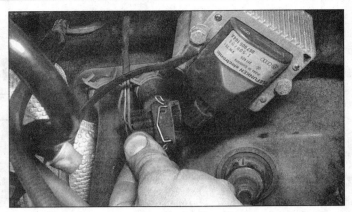

3.4 Disconnect the primary (low voltage) connector from the ignition coil

fires several times per ignition stroke, until the engine starts. This greatly improves the engines cold starting performance.

It should be noted that comprehensive fault diagnosis of all the engine management systems described in this Chapter is only possible with dedicated electronic test equipment. Problems with the systems operation that cannot be pinpointed by following the basic guidelines in Section 2 should therefore be referred to a VW dealer for assessment. Once the fault has been identified, the removal/installation sequences detailed in the following Sections will then allow the appropriate component(s) to replaced as required.

2   Ignition system - testing

Warning: *Extreme care must be taken when working on the system with the ignition switched on; it is possible to get a substantial electric shock from a vehicle's ignition system. Persons with cardiac pacemaker devices should keep well clear of the ignition circuits, components and test equipment. Always switch off the ignition before disconnecting or connecting any component and when using a multi-meter to check resistances.*

General

1   Most ignition system faults are likely to be due to loose or dirty connections or to 'tracking' (unintentional grounding) of secondary voltage due to dirt, dampness or damaged insulation, rather than by the failure of any of the system's components. **Always** check all wiring thoroughly before condemning an electrical component and work methodically to eliminate all other possibilities before deciding that a particular component is faulty.

2   The old practice of checking for a spark by holding the live end of a spark plug wire a short distance away from the engine is not recommended; not only is there a high risk of an electric shock, but the ignition coil could be damaged. Similarly, never try to 'diagnose' misfires by pulling off one spark plug wire at a time.

Engine will not start

3   If the engine either will not turn over at all, or only turns very slowly, check the battery and starter motor. Connect a voltmeter across the battery terminals (meter positive probe to battery positive terminal), disconnect the ignition coil wire from the distributor cap and ground it, then note the voltage reading obtained while turning over the engine on the starter for (no more than) ten seconds. If the reading obtained is less than approximately 9.5 volts, first check the battery, starter motor and charging systems (see Chapter 5A).

4   If the engine turns over at normal speed but will not start, check the secondary circuit by connecting a timing light (following the manufacturer's instructions) and turning the engine over on the starter motor; if the light flashes, voltage is reaching the spark plugs, so these should be checked first. If the light does not flash, check the spark plug wires themselves followed by the distributor cap, carbon brush

and rotor using the information given in Chapter 1.

5   If there is a spark, check the fuel system for faults referring to the relevant part of Chapter 4 for further information.

6   If there is still no spark, then the problem must lie within the engine management system. In these cases, the vehicle should be referred to a VW dealer or other qualified repair shop for assessment.

Engine misfires

7   An irregular misfire suggests either a loose connection or intermittent fault on the primary circuit, or a fault on the coil side of the rotor.

8   With the ignition switched off, check carefully through the system ensuring that all connections are clean and securely fastened. If the equipment is available, check the primary circuit as described above.

9   Check that the coil, the distributor cap and the spark plug wires are clean and dry. Check the leads themselves and the spark plugs (by substitution, if necessary), then check the distributor cap, carbon brush and rotor as described in Chapter 1.

10   Regular misfiring is almost certainly due to a fault in the distributor cap, spark plug wires or spark plugs. Use a timing light (Step 4 above) to check whether secondary voltage is present at all leads.

11   If voltage is not present on one particular lead, the fault will be in that lead or in the distributor cap. If voltage is present on all leads, the fault will be in the spark plugs; check and replace them if there is any doubt about their condition.

12   If no voltage is present, check the coil; its secondary windings may be breaking down under load.

Other problems

13   Problems with the system's operation that cannot be pinpointed by following the guidelines in the preceding Steps should be referred to a VW dealer or other qualified repair shop for assessment.

3   Ignition coil - removal and installation

Removal

Refer to illustrations 3.3 and 3.4

1   On all models, the ignition coil is mounted on the firewall.

2   Disconnect the battery negative cable and position it away from the terminal. **Caution:** *If the stereo in your vehicle is equipped with an anti-theft system, make sure you have the correct activation code before disconnecting the battery.*

3   Unplug the coil wire from the ignition coil **(see illustration)**.

4   Disconnect the primary (low voltage) electrical connector from the ignition coil **(see illustration)**.

5   Remove the mounting screws and remove the ignition coil.

Installation

6   Installation is a reversal of removal.

4.5 Unplug the Hall sensor from the distributor body

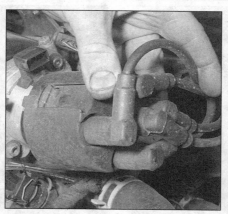

4.6a Pry off the clips and remove the distributor cap

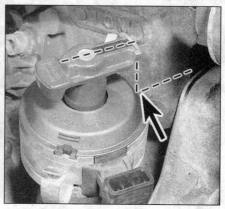

4.6b Check that the center of the rotor electrode is aligned with the cylinder No 1 mark (arrow) on the distributor body

4   Distributor - removal and installation

Removal

Refer to illustrations 4.5, 4.6a, 4.6b and 4.8

1   Disconnect the battery negative cable and position it away from the terminal. **Caution:** *If the stereo in your vehicle is equipped with an anti-theft system, make sure you have the correct activation code before disconnecting the battery.*

2   Set the engine to TDC on cylinder No 1, referring to Section 2 of Chapter 2A for guidance.

3   If required, unplug all four spark plug wires and the coil wire from the distributor cap, labeling them to aid installation later.

4   Where applicable, remove the screws and lift off the distributor shield.

5   Unplug the Hall sensor electrical connector from the distributor body **(see illustration)**.

6   Remove the screws/pry off the retaining clips (as applicable), then lift off the distributor cap. Check at this point that the center of the rotor electrode is aligned with the cylinder No 1 marking on the distributor body **(see illustrations)**.

7   Mark the relationship between the distributor body and the drive gear case flange by scribing arrows on each.

8   Loosen and remove the bolt, then lift off the clamp plate and withdraw the distributor body from the cylinder block **(see illustration)**. Recover the O-ring seal.

Inspection

9   Recover the O-ring seal(s) from the bottom of the distributor and inspect them. Replace them if they appear at all worn or damaged.

10   Inspect the teeth of the distributor drive gear for signs of wear or

4.8 Loosen and remove the bolt, then lift off the clamp plate

damage. Any slack in the distributor drive train will affect ignition timing. Replace the distributor if the teeth of the drive gear appear worn or chipped.

Installation

Refer to illustrations 4.12 and 4.13

11   Before progressing, check that the engine is still set to TDC on cylinder No 1.

12   If you're working on a 1.8L engine, check at this point that the oil pump shaft drive tongue is aligned with the threaded hole, adjacent to the distributor aperture **(see illustration)**.

13   If you're working on a 2.0L engine, check at this point that the oil

5B

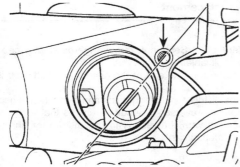

4.12 Oil pump shaft drive tongue is aligned with the threaded hole (arrow), adjacent to the distributor aperture (1.8L engine)

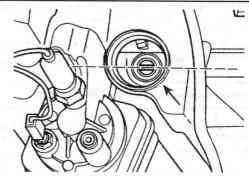

4.13 Oil pump shaft drive tongue is aligned with the axis of the crankshaft (2.0L engine)

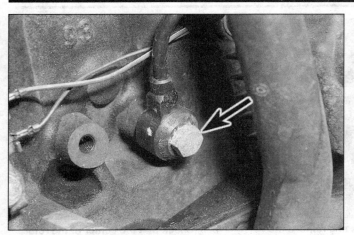

6.6 Unscrew the mounting bolt (arrow) and lift off the knock sensor

7.2 Pull the rotor from the end of the distributor shaft

pump shaft drive tongue is aligned with the axis of the crankshaft **(see illustration)**.

14   Install the distributor, then loosely install the clamp plate and securing bolt; it may be necessary to rotate the shaft slightly to allow it to engage with the intermediate shaft drive gear. Rotate the distributor body such that the alignment marks made during removal line up.

15   The shaft is engaged at the correct angle when the center of the rotor electrode is pointing directly at the No 1 cylinder mark on the distributor body - it may take a few attempts to get this right, as the helical drive gears make the alignment difficult to judge. Tighten the distributor clamp bolt to its specified torque. **Note:** *If alignment proves impossible, check that the intermediate shaft sprocket is correctly aligned with the crankshaft pulley - refer to Chapter 2A for further guidance.*

16   Install the distributor cap, pressing the retaining clips firmly into place/ tightening the retaining screws (as applicable).

17   Reconnect the Hall sensor cabling to the distributor.

18   Where applicable, install the distributor shield, tightening the screws securely.

19   Working from the No 1 terminal, connect the spark plug wires between the spark plugs and the distributor cap. Note that the firing order is 1-3-4-2.

20   Install the coil wire between the coil and the center terminal on the distributor cap.

21   If you are working on a 1.8L engine, it will be necessary to have the ignition timing checked and if necessary adjusted (see Section 5).

5   Ignition timing - checking and adjusting

The ignition timing is under the control of the engine management system ECU and is not manually adjustable without access to dedicated electronic test equipment. A basic setting cannot be quoted because the ignition timing is constantly being altered to control engine idle speed (see Section 1 for details).

The vehicle must be taken to a VW dealer or other qualified repair shop if the ignition timing requires checking or adjustment.

6   Ignition system sensors - removal and installation

1   Many of the engine management system sensors provide signals for both the fuel injection and ignition systems. Those which are specific to the ignition system are detailed in this Section.

2   Those sensors that are common to both systems are detailed in Chapter 4A or B as applicable. These include the coolant temperature sensor, the inlet air sensor, the air flow/mass meter, the engine speed/TDC sensor, the throttle potentiometer, the idle switch and the intake manifold depression sensor.

Knock sensor (2.0L engines only)

Refer to illustration 6.6

Removal

3   Disconnect the battery negative cable and position it away from the terminal. **Caution:** *If the stereo in your vehicle is equipped with an anti-theft system, make sure you have the correct activation code before disconnecting the battery.*

4   The knock sensor is located on the front of the cylinder block, below spark plug No 2.

5   Unplug the electrical connector from the sensor.

6   Loosen and withdraw the mounting bolt and lift off the sensor **(see illustration)**.

Installation

7   Installation is a reversal of removal, but note that the sensor operation will be affected if its mounting bolt is not tightened to exactly the right torque.

No 1 cylinder Hall-effect sensor

8   This sensor is an integral part of the distributor assembly. It can be removed and replaced separately, but disassembly of the distributor will be necessary. It is therefore recommended that this operation is entrusted to a VW dealer service department or other qualified repair shop.

7   Rotor - replacement

Refer to illustration 7.2

1   With reference to Section 4, remove the distributor cap and its shield (where applicable).

2   Pull the rotor from the end of the distributor shaft **(see illustration)**.

3   Inspect the distributor shaft contacts and clean them if necessary.

4   Installation is a reversal of removal - ensure that the rotor alignment lug engages with the recess in the distributor shaft before installing the distributor cap.

Chapter 5 Part C
Pre-heating system - diesel-engine models

Contents

Specifications

Glow plugs
Electrical resistance.. 1.5 Ω (approx.)
Current consumption... 8 amps (per glow plug)

Torque specifications
Glow plug-to-cylinder head... 18

1 General information

To assist cold starting, diesel engine models are equipped with a pre-heating system, which comprises four glow plugs, a glow plug control unit, a dash mounted warning lamp and the associated electrical wiring.

The glow plugs are miniature electric heating elements, encapsulated in a metal case with a probe at one end and electrical connection at the other. Each swirl chamber/intake tract has a glow plug threaded into it, the glow plug probe is positioned directly in line with incoming spray of fuel. When the glow plug is energized, the fuel passing over it is heated, allowing its optimum combustion temperature to be achieved more readily when it reaches the cylinder.

The duration of the pre-heating period is governed by the glow plug control unit, which monitors the temperature of the engine via the coolant temperature sensor and alters the pre-heating time to suit the conditions.

A dash mounted warning lamp informs the driver that pre-heating is taking place. The lamp extinguishes when sufficient pre-heating has taken place to allow the engine to be started, but power will still be supplied to the glow plugs for a further period until the engine is started. If no attempt is made to start the engine, the power supply to the glow plugs is switched off to prevent battery drain and glow plug burn-out. Note that on certain models, the warning lamp will also illuminate during normal driving if a pre-heating system malfunction occurs.

Generally, pre-heating is triggered by the ignition key being turned to the second position. However, certain models are equipped with a pre-heating system that activates when the drivers door is opened. Refer to the vehicles handbook for further information.

After the engine has been started, the glow plugs continue to operate for a further period of time. This helps to improve fuel combustion while the engine is warming up, resulting in quieter, smoother running and reduced exhaust emissions.

2 Glow plug control unit - removal and installation

Removal

1 The glow plug control unit is located behind the dash, above the main relay box - refer to Chapter 11 and remove the relevant sections of trim to gain access.
2 Disconnect the battery negative cable and position it away from the terminal. **Caution:** *If the stereo in your vehicle is equipped with an anti-theft system, make sure you have the correct activation code before disconnecting the battery.*
3 Unplug the wiring harness from the control unit at the connector.
4 Remove the retaining screws lift the control unit from its mounting bracket.

Installation

5 Installation is a reversal of removal.

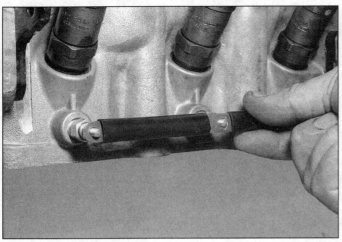

3.10 Remove the nuts and washers from the glow plug terminal. Lift off the bus bar

3.11 Loosen and withdraw the glow plug

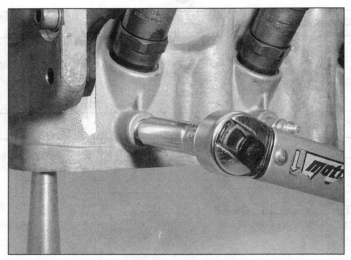

3.13 Tighten the glow plug to the specified torque

3   Glow plugs - testing, removal and installation

Testing

1   If the system malfunctions, testing is ultimately by substitution of known good units, but some preliminary checks may be made as described in the following paragraphs.

2   Connect a voltmeter or 12 volt test lamp to between the glow plug supply cable and a good ground point on the engine. **Caution:** *Make sure that the live connection is kept well clear of the engine and bodywork.*

3   Have an assistant activate the pre-heating system (either using the ignition key, or opening the drivers door as applicable) and check that a battery voltage is applied to the glow plug electrical connection. (Note that the voltage will drop to zero when the pre-heating period ends.)

4   If no supply voltage can be detected at the glow plug, then either the glow plug relay (where applicable) or the supply cabling must be faulty.

5   To locate a faulty glow plug, first disconnect the battery negative cable and position it away from the terminal. **Caution:** *If the stereo in your vehicle is equipped with an anti-theft system, make sure you have the correct activation code before disconnecting the battery.*

6   Refer to the next sub-Section and remove the supply cabling from the glow plug terminal. Measure the electrical resistance between the glow plug terminal and the engine ground. A reading of anything more than a few ohms indicates that the plug is defective.

7   If a suitable ammeter is available, connect it between the glow plug and its supply cable and measure the steady state current consumption (ignore the initial current surge which will be about 50% higher). Compare the result with the Specifications - high current consumption (or no current draw at all) indicates a faulty glow plug.

8   As a final check, remove the glow plugs and inspect them visually, as described in the next sub-Section.

Removal

Refer to illustrations 3.10 and 3.11

9   Disconnect the battery negative cable and position it away from the terminal. **Caution:** *If the stereo in your vehicle is equipped with an anti-theft system, make sure you have the correct activation code before disconnecting the battery.*

10   Remove the nuts and washers from the glow plug terminal. Lift off the bus bar **(see illustration)**.

11   Loosen and withdraw the glow plug **(see illustration)**.

12   Inspect the glow plug probe for signs of damage. A badly burned or charred probe is usually an indication of a faulty fuel injector; refer to Chapter 4C for greater detail.

Installation

Refer to illustration 3.13

13   Installation is a reversal of removal; tighten the glow plug to the specified torque **(see illustration)**.

Chapter 6
Emissions control and exhaust systems

Contents

6

Specifications

Turbocharger

Type	Garrett or KKK
Maximum boost pressure	8.7 to 12.0 psi

Torque specifications

	Ft-lbs
EGR valve-to-exhaust manifold connecting pipe	18
Turbocharger-to-exhaust manifold bolts	33
Exhaust downpipe-to-turbocharger nuts	18
Turbocharger oil supply line, turbocharger fitting	18
Turbocharger oil return line, turbocharger fitting	30
Turbocharger oil return line, cylinder block banjo bolt	37
Exhaust system clamp bolts	18
Exhaust mounting bracket-to-body bolts	30

1   General information

Emission control systems

All gasoline engine models have the ability to use unleaded gasoline and are controlled by engine management systems that are "tuned" to give the best compromise between driveability, fuel consumption and exhaust emission production. In addition, a number of systems are installed that help to minimize other harmful emissions: a crankcase emission-control system that reduces the release of pollutants from the engines lubrication system is installed to all models, catalytic converters that reduce exhaust gas pollutants are attached to most models and an evaporative loss emission control system that reduces the release of gaseous hydrocarbons from the fuel tank is attached to all models.

All diesel engine models are also equipped with a crankcase emission control system. In addition, all models have a catalytic converter and an Exhaust Gas Recirculation (EGR) system to reduce exhaust emissions.

Crankcase emission control

To reduce the emission of unburned hydrocarbons from the crankcase into the atmosphere, the engine is sealed and the blow-by gases and oil vapor are drawn from inside the crankcase, through a wire mesh oil separator, into the intake tract to be burned by the engine during normal combustion.

Under conditions of high manifold vacuum (idling, deceleration) the gases will be sucked positively out of the crankcase. Under conditions of low manifold vacuum (acceleration, full-throttle running) the gases are forced out of the crankcase by the (relatively) higher crankcase pressure; if the engine is worn, the raised crankcase pressure (due to increased blow-by) will cause some of the flow to return under all manifold conditions. On certain engines, a pressure regulating valve (mounted on the camshaft cover) controls the flow of gases from the crankcase.

Exhaust emission control - gasoline models

To minimize the amount of pollutants which escape into the atmosphere, most models have a catalytic converter in the exhaust system. On all models where a catalytic converter is attached, the fuel system is of the closed-loop type, in which a oxygen sensor in the exhaust system provides the engine management system ECU with constant feedback, enabling the ECU to adjust the air/fuel mixture to optimize combustion.

The oxygen sensor has a heating element built-in that is controlled by the ECU through the oxygen sensor relay to quickly bring the sensor's tip to its optimum operating temperature. The sensor's tip is sensitive to oxygen and relays a voltage signal to the ECU that varies according on the amount of oxygen in the exhaust gas. If the intake air/fuel mixture is too rich, the exhaust gases are low in oxygen so the sensor sends a low-voltage signal, the voltage rising as the mixture weakens and the amount of oxygen rises in the exhaust gases. Peak conversion efficiency of all major pollutants occurs if the intake air/fuel mixture is maintained at the chemically-correct ratio for the complete combustion of gasoline of 14.7 parts (by weight) of air to 1 part of fuel (the `stoichiometric' ratio). The sensor output voltage alters in a large step at this point, the ECU using the signal change as a reference point and correcting the intake air/fuel mixture accordingly by altering the fuel injector pulse width. Details of the oxygen sensor removal and installation are given in Chapter 4 Part A or B as applicable.

Exhaust emission control - diesel models

An oxidation catalyst is attached in the line with the exhaust system of all diesel engine models. This has the effect of removing a large proportion of the gaseous hydrocarbons, carbon monoxide and particulates present in the exhaust gas.

An Exhaust Gas Recirculation (EGR) system is attached to all diesel engine models. This reduces the level of nitrogen oxides produced during combustion by introducing a proportion of the exhaust gas back into the intake manifold, under certain engine operating conditions, via a plunger valve. The system is controlled electronically by the glow plug control module.

Evaporative emission control - gasoline models

To minimize the escape of unburned hydrocarbons into the atmosphere, an evaporative loss emission control system is attached to all gasoline models. The fuel tank filler cap is sealed and a charcoal canister is mounted underneath the right-hand fender to collect the gasoline vapors released from the fuel contained in the fuel tank. It stores them until they can be drawn from the canister (under the control of the fuel-injection/ignition system ECU) via the purge valve(s) into the intake tract, where they are then burned by the engine during normal combustion.

To ensure that the engine runs correctly when it is cold and/or idling and to protect the catalytic converter from the effects of an over-rich mixture, the purge control valve(s) are not opened by the ECU until the engine has warmed up, and the engine is under load; the valve solenoid is then modulated on and off to allow the stored vapor to pass into the intake tract.

Exhaust systems

The exhaust system comprises the exhaust manifold, one or two muffler units (depending on model and specification), a catalytic converter (where installed), a number of mounting brackets and a series of connecting pipes.

On the diesel engine, a turbocharger is attached to the exhaust manifold - refer to Section 6 for further details.

2   Evaporative loss emission control system - general information and component replacement

Refer to illustration 2.3

General information

1   The evaporative loss emission control system consists of the purge valve, the activated charcoal filter canister and a series of connecting vacuum hoses.
2   The purge valve is mounted on a bracket behind the air cleaner housing, and the charcoal canister is mounted on a bracket inside the right hand front wheel housing.

Component replacement
Purge valve

3   Ensure that the ignition is switched off, then unplug the wiring harness from the purge valve at the connector **(see illustration)**.

2.3 Unplug the wiring harness from the purge valve at the connector (arrow)

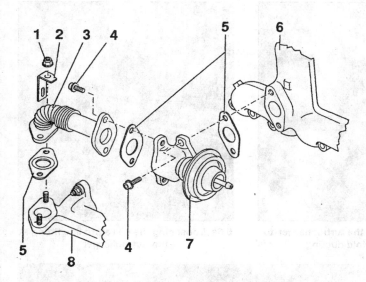

4.5 EGR valve and associated components: engine code 1Z

1	*Nut*	5	*Gaskets*
2	*Oil supply pipe bracket*	6	*Intake manifold*
3	*Semi-flexible pipe*	7	*EGR valve*
4	*Screw*	8	*Exhaust manifold*

4   Loosen the clips and pull the vacuum hoses off the purge valve ports. Make a note of their orientation to aid installation later.
5   Slide the purge valve out of its retaining ring and remove it from the engine bay.
6   Installation is a reversal of removal.

Charcoal canister

7   Locate the canister in the wheel housing. Disconnect the vacuum hoses from it, noting which ports they connect to. Depress the locking tab on the side of the securing strap and lift the canister out of the wheel housing.
8   Installation is a reversal of removal.

3   Crankcase emission system - general information

The crankcase emission control system consists of a series of hoses that connect the crankcase vent to the camshaft cover vent and the air intake, a pressure regulating valve (where applicable) and an oil separator unit.

The components of this system require no attention other than to check at regular intervals that the hose(s) are free of blockages and undamaged.

4   Exhaust Gas Recirculation (EGR) system - general information and component replacement

General information

1   The EGR system consists of the EGR valve, the EGR solenoid valve and a series of connecting vacuum hoses.
2   The EGR valve is mounted on a flange joint at the intake manifold and is connected to a second flange joint at the exhaust manifold by a semi-flexible pipe.
3   The EGR solenoid valve/modulator valve is mounted on a bracket on the right hand front suspension tower, behind the air cleaner.

Component replacement

Refer to illustration 4.5

EGR valve

4   Disconnect the vacuum hose from the port at the top of the EGR valve.
5   Loosen and withdraw the bolts that secure the semi-flexible connecting pipe to the EGR valve flange **(see illustration)**. Recover and discard the gasket from the joint.
6   Remove the bolts that secure the EGR valve to the intake manifold flange and lift off the EGR valve. Recover and discard the gasket.
7   Installation is a reversal of removal, noting the following points:
(a) *Use new flange joint gaskets and self-locking nuts.*
(b) *When reconnecting the semi-flexible pipe, attach the retaining bolts loosely and ensure that the pipe is unstressed before tightening the bolts to the specified torque.*

EGR solenoid valve/Modulator valve

8   Ensure that the ignition is switched off, then unplug the wiring harness from the valve at the connector.
9   Loosen the clips and pull the vacuum hoses off the valve ports. Make a *careful* note of their orientation to aid installation later.
10   Remove the retaining screws and lift off the valve.
11   Installation is a reversal of removal.
Caution: *Ensure that the vacuum hoses are installed correctly; combustion and exhaust smoke production can be drastically affected by an incorrectly operating EGR system.*

5   Exhaust manifold - removal and installation

The exhaust manifold removal is described as part of the cylinder head disassembly sequence; refer to Chapter 2 Part A or B as applicable.

6   Turbocharger - removal and installation

General information

1   A turbocharger is installed on the diesel engine and is mounted directly on the exhaust manifold. Lubrication is provided by a dedicated oil supply line that runs from the engine oil filter mounting. Oil is returned to the oil pan via a return line that connects to the side of the cylinder block. The turbocharger unit has an integral wastegate valve and vacuum actuator diaphragm, which is used to control the boost pressure applied to the intake manifold.
2   The turbocharger's internal components rotate at very high speed and as such are very sensitive to contamination; a great deal of damage can be caused by small particles of dirt, particularly if they strike the delicate turbine blades. Refer to the *Caution* and *Warning* notes given below before working on or removing the turbocharger unit.
Caution: *Thoroughly clean the area around all oil pipe fittings before disconnecting them, to prevent the ingress of dirt. Store disassembled components in a sealed container to prevent contamination. Cover the turbocharger air intake ducts to prevent debris entering and clean using lint-free cloths only.*
Warning: *Do not run the engine with the turbocharger air intake hose disconnected; the vacuum at the intake can build up very suddenly if the engine speed is raised and there is the risk of foreign objects being sucked in and ejected at very high speed.*

Removal

Refer to illustrations 6.4a, 6.4b, 6.6a, 6.6b, 6.7a and 6.7b
3   Disconnect the battery negative cable and position it away from the terminal. **Caution:** *If the stereo in your vehicle is equipped with an anti-theft system, make sure you have the correct activation code before disconnecting the battery.*

6

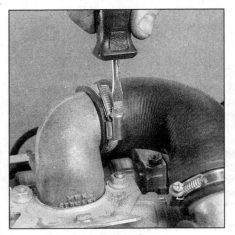

6.4a Loosen the clamps . . .   **6.4b . . . and remove the turbocharger-to-intake manifold ducting**   **6.6a Loosening the oil return line fitting at the turbocharger**

4   Loosen the clamps and remove the turbocharger-to-intake mani-fold **(see illustrations)** and air cleaner-to-turbocharger ducting.

5   Disconnect the vacuum hoses from the wastegate actuator diaphragm housing; note their order of connection and color coding to aid correct installation later.

6   Loosen the fittings and disconnect the oil supply and return lines from the turbocharger unit **(see illustrations)**. Recover the sealing washers and discard them - new items must be used on installation. Free the supply line from the clip on the intake manifold.

7   Remove the nuts and disconnect the exhaust downpipe from the turbocharger outlet. Recover and discard the gasket - a new item must be used on installation **(see illustrations)**. Loosen the mountings and remove the downpipe from the exhaust manifold support bracket.

8   Loosen and withdraw the turbocharger-to-exhaust manifold bolts. **Note:** *Access to the lowest bolt is restricted; a universal-joint extension bar will ease its removal.* Discard the bolts as new items must be used on installation.

9   Lift the turbocharger unit away from the exhaust manifold.

Installation

Refer to illustration 6.10

10   Install the turbocharger by following the removal procedure in reverse, noting the following points:

(a)   *Install the turbocharger to the exhaust manifold, then attach and hand tighten the exhaust downpipe nuts.*

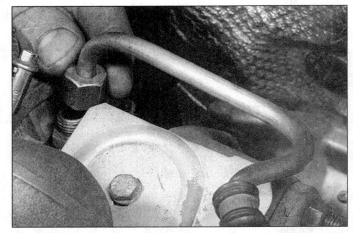

6.6b Disconnecting the oil supply line at the turbocharger

(b)   *Apply high temperature grease to the threads and heads of the new turbocharger-to-exhaust manifold bolts, then attach and tighten them to the specified torque.*

(c)   *Tighten the exhaust downpipe nuts to the specified torque.*

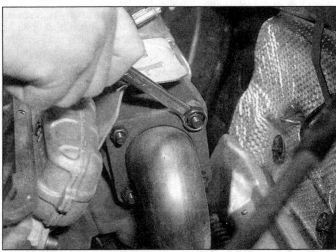
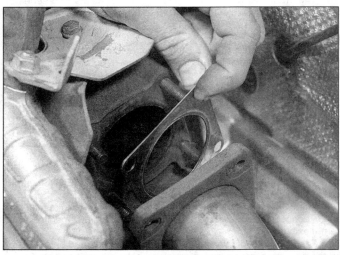

6.7a Remove the nuts and disconnect the exhaust downpipe from the turbocharger outlet   **6.7b Recover and discard the gasket**

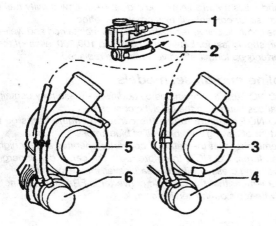

6.10 Wastegate vacuum hose connections

1	Two-way valve	4	Wastegate actuator
2	To vacuum pump	5	Turbocharger (Garrett)
3	Turbocharger (KKK)	6	Wastegate actuator

(d) *Prime the oil supply line and turbocharger oil intake port with clean engine oil before reconnecting the fitting and tightening it to the specified torque.*

(e) *Tighten the oil return fitting to the specified torque.*

(f) *Reconnect the wastegate actuator vacuum hoses according to the notes made during removal (see illustration).*

(g) *When the engine is started after installation, allow it idle for approximately one minute to give the oil time to circulate around the turbine shaft bearings.*

7 Exhaust system - general information and component replacement

General information

1   The exhaust system is made up of the downpipe, the front muffler (or catalytic converter, depending on specification), the intermediate pipe, and the tail section which contains the intermediate and rear silencers.

2   On diesel-engine models, a flexible coupling is attached in the downpipe, downstream of the exhaust manifold flange.

3   On all models, the system is suspended throughout its entire length by rubber mountings, which are secured to the underside of the vehicle by metal brackets.

Removal

4   Each exhaust section can be removed individually or, alternatively, the complete system can be removed as a unit.

5   To remove the system or part of the system, first jack up the front or rear of the car and support it on axle stands (see "*Jacking and Vehicle Support*"). Alternatively position the car over an inspection pit or on car ramps.

Downpipe

6   Place blocks of wood under the catalytic converter/front muffler to act as a support. Where applicable, refer to Chapter 4 Part A or B and remove the oxygen sensor from the exhaust pipe.

7   Loosen and remove the nuts securing the downpipe to the catalytic converter/front muffler (as applicable). Remove the bolts and recover the sealing ring from the joint.

8   Undo the nuts and separate the downpipe from the exhaust manifold. Recover the gasket then withdraw the downpipe from underneath the vehicle.

Catalytic converter

9   Loosen and remove the nuts securing the downpipe to the catalytic converter. Remove the bolts and recover the sealing ring from the joint.

10   Loosen the catalytic converter to intermediate pipe clamping ring bolts.

11   Free the catalytic converter from the intermediate pipe then withdraw it from underneath the vehicle.

Intermediate pipe

12   Loosen the clamping ring bolts and disengage the clamp from the intermediate pipe-to-tailpipe joint and the intermediate pipe-to-catalytic converter/front muffler joint.

13   Disengage the intermediate pipe from the tailpipe and the catalytic converter/front muffler and remove it from underneath the vehicle.

Tailpipe

14   Loosen the clamping ring bolts and disengage the tailpipe at the joint.

15   Unhook the tailpipe from its mounting rubbers and remove it from the vehicle. **Note:** *Where applicable, mufflers in the tail section can be carefully cut from the exhaust system using a hacksaw and replaced individually; refer to a VW dealer or an exhaust specialist for further advice.*

Complete system

16   Disconnect the front pipe from the manifold as described in paragraphs 6, 7 and 8.

17   With the aid of an assistant, free the system from all its mounting rubbers and maneuver it out from underneath the vehicle.

Heatshield(s)

18   The heatshields are secured to the underside of the body by a mixture of nuts, bolts and clips. Each shield can be removed once the relevant exhaust section has been removed. Note that if the shield is being removed to gain access to a component located behind it, in some cases it may prove sufficient to remove the retaining nuts and/or bolts and simply lower the shield, removing the need to disturb the exhaust system.

Installation

19   Each section is installed by a reverse of the removal sequence, noting the following points.

(a) *Ensure that all traces of corrosion have been removed from the flanges and replace all necessary gaskets.*

(b) *Inspect the rubber mountings for signs of damage or deterioration and replace as necessary.*

(c) *Replace the sealing ring in the catalytic converter/front muffler-to-downpipe joint.*

(d) *On joints which are secured by clamping rings, apply a smear of exhaust system jointing paste to the joint mating surfaces to ensure an air-tight seal. Tighten the clamping ring nuts evenly and progressively to the specified torque so that the clearance between the clamp halves is equal on either side.*

(e) *Prior to tightening the exhaust system fasteners, ensure that all rubber mountings are correctly located and that there is adequate clearance between the exhaust system and vehicle underbody.*

8 Catalytic converter - general information and precautions

1   The catalytic converter is a reliable and simple device which needs no maintenance in itself, but there are some facts of which an owner should be aware if the converter is to function properly for its full service life.

6

Gasoline models

(a) *DO NOT use leaded gasoline in a car equipped with a catalytic converter - the lead will coat the precious metals reagents, reducing their converting efficiency and will eventually destroy the converter.*

(b) *Always keep the ignition and fuel systems well-maintained in accordance with the manufacturer's schedule.*

(c) *If the engine develops a misfire, do not drive the car at all (or at least as little as possible) until the fault is cured.*

(d) *DO NOT push- or tow-start the car - this will soak the catalytic converter in unburned fuel, causing it to overheat when the engine does start.*

(e) *DO NOT switch off the ignition at high engine speeds.*

(f) *In some cases a sulfurous smell (like that of rotten eggs) may be noticed from the exhaust. This is common to many catalytic converter-equipped cars and once the car has covered a few thousand miles the problem should disappear. Low quality fuel with a high sulfur content will exacerbate this effect.*

(g) *The catalytic converter, used on a well-maintained and well-driven car, should last for between 50,000 and 100,000 miles - if the converter is no longer effective it must be replaced.*

Gasoline and diesel models

(h) *DO NOT use fuel or engine oil additives - these may contain substances harmful to the catalytic converter.*

(i) *DO NOT continue to use the car if the engine burns oil to the extent of leaving a visible trail of blue smoke.*

(j) *Remember that the catalytic converter operates at very high temperatures. DO NOT, therefore, park the car in dry undergrowth, over long grass or piles of dead leaves after a long run.*

(k) *Remember that the catalytic converter is FRAGILE - do not strike it with tools during servicing work.*

Chapter 7 Part A
Manual transaxle

Contents

Specifications

General
Lubricant capacity and type .... See Chapter 1

Torque specifications
Ft-lbs (unless otherwise indicated)

Clutch shield plate	132 in-lbs
Engine front crossmember-to-body bolts	37
Transmission bellhousing-to-engine bolts, M10	44
Transmission bellhousing-to-engine bolts, M12	59

1 General information

The manual transaxle is mounted transversely in the engine bay, bolted directly to the engine. This layout has the advantage of providing the shortest possible drive path to the front wheels, as well as locating the transaxle in the airflow through engine bay, optimizing cooling. The unit is cased in aluminum alloy.

Power from the crankshaft is transmitted via the clutch to the gearbox input shaft, which is splined to accept the clutch friction plate.

All forward gears are equipped with syncromeshes. When a gear is selected, the movement of the cabin floor-mounted gear lever is communicated to the gearbox either by a selector rod, or selector and shift cables, depending on the transaxle type. This in turn actuates a series of selector forks inside the gearbox which are slotted onto the synchromesh sleeves. The sleeves, which are locked to the gearbox shafts but can slide axially by means of splined hubs, press balk rings into contact with the respective gear/pinion. The coned surfaces between the balk rings and the pinion/gear act as a friction clutch, that progressively matches the speed of the synchromesh sleeve (and hence the gearbox shaft) with that of the gear/pinion. The dog teeth on the outside of the balk ring prevent the synchromesh sleeve ring from

meshing with the gear/pinion until their speeds are exactly matched; this allows gear changes to be carried out smoothly and greatly reduces the noise and mechanical wear caused by rapid gear changes.

Power is transmitted to the differential ring gear, which rotates the differential case and planetary gears, thus driving the sun gears and driveaxles. The rotation of the planetary gears on their shaft allows the inner wheel to rotate at a slower speed than the outer wheel during cornering.

2 Shift linkage - adjustment

Refer to illustration 2.4

1   If the gear change quality proves to be unsatisfactory following transaxle installation, proceed as described in the following paragraphs.

2   Refer to Chapter 11 and remove the knob and boot from the gear change lever, to expose the adjustment collar.

3   Select first gear, then take up the play in the gear change mechanism by gently pressing the gear change lever to the left.

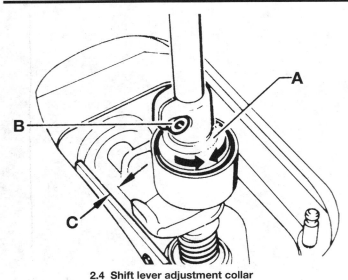

2.4 Shift lever adjustment collar

A   *Adjustment collar*
B   *Clamp bolt*
C   *Clearance less than 1/16-inch*

4   Measure the clearance between the gear change lever stop and the side of the lever housing **(see illustration)**.
5   If the clearance is greater than 1/16-inch, loosen the adjustment collar clamping bolt and rotate the collar until the correct clearance is achieved **(refer to illustration 2.4)**.
6   On completion, tighten the clamping bolt. Install the gear change lever boot and knob.

3   Manual transaxle - removal and installation

Refer to illustrations 3.10, 3.11, 3.12, 3.13, 3.19a and 3.19b

Removal

1   Select a solid, level surface to park the vehicle upon. Give yourself enough space to move around it easily. Apply the parking brake and block the rear wheels.
2   Raise the front of the vehicle and rest it securely on axle stands.
3   Refer to Chapter 11 and remove the hood from its hinges.
4   Disconnect the battery negative cable and position It away from

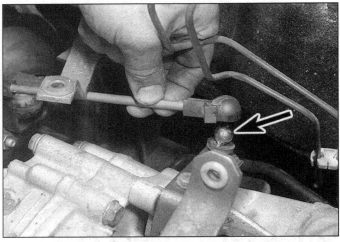

3.10 Disconnect the gearshift selector rod from the relay lever (arrows)

the terminal. **Caution:** *If the stereo in your vehicle is equipped with an anti-theft system, make sure you have the correct activation code before disconnecting the battery.*
5   The "lock carrier" is a panel assembly comprising the front bumper molding, radiator and grille, cooling fan(s) headlight units, front valence and hood lock mechanism. Although its removal is not essential, its does give greatly improved access to the engine. Its removal is relatively simple and is described at the beginning of the engine removal procedure - refer to Chapter 2C for details.
6   Refer to Chapter 8 and disconnect the clutch cable from the transaxle release lever.
7   Unbolt the ground strap from the transaxle.
8   With reference to Sections 5 and 6, disconnect the harness cabling from the speedometer transducer and back-up light switch.
9   Position a floor jack underneath the transaxle and raise it to just take the weight of the unit.
10   Disconnect the longer of the two gearshift selector rods from the relay lever; pry open the plastic clip and pull off the balljoint **(see illustration)**.
11   Disconnect the shorter of the two gearshift selector rods with the damper weight from the lever at the transaxle and the gear shift shaft; pull out the locking pins to separate the joints **(see illustration)**.
12   Refer to Chapter 2A or B as applicable and carry out the following:

3.11 Pull out the locking pins to separate the selector shaft joints

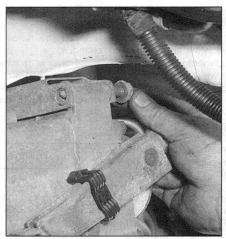

3.12 Unbolting the rear left hand engine mounting bracket from the transaxle casing

3.13 Unbolt and remove the relay lever support bracket

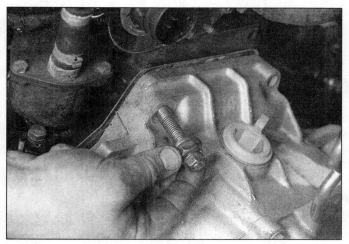

3.19a Remove the last bellhousing bolt . . .

3.19b . . . and pull the transaxle away from the engine

a) *Unbolt and remove the front engine mounting bracket from the transaxle bellhousing (support the engine with a lifting beam or hoist).*

b) *Unbolt and remove the rear left hand engine mounting bracket from the transaxle casing and the bodywork* (see illustration).

13   Unbolt and remove the relay lever support bracket from the rear of the transaxle casing (see illustration)

14   Where applicable, unbolt the clutch shield plate from the underside of the transaxle bellhousing.

15   Refer to Chapter 8 and unbolt the driveaxles from the transaxle output shafts. Suspend the driveaxles as high as possible inside the engine using cable-ties or wire. Turn the steering to full left lock.

16   Starting at the bottom, work around the transaxle bellhousing and remove all except the uppermost retaining bolts.

17   Loosen and withdraw the bolts, then lower the front engine mounting crossmember away from the vehicle.

18   Check that nothing remains connected to the transaxle, before attempting to separate it from the engine.

19   Remove the last retaining bolt from the top of the bellhousing and pull the transaxle away from the engine (see illustrations). **Warning:** *Maintain firm support of the transaxle to ensure that it remains steady on the jack head. Keep the transaxle level until the input shaft is fully withdrawn from the clutch friction plate* (see illustrations).

20   When all the locating dowels are clear of their mounting holes, lower the transaxle out of the engine bay using the jack.

Installation

21   Installation the transaxle is essentially a reversal of the removal procedure, but note the following points:

a) *Apply a smear of high-melting-point grease to the clutch friction plate splines; take care to avoid contaminating the friction surfaces.*

b) *When installing the engine front crossmember, tighten the bolts to the correct torque before the weight of the engine is allowed to rest upon it.*

c) *Tighten the bellhousing bolts to the specified torque*

d) *Refer to Chapter 2A or B (as applicable) and tighten the engine mounting bolts to the correct torque.*

e) *Refer to Chapter 8 and install the clutch cable.*

f) *On completion, refer to Section 2 and check the shift linkage adjustment.*

---

4   Manual transaxle overhaul - general information

The overhaul of a manual transaxle is a complex (and often expensive) task for the home mechanic to undertake and requires

access to specialist equipment. It involves disassembly and reassembly of many small components, measuring clearances precisely and if necessary, adjusting them by the selection shims and spacers. Internal transaxle components are also often difficult to obtain and in many instances, extremely expensive. Because of this, if the transaxle develops a fault or becomes noisy, the best course of action is to have the unit overhauled by a transmission repair specialist or to obtain an exchange reconditioned unit.

Nevertheless, it is not impossible for the more experienced mechanic to overhaul the transaxle if the special tools are available and the job is carried out in a deliberate step-by-step manner, to ensure that nothing is overlooked.

The tools necessary for an overhaul include internal and external snap-ring pliers, bearing pullers, a slide hammer, a set of pin punches, a dial indicator and possibly a hydraulic press. In addition, a large, sturdy workbench and a vise will be required.

During disassembly of the transaxle, make careful notes of how each component is assembled to make reassembly easier and accurate.

Before disassembling the transaxle, it will help if you have some idea of where the problem lies. Certain problems can be closely related to specific areas in the transaxle which can make component examination and replacement easier. Refer to the *Troubleshooting* section in this manual for more information.

---

5   Back-up light switch - testing, removal and installation

Testing

1   Ensure that the ignition switch is turned to the 'OFF' position.

2   Unplug the wiring harness from the back-up light switch at the connector. The switch is located on the top of the transaxle casing.

3   Connect the probes of a continuity tester, or multimeter set to the resistance measurement function, across the terminals of the reverse lamp switch.

4   The switch contacts are normally open, so with any gear other than reverse selected, the tester/meter should indicate an open circuit. When reverse gear is then selected, the switch contacts should close, causing the tester/meter to indicate a short circuit.

5   If the switch appears to be constantly open or short circuit, or is intermittent in its operation, it should be replaced.

Removal

Refer to illustrations 5.7 and 5.8

6   Ensure that the ignition switch is turned to the 'OFF' position.

7   Unplug the electrical connector from the back-up light switch **(see illustration)**.

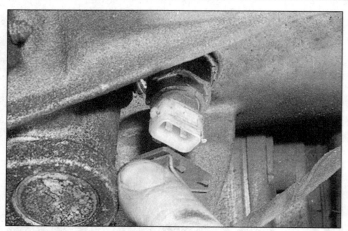

5.7 Unplug the electrical connector from the back-up light switch

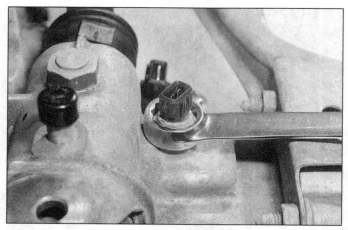

5.8 Loosen the switch body using a box-end wrench and remove it from the transaxle case

8   Loosen the switch body using a box-end wrench and withdraw it from the transaxle casing. Recover the sealing ring **(see illustration)**.

Installation

9   Install the switch by reversing the removal procedure.

---

6   Speedometer drive - removal and installation

General information

1   All transaxles are equipped with an electronic speedometer transducer. This device measures the rotational speed of the transaxle final drive and converts the information into an electronic signal, which is then sent to the speedometer module in the instrument panel. On cer-

tain models, the signal is also used as an input by the engine management system ECU.

Removal

Refer to illustrations 6.4a and 6.4b

2   Ensure that the ignition switch is turned to the 'OFF' position.
3   Locate the speed transducer, at the top of the transaxle casing. Unplug the wiring harness from the transducer.
4   Remove the transducer retaining screw using an Allen key and withdraw the unit from the transaxle casing **(see illustrations)**.
5   Recover the sealing ring.

Installation

6   Install the transducer by following the removal procedure in reverse.

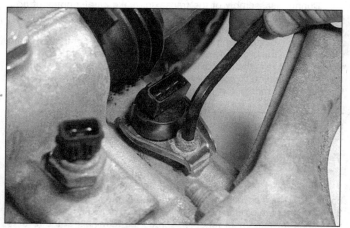

6.4a Remove the transducer retaining screw using an Allen key . . .

6.4b . . . and withdraw the unit from the transaxle case

Chapter 7 Part B
Automatic transaxle

Contents

Specifications

General
Designation
1993 and 1994 models.. 096
1995 and later models.. 01M
Automatic transmission fluid capacity and type.................... See Chapter 1

Torque wrench settings
Ft-lbs (unless otherwise indicated)
Torque converter shield plate...................................... 132 in-lbs
Torque converter-to-driveplate bolts............................. 44
Transaxle bellhousing-to-engine bolts, M10..................... 44
Transaxle bellhousing-to-engine bolts, M12..................... 59

1 General information

The automatic transaxle is a four speed unit, incorporating a hydrokinetic torque converter with and a planetary gearbox.

Gear selection is achieved by means of a cabin floor mounted, seven position selector lever. The transaxle operates in different modes, depending on the position of the selector lever.

The overall operation of the transaxle is managed by an electronic control unit (ECU) and as a result there are no manual adjustments. Comprehensive fault diagnosis can therefore only be carried out using dedicated electronic test equipment.

Due to the complexity of the transaxle and its control system, major repairs and overhaul operations should be left to a VW dealer, who will be equipped with the necessary equipment for fault diagnosis and repair. The information in this Chapter is therefore limited to a description of the removal and installation of the transaxle as a complete unit. The removal, installation and adjustment of the selector cable is also described.

2 Automatic transaxle - removal and installation

7B

Removal
Refer to illustrations 2.8, 2.9 and 2.16

1 Select a solid, level surface to park the vehicle upon. Give yourself enough space to move around it easily. Apply the parking brake and chock the rear wheels.

2 Raise the front of the vehicle and rest it securely on axle stands (see *"Jacking and Vehicle Support"*).

3 Refer to Chapter 11 and remove the hood from its hinges.

4 Disconnect the battery negative cable and position It away from the terminal. **Note:** *If the vehicle has a security coded radio, check that you have a copy of the code number before disconnecting the battery cable; refer to Chapter 12 for details.*

5 The "lock carrier" is a panel assembly comprising the front bumper molding, radiator and grille, cooling fan(s) headlight units, front valence and hood lock mechanism. Although its removal is not essen-

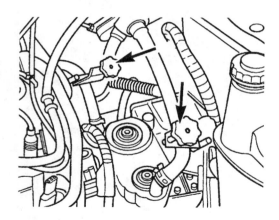

2.8 Clamp off the coolant hoses leading to and from the transmission fluid cooler unit (arrows)

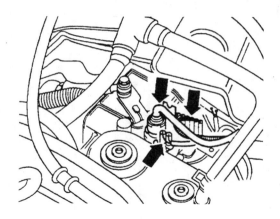

2.9 Disconnect the wiring harness from the transaxle at the multi-way connectors (arrows)

tial, its does give greatly improved access to the engine. Its removal is relatively simple and is described at the beginning of the engine removal procedure - refer to Chapter 2C for details.

6   Refer to Section 4 and disconnect the selector cable from the transaxle selector shaft.

7   Unbolt the ground strap from the transaxle.

8   Clamp off the coolant hoses leading to and from the transmission fluid cooler unit, then loosen the hose clamps and pull off the hoses **(see illustration)**. Absorb any coolant that escapes with old rags.

9   Disconnect the wiring harness from the transaxle at the multi-way connectors, labeling each one to aid installation later **(see illustration)**.

10   Unbolt the protective plate from the underside of the transaxle oil pan.

11   Refer to Chapter 5A and remove the starter motor.

12   Position a floor jack underneath the transaxle and raise it to just take the weight of the unit.

13   Refer to Chapter 2A or 2B as applicable and carry out the following:

   a) *Unbolt and remove the front engine mounting bracket from the transaxle bellhousing (support the engine with a lifting beam or hoist).*

   b) *Unbolt and remove the rear left hand engine mounting bracket from the transaxle casing and the engine mounting block.*

14   Unbolt the torque converter shield plate from the underside of the transaxle bellhousing.

15   Refer to Chapter 8 and unbolt the driveshafts from the transaxle output shafts.

16   Refer to Chapter 10 and separate the left hand suspension lower arm from the lower balljoint. Pass the left hand driveshaft over the suspension arm and swing it towards the rear of the vehicle, out of the way **(see illustration)**.

17   Suspend the right hand driveshaft as high as possible inside the engine compartment using cable-ties or wire. Turn the steering to full left lock.

18   Working through the starter motor aperture, loosen and withdraw each torque converter-to-driveplate bolt in turn. As each bolt is removed, rotate the crankshaft using a wrench and socket on the crankshaft sprocket to expose the next bolt. Repeat until all the bolts are removed.

19   Loosen and withdraw the bolts, then lower the front engine mounting crossmember away from the vehicle.

20   Starting at the bottom, work around the transaxle bellhousing and remove all except the uppermost retaining bolts.

21   Check that nothing remains connected to the transaxle, before attempting to separate it from the engine.

22   Remove the last retaining bolt from the top of the bellhousing and

pull the transaxle away from the engine. **Warning:** *Maintain firm support of the transaxle to ensure that it remains steady on the jack head. Take care to prevent the torque converter from falling out as the transaxle is removed.*

23   When all the locating dowels are clear of their mounting holes, lower the transaxle out of the engine bay using the jack. Strap a restraining bar across the front of the bellhousing to keep the torque converter in position.

Installation

24   Installation of the transaxle is essentially a reversal of the removal procedure, but note the following points:

   a) *As the torque converter is installed, ensure that the drive pins at the center of the torque converter hub engage with the recesses in the automatic transmission fluid pump inner wheel.*

   b) *Tighten the bellhousing bolts and torque converter-to-driveplate bolts to the specified torque.*

   c) *When installing the engine front crossmember, tighten the bolts to the correct torque before the weight of the engine is allowed to rest upon it.*

   d) *Refer to Chapter 2A or B (as applicable) and tighten the engine mounting bolts to the correct torque.*

   e) *On completion, refer to Section 4 and check the gear selector cable adjustment.*

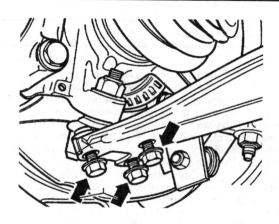

2.16 Suspension lower arm-to-balljoint bolts (arrows)

3 Automatic transaxle overhaul - general information

In the event of a fault occurring, it will be necessary to establish whether the fault is electrical, mechanical or hydraulic in nature, before repair work can be contemplated. Diagnosis requires detailed knowledge of the transaxle's operation and construction, as well as access to specialized test equipment, and so is deemed to be beyond the scope of this manual. It is therefore essential that problems with the automatic transaxle are referred to a VW dealer for assessment.

Note that a faulty transaxle should not be removed before the vehicle has been assessed by a dealer, as fault diagnosis is carried out with the transaxle in place.

4 Selector cable - removal, installation and adjustment

Removal

Refer to illustration 4.4

1   Move the selector lever to the 'P' position

2   At the transaxle end of the cable, undo the cable clamp nut and detach the cable from the selector lever cable bracket.

3   Working inside the car, remove the retaining screws securing the selector lever cover to the console, pull the cover up along the lever and move it to one side.

4   Pry free the snap-ring and detach the cable from the shift mechanism **(see illustration)**.

5   Withdraw the cable from the selector lever housing, then working along its length, release the cable from the securing clips and remove it from the vehicle.

Installation

6   Install the selector cable by following the removal procedure in reverse. When connecting the cable to the selector lever, use a new snap-ring.

7   Before tightening the connection at the transaxle selector shaft, refer to the adjustment procedure in the following sub-Section.

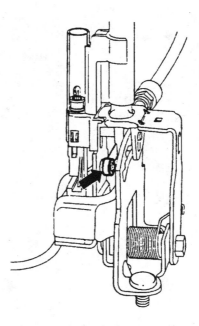

4.4 Pry free the snap-ring (arrow) and detach the cable from the shift mechanism

Adjustment

8   Move the selector lever to the 'P' position.

9   At the transaxle, loosen the cable locking bolt on the side of the selector shaft lever. Push the selector shaft up against its end stop, corresponding to the 'P' position, then tighten the locking bolt.

10   Verify the operation of the selector lever by shifting through all gear positions and checking that every gear can be selected smoothly and without delay.

Notes

Chapter 8
Clutch and driveaxles

Contents

Specifications

Torque specifications

	Ft-lbs
Flywheel-to-pressure plate bolts...	15
Pressure plate-to-crankshaft bolts	
Stage 1	44
Stage 2	Rotate an additional 90-degrees (1/4-turn)
Driveaxle retaining nut	
Models with Plus suspension	
Stage 1	66
Stage 2	Rotate an additional 45-degrees (1/8-turn)
Models with base suspension..	195
Inner constant velocity joint retaining bolts..............................	33

1   General information

Models with manual transaxles are equipped with a pedal-oper-ated, single dry plate clutch system. When the clutch pedal is depressed, effort is transmitted to the clutch release mechanism by means of a cable. The release mechanism transfers effort to the pres-sure plate diaphragm spring, which withdraws from the flywheel and releases the driven plate.

The clutch system comprises the clutch pedal, a self-adjusting clutch cable, the clutch release components, the pressure plate, the driven plate and the flywheel. Note that the clutch pressure plate is bolted directly to the crankshaft flange - the dished flywheel is then mounted on the pressure plate.

Power is transmitted from the transaxle to the front wheels by means of two solid-steel driveaxles of unequal length. The right-hand driveaxle is longer than the left-hand, due to the position of the transaxle.

Both driveaxles are splined at their outer ends to accept the wheel hubs, and are threaded so that each hub can be fastened by a large nut. The inner end of each driveaxle is bolted to the transaxle drive flanges.

Constant velocity (CV) joints are installed on each end of the dri-veaxles. On manual transaxle models, both inner and outer constant velocity joints are of the ball-and-cage type. On automatic transaxle models, the outer joint is of the ball-and-cage type, but the inner joint is of the tripod type.

2   Clutch cable - removal and installation

Refer to illustrations 2.2 and 2.12

Removal

1   Depress the clutch pedal several times to settle the automatic adjustment mechanism.
2   Working in the engine bay, slide the locking strap down to the top of the adjustment mechanism protective boot. Grasp the top and bot-tom of the adjustment mechanism and compress it - at the same time, hook the ends of the locking strap over the lugs protruding from the side of the adjustment mechanism **(see illustration)**. **Note:** *If the lock-ing strap is no longer attached to the clutch cable, a homemade strap can be fabricated using nylon cable-ties or a length of wire.*
3   Lift the clutch release lever up and extract the inner clutch cable from it, together with the locking plates and rubber damper.
4   Working inside the vehicle, refer to Chapter 11 and remove the trim panels from the area underneath the steering column.
5   Unhook the inner cable from the recess at the top of clutch pedal.
6   Push the clutch cable out through the firewall grommet, into the engine bay.
7   Disconnect the clutch cable from its securing clips and lift it out of the engine bay.

**2.2 Compressing the adjustment mechanism using
the locking strap**

8

2.12 Install the base of the adjustment mechanism into the retaining bracket on the transaxle case

3.2 Homemade flywheel locking tool in use

3.5 Lock the pressure plate in position by bolting a piece of scrap metal between it and one of the bellhousing mounting bolt holes

Installation

8   Feed the clutch cable through the firewall grommet, into the cabin area.

9   Apply multi-purpose grease to the inner clutch cable end, then install the cable end into the recess at the top of the clutch pedal. Install the trim panels.

10   Pre-tension the adjustment mechanism by depressing the clutch pedal while an assistant pulls on the inner clutch cable and simultaneously compresses the adjustment mechanism. Use the locking strap to keep the adjustment mechanism compressed, as described in Step 2.

11   Working in the engine compartment, guide the inner cable through the release lever and install the retaining plates and rubber damper.

12   Insert the base of the adjustment mechanism into the retaining bracket on the transaxle casing **(see illustration)**.

13   Unhook the locking strap from the lugs at the side of the adjustment mechanism, then depress the clutch pedal several times, until the cable tension is set.

14   On completion, assess the feel of the clutch pedal before bringing the vehicle back into service. If it exhibits any stiffness or shows signs of binding, check the routing of the cable and ensure that there are no sharp bends or kinks along its length.

15   Finally, road test the vehicle and check the operation of the clutch system while shifting up and down through the gears, while pulling away from a standstill and from a hill start.

3   Clutch components - removal and installation

Warning: *Dust created by clutch wear and deposited on the clutch components may contain asbestos, which is a health hazard. DO NOT blow it out with compressed air or inhale any of it. DO NOT use gasoline or petroleum-based solvents to clean off the dust. Brake system cleaner should be used to flush the dust into a suitable receptacle. After the clutch components are wiped clean with clean rags, dispose of the contaminated rags and cleaner in a sealed, marked container.*
Note: *Although some friction materials may no longer contain asbestos, it is safest to assume that they do and to take precautions accordingly.*

Removal

Refer to illustrations 3.2 and 3.5

1   Refer to Chapter 7A and remove the transaxle from the engine.

2   Before the flywheel bolts can be removed, the flywheel must be locked in position - a homemade flywheel locking tool can easily be fabricated from scrap metal **(see illustration)**.

3   Loosen the flywheel bolts progressively, then lift the flywheel away from the clutch pressure plate and remove the friction plate.

4   Pry off the spring clip and remove the clutch release plate.

5   Lock the pressure plate in position by bolting a piece of scrap metal between it and one of the bellhousing mounting bolt holes

3.8a Lift the pressure plate up to the crankshaft flange . . .   **3.8b . . . together with the intermediate plate . . .**

3.10a Install the release plate . . .

3.10b . . . and secure it in position with the spring clip

3.12a Install the clutch friction plate, with the hub springs facing outwards, then install the flywheel

(see illustration).

6   Progressively loosen the pressure plate bolts until they can be removed by hand. Remove the intermediate plate.

7   Lift the pressure plate away from the crankshaft flange.

Installation

Refer to illustrations 3.8a, 3.8b, 3.10a, 3.10b, 3.12a, 3.12b, 3.12c and 3.13

8   If a new pressure plate is to be installed, first wipe the protective grease from the friction surface only. Clean the friction surface with brake system cleaner or acetone. Lift the pressure plate up to the crankshaft flange together with the intermediate plate, then insert a set of new retaining bolts **(see illustrations)**. **Note:** *Coat the bolt threads with a non-hardening locking compound, if they are not supplied already coated.*

9   Hold the pressure plate still using the method described in Step 5 and tighten the retaining bolts progressively to the torque listed in this Chapter's Specifications.

10   Install the release plate and secure it in position with the spring clip **(see illustrations)**. Apply a film of high-temperature grease to the center of the release plate.

11   Smear a film of high-temperature grease on the splines at the center of the clutch friction plate - take care to avoid contaminating the friction surfaces.

12   Hold the friction plate up to the pressure plate, with the spring

loaded boss facing outwards, then install the flywheel, ensuring that the locating dowels engage with the recess on the edge of the pressure plate **(see illustrations)**. Insert a set of new flywheel retaining bolts - hand tighten them only at this stage.

13   Center the friction plate using vernier calipers; ensure that there is uniform gap between outer edge of the friction plate and the inner edge of the flywheel, around the whole circumference **(see illustration)**.

14   Lock the flywheel as described in Step 2, then tighten the flywheel retaining bolts diagonally and progressively to the torque listed in this Chapter's Specifications. Re-check the friction plate centralization.

15   Refer to Chapter 7A and install the transaxle.

4   Driveaxle - removal and installation

Refer to illustrations 4.1, 4.3a and 4.3b
Note: *A new driveaxle retaining nut will be required on installation.*

Removal

Note: *On automatic transaxle models, it may be necessary to unbolt the front and rear engine/transaxle mounts and lift the engine slightly in order to gain the necessary clearance required to withdraw the left-hand driveaxle (refer to Chapter 2 for details).*

1   Remove the wheel trim/hub cap (as applicable) and loosen the

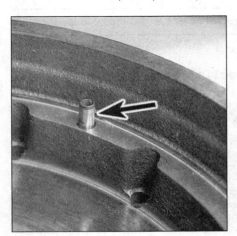

3.12b Make sure the locating dowels (arrow) engage with . . .

3.12c . . . the recesses on the edge of the pressure plate (arrow)

3.13 Center the friction plate using vernier calipers, then tighten the flywheel bolts to the torque listed in this Chapter's Specifications

8

4.1 Remove the trim/hub cap and loosen the driveaxle retaining nut, then loosen the wheel lug bolts

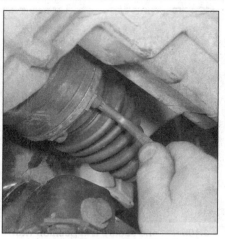

4.3a Loosen the inner CV joint retaining bolts . . .

4.3b . . . and remove them along with their retaining plates (arrow)

driveaxle retaining nut with the vehicle resting on its wheels **(see illustration)**. Also loosen the wheel bolts.

2   Block the rear wheels of the car, firmly apply the parking brake, then raise the front of the vehicle and support it securely on jackstands. Remove the front wheel.

3   Unscrew the bolts securing the inner driveaxle joint to the transaxle flange and, if applicable, remove the retaining plates from underneath the bolts **(see illustrations)**. Support the driveaxle by suspending it with wire or string - do not allow it to hang under its weight, or the joint may be damaged.

4   Using a marking pen, draw around the end of the suspension control arm, marking the correct installed position of the balljoint. Unscrew the balljoint retaining bolts and remove the retaining plate from the top of the control arm (see Chapter 10, Section 2). **Note:** *On some models the balljoint inner retaining bolt hole is slotted; on these models the inner retaining bolt can be loosened, leaving the retaining plate and bolt in position in the arm, and the balljoint disengaged from the bolt.*

5   Unscrew the driveaxle retaining nut and (where necessary) remove its washer.

6   Carefully pull the hub assembly outwards and withdraw the driveaxle outer constant velocity joint from the hub assembly. If the splines of the outer joint are stuck in the hub, tap the joint out of the hub using a soft-faced mallet. If this fails to free it from the hub, the joint will have to be pressed out using a puller which is bolted to the hub.

7   Maneuver the driveaxle out from underneath the vehicle and recover the gasket from the end of the inner constant velocity joint. Discard the gasket - a new one should be used on installation.

8   Don't allow the vehicle to rest on its wheels with one or both driveaxle(s) removed, as damage to the wheel bearing(s) may result. If moving the vehicle is unavoidable, temporarily insert the outer end of the driveaxle(s) in the hub(s) and tighten the driveaxle retaining nut(s); in this case, the inner end(s) of the driveaxle(s) must be supported, for example by suspending with string from the vehicle underbody. Do not allow the driveaxle to hang down under its weight, or the joint may be damaged.

Installation

9   Ensure that the transaxle flange and inner joint mating surfaces are clean and dry. Install a new gasket to the joint by peeling off its backing foil and sticking it in position.

10   Ensure that the outer joint and hub splines and threads are clean and dry. On models with Plus suspension, remove all traces of sealant from both sets of splines and apply a 1/8-inch bead of sealant to the outer joint splines (VW recommends the use of part number D 185 400 A2 - available from a VW dealer parts department).

11   Maneuver the driveaxle into position and engage the outer joint with the hub. Ensure that the threads are clean and install the washer (if so equipped) and nut and use it to draw the joint fully into position. **Note:** *Models with the base suspension use a six-point nut and a washer; models with Plus suspension use a twelve-point nut with no washer.*

12   Install the suspension control arm balljoint retaining bolts, tightening them to the torque listed in the Chapter 10 Specifications, using the marks made on removal to ensure that the balljoint is correctly positioned.

13   Align the driveaxle inner joint with the transaxle flange and install the retaining bolts and (if equipped) plates. Tighten the retaining bolts to the torque listed in this Chapter's Specifications.

14   Ensure that the outer joint is drawn fully into position, then install the wheel and lower the vehicle to the ground.

15   On models with Plus suspension, tighten the driveaxle nut to the stage 1 torque listed in this Chapter's Specifications, then rotate the nut through the specified stage 2 angle.

17   On models with the base suspension, tighten the driveaxle nut to the torque listed in this Chapter's Specifications.

18   Once the driveaxle nut is correctly tightened, tighten the wheel bolts to the torque listed in the Chapter 1 Specifications and install the wheel trim/hub cap.

5   Driveaxle boot replacement and CV joint inspection

1   Remove the driveaxle from the car, as described in Section 4.

Outer CV joint boot (all models)

Refer to illustrations 5.5a, 5.5b, 5.8, 5.9, 5.10, 5.17 and 5.22

2   Secure the driveaxle in a vise equipped with soft jaws, then loosen the two outer joint boot retaining clamps. If necessary, the clamps can be cut off.

3   Slide the boot down the shaft to expose the constant velocity (CV) joint and wipe off as much grease as possible.

4   Using a soft-faced mallet, tap the joint off the end of the driveaxle.

5   Remove the snap-ring from the driveaxle groove, then slide off the thrust washer and dished washer, noting which way it is installed **(see illustrations)**.

6   Slide the boot off the driveaxle and discard it.

7   Clean the outer CV joint assembly to remove as much grease as possible. Mark the relative position of the bearing cage, inner race and housing.

8   Mount the outer CV joint in a vise equipped with soft jaws. Push down on one side of the cage and remove the ball bearing from the

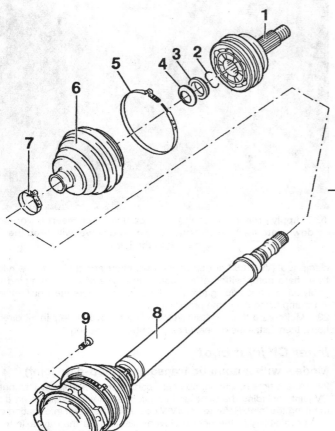

5.5a Exploded view of the driveaxle - models with automatic
transaxle (tripod-type inner CV joint)

1	Outer constant velocity (CV) joint	6	Boot
2	Snap-ring	7	Boot clamp
3	Thrust washer	8	Driveaxle and inner constant velocity (CV) joint assembly
4	Dished washer		
5	Boot clamp	9	Retaining bolt

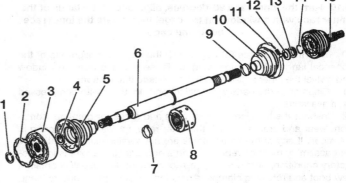

5.5b Exploded view of the driveaxle - models with manual
transaxle (ball-and-cage type inner CV joint)

1	Snap-ring	9	Boot clamp
2	Gasket	10	Boot
3	Inner CV joint	11	Boot clamp
4	Dished washer	12	Dished washer
5	Boot	13	Thrust washer
6	Driveaxle shaft	14	Snap-ring
7	Boot clamp	15	Outer CV joint
8	Vibration damper		

5.8 Tilt the cage and inner race to remove the balls

opposite side. Repeat this procedure in a criss-cross pattern until all of
the balls are removed **(see illustration)**. If the joint is tight, tap on the
inner race (not the cage) with a hammer and brass punch.
9   Remove the cage and inner race assembly from the housing by
tilting it vertically and aligning two opposing cage windows in the area
between the ball grooves **(see illustration)**.

**5.9 With the inner race and cage vertical, align the windows in the
cage with the lands of the housing, then rotate the inner race and
cage out of the housing**

8

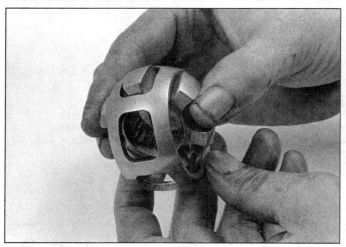

5.10 Turn the inner race 90-degrees, align one of the lands of the inner race with a window on the cage, then rotate the inner race out of the cage

5.17 Apply grease through the splined hole, then insert a wooden dowel into the hole and push down - the dowel will force the grease into the joint

10   Turn the inner race 90-degrees to the cage and align one of the spherical lands with a cage window. Raise the land into the window and swivel the inner race out of the cage **(see illustration).**

11   Clean all of the parts with solvent and dry them with compressed air, if available.

12   Inspect the housing, splines, balls and races for damage, corrosion, wear and cracks. Check the inner race, for wear and scoring in the races. If any of the components are not serviceable, the entire CV joint assembly must be replaced with a new one. If the joint is in satisfactory condition, obtain a boot replacement kit; kits usually contain a new boot and retaining clamps, a constant velocity joint snap-ring and the correct type of grease. If grease isn't included in the kit, be sure to obtain some CV joint grease.

13   Coat all of the CV joint components with CV joint grease before beginning reassembly.

14   Install the inner race in the cage and align the marks made in Step 7.

15   Install the inner race and cage assembly into the CV joint housing, aligning the marks on the inner race and cage assembly with the mark on the housing.

16   Install the balls into the holes, one at a time, until they are all in place.

17   Apply CV joint grease through the hole in the inner race, then force a wooden dowel down through the hole **(see illustration).** This will force the grease into the joint. Continue this procedure until the joint is completely packed.

18   Place the axleshaft in the vise. Clean the end of the axleshaft, then slide the new clamp and boot into place. **Note:** *It's a good idea to wrap the axleshaft splines with electrical tape to prevent damage to the boot.* Apply the remainder of the grease from the kit into the CV joint boot.

19   Remove the protective tape from the driveaxle splines. Slide on the dished washer, convex side first, followed by the thrust washer.

20   Install a new snap-ring in the groove on the driveaxle, then tap the joint onto the driveaxle until the snap-ring engages with the groove in the inner race. Make sure that the joint is securely retained by the snap-ring.

21   Ease the boot over the joint, making sure that the boot lips are correctly located on both the driveaxle and constant velocity joint. Lift the outer sealing lip of the boot to equalize air pressure within the boot.

22   Install the large retaining clamp on the boot. Pull the clamp as tight as possible and locate the hooks on the clamp in their slots. Tighten the clamp by crimping the raised area with a special boot

clamp tool (VW 1682 or equivalent) **(see illustration).** Due to the relatively hard composition of the boots, this type of tool is required to apply adequate crimping force on the clamps. Secure the small retaining clamp using the same procedure.

23   Make sure the constant velocity joint moves freely in all directions, then install the driveaxle as described in Section 4.

Inner CV joint boot

Models with automatic transaxle (tripod-type joint)

24   At the time of writing, no spare parts were available for this inner CV joint, including the boot, and no information was available on dismantling the joint. Refer to your VW dealer for the latest information on parts availability. If the boot is now available, take the driveaxle to a VW dealer service department or other qualified repair shop who will be able to install it for a small charge.

Models with manual transaxle (ball-and-cage type joint)

25   A hydraulic press and several special tools are required to remove and install the inner CV joint. Therefore it is recommended that boot replacement be left to a VW dealer service department or other qualified repair shop.

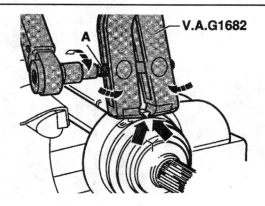

5.22 This tool (or an equivalent) is required to tighten the CV joint boot clamps. Be sure to lubricate the threads of the tool (A), then tighten the clamping bolt to 18 ft-lbs

Chapter 9 Brakes

Contents

Specifications

Front brakes
Disc thickness (minimum)	Refer to dimension on disc
Maximum disc runout	0.004 inch
Brake pad minimum thickness (all models)	See Chapter 1

Rear drum brakes
Drum diameter (maximum)	Refer to dimension on drum
Maximum drum out-of-round	0.004 inch
Brake shoe friction material thickness	See Chapter 1

Rear disc brakes
Disc thickness (minimum)	Refer to dimension on disc
Maximum disc runout	0.004 inch
Brake pad thickness	See Chapter 1

Torque specifications

	Ft-lbs (unless otherwise indicated)
ABS wheel sensor retaining bolts	84 in-lbs
Front brake caliper:	
VW caliper mounting bolts	18
Girling caliper	
Guide pin bolts	26
Mounting bracket bolts	92
Master cylinder mounting nuts	15
Rear brake caliper	
Guide pin bolts	26
Mounting bracket bolts	41
Rear brake wheel cylinder bolts	84 in-lbs
Power brake booster mounting nuts	15
Wheel lug bolts	See Chapter 1

9

1   General information

The brake system is of the power-assisted, dual-circuit hydraulic type. The arrangement of the hydraulic system is such that each circuit operates one front and one rear brake from a tandem master cylinder. Under normal circumstances, both circuits operate in unison. However, if there is hydraulic failure in one circuit, full braking force will still be available at two wheels.

Some models have disc brakes at all four wheels as standard equipment; all other models are equipped with front disc brakes and rear drum brakes. An Anti-lock Braking System (ABS) is standard equipment on some models and was offered as an option on most other models (refer to Section 22 for more information on the ABS system).

The front disc brakes are actuated by single-piston sliding type calipers, which ensure that equal pressure is applied to each disc pad.

On models with rear drum brakes, the rear brakes incorporate leading and trailing shoes, which are actuated by twin-piston wheel cylinders. A self-adjusting mechanism is incorporated to automatically compensate for brake shoe wear.

On models with rear disc brakes, the brakes are actuated by single-piston sliding calipers which incorporate mechanical parking brake mechanisms.

A pressure-regulating set-up is incorporated in the braking system; this helps to prevent rear wheel lock-up during emergency braking. The system is controlled by a load-dependent valve which is linked to the rear axle.

The cable-actuated parking brake provides an independent mechanical means of rear brake application.

Note: *When servicing any part of the system, work carefully and methodically; also observe scrupulous cleanliness when overhauling any part of the hydraulic system. Always replace components (in axle sets, where applicable) if in doubt about their condition. Note the warnings given in* Safety first *at the front of this manual and at relevant points in this Chapter concerning the dangers of asbestos dust and brake fluid.*

2   Hydraulic system - bleeding

Warning: *Brake fluid is poisonous; wash it off immediately and thoroughly in the case of skin contact, and seek immediate medical advice if any fluid is swallowed or gets into the eyes. Certain types of brake fluid are flammable, and may ignite when allowed into contact with hot components; when servicing any hydraulic system, it is safest to assume that the fluid IS flammable and to take precautions against the risk of fire as though it is gasoline that is being handled. Brake fluid is also an effective paint stripper, and will attack some plastics; if any is spilled it should be washed off immediately using copious quantities of water. Finally, it is hygroscopic (it absorbs moisture from the air) - old fluid may be contaminated and unfit for further use. When topping-up or replacing the fluid, always use the recommended type (see Chapter 1) and ensure that it comes from a freshly opened sealed container.*

General

1   The correct operation of any hydraulic system is only possible after removing all air from the components and circuit; this is achieved by bleeding the system.

2   During the bleeding procedure, add only clean, unused brake fluid of the recommended type; never re-use fluid that has already been bled from the system. Ensure that sufficient fluid is available before starting work.

3   If there is any possibility of incorrect fluid being in the system, the brake components and circuit must be flushed completely with uncontaminated fluid of the proper type, and new seals should be installed on the all of the hydraulic components.

4   If brake fluid has been lost from the system, or air has entered

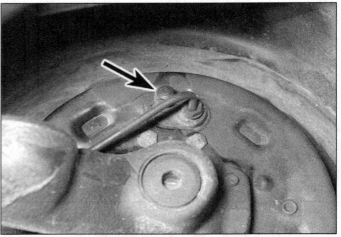

2.14a   Dust cap (arrow) over the bleed screw on a rear brake wheel cylinder - models with rear drum brakes

because of a leak, ensure that the fault is cured before continuing further.

5   Park the vehicle on level ground, switch off the engine and select first or reverse gear, then block the wheels and release the parking brake.

6   Check that all lines and hoses are secure, fittings are tight and bleed screws closed. Clean any dirt from around the bleed screws.

7   Unscrew the master cylinder reservoir cap and fill the master cylinder reservoir up to the "MAX" level line; install the cap loosely, and remember to maintain the fluid level at least above the "MIN" level line throughout the procedure, or there is a risk of further air entering the system.

8   There are a number of one-man, do-it-yourself brake bleeding kits currently available from auto parts stores. It is recommended that one of these kits is used whenever possible, as they greatly simplify the bleeding operation, and reduce the risk of expelled air and fluid being drawn back into the system. If such a kit is not available, the basic (two-man) method must be used, which is described in detail below.

9   If a kit is to be used, prepare the vehicle as described previously, and follow the kit manufacturer's instructions, as the procedure may vary slightly according to the type being used; generally, they are as outlined below in the relevant sub-section.

10   Whichever method is used, the same sequence must be followed (Steps 11 and 12) to ensure the removal of all air from the system.

Bleeding sequence

11   If the system has been only partially disconnected, and precautions were taken to minimize fluid loss, it should be necessary only to bleed that part of the system (i.e. the primary or secondary circuit).

12   If the complete system is to be bled, then it should be done working in the following sequence:

a) *Right rear brake.*
b) *Left rear brake.*
c) *Right front brake.*
d) *Left front brake.*

Warning: *On models with ABS, under no circumstances should the hydraulic unit bleed screws be opened.*

Bleeding - basic (two-man) method

Refer to illustrations 2.14a and 2.14b

13   Obtain a clean glass jar, a length of plastic or rubber tubing which is a tight fit over the bleed screw, and a box-end wrench to fit the screw. The help of an assistant will also be required.

14   Remove the dust cap from the first screw in the sequence **(see illustration)**. Place the wrench and tube on the screw, place the other end of the tube in the jar, and pour in sufficient fluid to cover the end of

2.14b When bleeding the brakes, a hose is connected to the bleed screw at the caliper or wheel cylinder and then submerged in brake fluid - air will be seen as bubbles in the tube and container (all air must be expelled before moving to the next wheel)

2.22 Bleeding a rear brake caliper using a one-way valve kit

the tube **(see illustration)**.
15   Ensure that the master cylinder reservoir fluid level is maintained at least above the "MIN" level line throughout the procedure.
16   Have the assistant fully depress the brake pedal several times to build up pressure, then maintain it on the final downstroke.
17   While pedal pressure is maintained, unscrew the bleed screw (approximately one turn) and allow the compressed fluid and air to flow into the jar. The assistant should maintain pedal pressure, following it down to the floor if necessary, and should not release it until instructed to do so. When the flow stops, tighten the bleed screw again, have the assistant release the pedal slowly, and recheck the reservoir fluid level.
18   Repeat the steps given in Steps 16 and 17 until the fluid emerging from the bleed screw is free from air bubbles. If the master cylinder has been drained and refilled, and air is being bled from the first screw in the sequence, allow approximately five seconds between cycles for the master cylinder passages to refill.
19   When no more air bubbles appear, tighten the bleed screw securely, remove the tube and wrench, and install the dust cap. Do not overtighten the bleed screw.
20   Repeat the procedure on the remaining screws in the sequence, until all air is removed from the system and the brake pedal feels firm again.

Bleeding - using a one-way valve kit

Refer to illustration 2.22
21   As the name implies, these kits consist of a length of tubing equipped with a one-way valve to prevent expelled air and fluid being drawn back into the system; some kits include a translucent container, which can be positioned so that the air bubbles can be more easily seen flowing from the end of the tube.
22   The kit is connected to the bleed screw, which is then opened. The user returns to the driver's seat, depresses the brake pedal with a smooth, steady stroke, and slowly releases it; this is repeated until the expelled fluid is clear of air bubbles **(see illustration)**.
23   Note that these kits simplify work so much that it is easy to forget the master cylinder reservoir fluid level; ensure that this is maintained at least above the "MIN" level line at all times.

Bleeding - using a pressure-bleeding kit

24   These kits are usually operated by the reservoir of pressurized air contained in the spare tire. However, note that it will probably be necessary to reduce the pressure to a lower level than normal; refer to the instructions supplied with the kit.
25   By connecting a pressurized, fluid-filled container to the master cylinder reservoir, bleeding can be carried out simply by opening each

screw in turn (in the specified sequence), and allowing the fluid to flow out until no more air bubbles can be seen in the expelled fluid.
26   This method has the advantage that the large reservoir of fluid provides an additional safeguard against air being drawn into the system during bleeding.
27   Pressure-bleeding is particularly effective when bleeding "difficult" systems, or when bleeding the complete system at the time of routine fluid replacement.

All methods

28   When bleeding is complete and firm pedal feel is restored, wash off any spilled fluid, tighten the bleed screws securely, and install their dust caps.
29   Check the brake fluid level in the master cylinder reservoir, adding fluid if necessary (see Chapter 1).
30   Discard any brake fluid that has been bled from the system; it is not fit for re-use.
31   Check the feel of the brake pedal. If it feels at all spongy, air must still be present in the system, and further bleeding is required. Failure to bleed satisfactorily after a reasonable repetition of the bleeding procedure may be due to worn master cylinder seals.

3   Brake lines and hoses - replacement

Warning: *Before starting work, refer to the Warning at the beginning of Section 2 concerning the dangers of brake fluid.*
1   If any line or hose is to be replaced, minimize fluid loss by first removing the master cylinder reservoir cap, then tightening it down onto a piece of cellophane to obtain an airtight seal. Hoses and lines should be plugged (if care is taken not to allow dirt into the system) or capped immediately they are disconnected. Place a wad of rag under any fitting that is to be disconnected, to catch any spilled fluid.
2   If a flexible hose is to be disconnected, unscrew the brake line fitting nut before removing the spring clip which secures the hose to its mounting bracket.
3   To unscrew the fitting nuts, it is preferable to obtain a flare-nut wrench of the correct size; these are available from most large auto parts stores. These wrenches wrap around the fitting nut, reducing the possibility of rounding-off the corners. Always clean the fitting and surrounding area before disconnecting it. If disconnecting a component with more than one fitting, make a careful note of the connections before disturbing any of them.
4   When replacing brake lines, be sure to use the correct parts. Don't use copper tubing for any brake system components. Purchase prefabricated steel *brake* lines, with the tube ends already flared and fittings installed, from a dealer parts department or auto parts store. These lines are also sometimes bent to the proper shapes, but if you

9

4.4 On models with VW calipers, unscrew the caliper mounting bolts and lift off the caliper

4.10a Install the anti-rattle springs, making sure they are correctly located . . .

purchase straight steel tubing, be sure to use a bending tool to make kink-free bends. Alternatively, most auto parts stores can make up brake lines from kits, but this requires very careful measurement of the original, to ensure that the replacement is of the correct length. The safest answer is usually to take the original to the shop as a pattern.

5   On installation, do not overtighten the fitting nuts. It is not necessary to exercise brute force to obtain a sound joint.

6   Ensure that the lines and hoses are correctly routed, with no kinks, and that they are secured in the clips or brackets provided and have plenty of clearance between moving or hot components. After installation, remove the cellophane from the reservoir, and bleed the hydraulic system as described in Section 2. Wash off any spilled fluid, and check carefully for fluid leaks.

4   Brake pads (front) - replacement

Warning: *Replace both sets of front brake pads at the same time - never replace the pads on only one wheel, as uneven braking may result. Note that the dust created by wear of the pads may contain asbestos, which is a health hazard. Never blow it out with compressed air, and do not inhale any of it. An approved filtering mask should be worn when working on the brakes. DO NOT use gasoline or petroleum-based solvents to clean brake parts; use brake system cleaner only.*

Note: *New pads will not give full braking efficiency until they have bedded-in. Be prepared for this, and avoid hard braking as far as possible for the first hundred miles or so after pad replacement.*

1   Loosen the wheel lug bolts, apply the parking brake, then raise the front of the vehicle and support it securely on jackstands. Remove the front wheels. Before starting work, place a drip pan under the brake and clean the caliper and surrounding area with brake system cleaner.

2   Trace the brake pad wear sensor wiring (if equipped) back from the pads, and disconnect it from the electrical connector. Note the routing of the wiring and free it from any retaining clips. Continue as described under the relevant sub-heading.

VW calipers

Refer to illustrations 4.4, 4.10a, 4.10b and 4.11

3   To improve access, unscrew the retaining bolts and remove the air deflector shield from the caliper.

4   Loosen and remove the two caliper mounting bolts, then lift the caliper off and tie it to the suspension strut using a piece of wire **(see illustration)**. Do not allow the caliper to hang unsupported on the flexible brake hose.

5   Withdraw the two brake pads and remove the anti-rattle springs, noting their correct installed locations. Note that the springs are different and are not interchangeable.

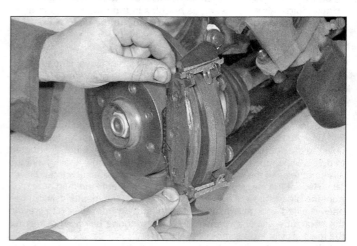

4.10b . . . and install the brake pads with their friction material facing the disc

6   Measure the thickness of each brake pad (including the backing plate). If either pad is worn at any point to the specified minimum thickness or less, all four pads must be replaced. Also, the pads should be replaced if any are fouled with oil or grease; there is no satisfactory way of degreasing friction material once contaminated.

7   If any of the brake pads are worn unevenly, or are fouled with oil or grease, trace and repair the cause before reassembly.

8   Prior to installing the pads, check that the spacers are free to slide easily in the caliper body bushings, and are a reasonably tight fit. Clean the caliper and pad mounting area with brake system cleaner. Inspect the dust seal around the piston for damage, and the piston for evidence of fluid leaks, corrosion or damage. If attention to any of these components is necessary, refer to Section 10.

9   The caliper piston must be pushed back into the cylinder to make room for the new pads. This can be accomplished with a C-clamp. Provided that the master cylinder reservoir has not been overfilled with brake fluid, there should be no spillage, but keep a careful watch on the fluid level while retracting the piston. If the fluid level rises above the "MAX" level line at any time, the surplus must be siphoned off or ejected through a plastic tube connected to the bleed screw (see Section 2). **Note:** *Do not siphon the fluid by mouth, as it is poisonous; use a syringe or an old poultry baster (if a baster is used, never again use it for preparing food).*

10   Install the new anti-rattle springs, making sure they are correctly positioned. Install the pads, ensuring that the friction material of each

4.11 Make sure the pads and springs are correctly located, and slide the caliper back into position

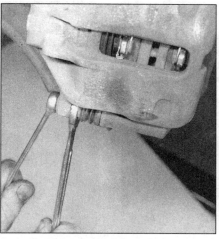

4.18 On Girling calipers, remove the lower guide pin bolt, holding the guide pin with an open-end wrench

4.19a Pivot the caliper upwards . . .

4.19b . . . then recover the shim from the caliper piston . . .

4.20 . . . and remove the pads from the caliper mounting bracket

pad is against the brake disc **(see illustrations)**. Note that, where necessary, the pad with the wear sensor wire should be installed as the inner pad.

11   Position the caliper over the pads, and pass the pad warning sensor wiring (where installed) through the caliper aperture **(see illustration)**.

12   Press the caliper into position sufficiently until it is possible to install caliper mounting bolts. Tighten the mounting bolts to the torque listed in this Chapter's Specifications. **Note:** *Do exert excess pressure on the caliper, as this will deform the pad springs, resulting in noisy operation of the brakes.*

13   Reconnect the wear sensor electrical connectors, ensuring that the wiring is correctly routed. Install the air deflector shield to the caliper, if so equipped.

14   Depress the brake pedal repeatedly until the pads are pressed into firm contact with the brake disc, and normal (non-assisted) pedal pressure is restored.

15   Repeat the above procedure on the remaining front brake caliper.

16   Install the wheels, then lower the vehicle to the ground and tighten the wheel bolts to the torque listed in the Chapter 1 Specifications.

17   Check the brake fluid level in the reservoir and add some, if necessary, to bring the fluid to the proper level (see Chapter 1).

Girling caliper

Refer to illustrations 4.18, 4.19a, 4.19b, 4.20 and 4.23

18   Loosen and remove the lower caliper guide pin bolt, using a slim open-ended wrench to prevent the guide pin itself from rotating **(see illustration)**. Discard the guide pin bolt - a new bolt must be used on installation.

19   With the lower guide pin bolt removed, pivot the caliper upwards until it is clear of the brake pads and mounting bracket. Remove the shim from the caliper piston **(see illustrations)**.

20   Remove the two brake pads from the caliper mounting bracket **(see illustration)**.

21   Examine the pads and caliper as described above in Steps 6 to 9, substituting "guide pins" for references to spacers and bushings.

22   Install the pads in the caliper mounting bracket, ensuring that the friction material of each pad is against the brake disc. Note the pad with the wear sensor wiring should be installed as the inner pad. Bottom the piston in the caliper bore as described in Step 9 (be sure to watch the fluid level in the master cylinder reservoir).

23   Install the shim to the caliper piston. Pivot the caliper down into position and pass the pad warning sensor wiring through the caliper aperture. If the threads of the new guide pin bolt are not already pre-coated with locking compound, apply a non-hardening thread-locking

9

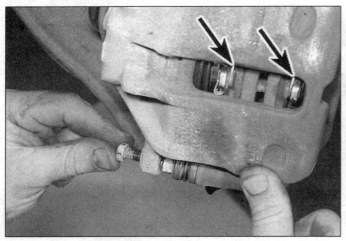

4.23   Make sure the anti-rattle springs (arrows) are correctly located, then install the new guide pin bolt

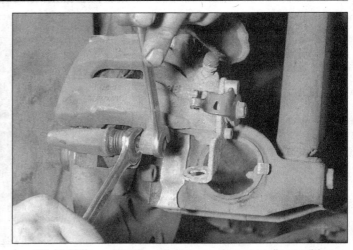

5.3   Hold the guide pin and unscrew the rear caliper guide pin bolts

5.4   Lift the caliper upwards and away . . .

5.5a   . . . and remove the pads . . .

5.5b   . . . and anti-rattle springs from the caliper mounting bracket

compound to them. Press the caliper into position while ensuring that the pad anti-rattle springs engage correctly with the caliper **(see illustration)**. Install the guide pin bolt, tightening it to the torque listed in this Chapter's Specifications while holding the guide pin with an open-ended wrench.

24   Reconnect the brake pad wear sensor wiring connectors (if equipped) ensuring that the wiring is correctly routed.

25   Depress the brake pedal repeatedly, until the pads are pressed into firm contact with the brake disc, and normal (non-assisted) pedal pressure is restored.

26   Repeat the above procedure on the remaining front brake caliper.

27   Install the wheels, then lower the vehicle to the ground and tighten the wheel bolts to the torque listed in the Chapter 1 Specifications.

28   Check the brake fluid level as described in Chapter 1.

5   Brake pads (rear) - replacement

Refer to illustrations 5.3, 5.4, 5.5a, 5.5b and 5.9

Warning: *Replace both sets of rear brake pads at the same time - never replace the pads on only one wheel, as uneven braking may result. Note that the dust created by wear of the pads may contain asbestos, which is a health hazard. Never blow it out with compressed air, and do not inhale any of it. An approved filtering mask should be worn when working on the brakes. DO NOT use gasoline or petroleum-based solvents to clean brake parts; use brake system cleaner only.*

Note: *New pads will not give full braking efficiency until they have bedded-in. Be prepared for this, and avoid hard braking as far as possible for the first hundred miles or so after pad replacement.*

1   Loosen the wheel lug bolts. Block the front wheels, then raise the rear of the vehicle and support it securely on jackstands. Remove the rear wheels. Before starting work, place a drip pan under the brake and clean the caliper and surrounding area with brake system cleaner.

2   Loosen the parking brake cable and detach it from the caliper as described in Section 19.

3   Loosen and remove the caliper guide pin bolts, using a slim open-ended wrench to prevent the guide pins from rotating **(see illustration)**. Discard the guide pin bolts - new bolts must be used on installation.

4   Lift the caliper away from the brake pads and tie it to the suspension strut using a piece of wire **(see illustration)**. Do not allow the caliper to hang unsupported on the flexible brake hose.

5   Withdraw the two brake pads from the caliper mounting bracket and recover the anti-rattle springs from the mounting bracket, noting their correct installed locations **(see illustrations)**.

6   Measure the thickness of each brake pad (including the backing plate). If either pad is worn at any point to the specified minimum thickness or less, all four pads must be replaced. Also, the pads should be replaced if any are fouled with oil or grease; there is no satisfactory way of degreasing friction material, once contaminated.

7   If any of the brake pads are worn unevenly, or fouled with oil or

5.9 In the absence of the special tool, the piston can be screwed back into the caliper using a pair of needle-nose or snap-ring pliers

6.7a Remove the spring cup . . .

6.7b . . . then lift off the spring . . .

grease, trace and rectify the cause before reassembly.

8   Prior to installing the pads, check that the guide pins are free to slide easily in the caliper bracket, and check that the guide pin dust boots are undamaged. Clean the caliper and mounting bracket with brake system cleaner. Inspect the dust seal around the piston for damage, and the piston for evidence of fluid leaks, corrosion or damage. If attention to any of these components is necessary, refer to Section 11.

9   If new brake pads are to be installed, it will be necessary to retract the piston fully into the caliper bore, by rotating it in a clockwise direction **(see illustration)**. Provided that the master cylinder reservoir has not been overfilled with brake fluid, there should be no spillage, but keep a careful watch on the fluid level while retracting the piston. If the fluid level rises above the "MAX" level line at any time, the surplus must be siphoned off, or ejected through a plastic tube connected to the bleed screw (see Section 2). **Note:** *Do not siphon the fluid by mouth, as it is poisonous; use a syringe or an old poultry baster (if a baster is used, it must never again be used for preparing food).*

10   Install the anti-rattle springs on the caliper mounting bracket, ensuring that they are correctly located. Install the pads in the mounting bracket, ensuring that each pad's friction material is against the brake disc.

11   Slide the caliper back into position over the pads.

12   If the threads of the new guide pin bolts are not already pre-coated with locking compound, apply a non-hardening thread-locking compound to them. Press the caliper into position, then install the bolts, tightening them to the torque listed in this Chapter's Specifications while holding the guide pin with an open-ended wrench.

13   Depress the brake pedal repeatedly until the pads are pressed into firm contact with the brake disc and normal (non-assisted) pedal pressure is restored.

14   Repeat the above procedure on the remaining rear brake caliper.

15   Reconnect the parking brake cables to the calipers, and adjust the parking brake as described in Section 17.

16   Install the wheels, then lower the vehicle to the ground and tighten the wheel bolts to the torque listed in the Chapter 1 Specifications.

17   Check the brake fluid level as described in Chapter 1.

6   Rear brake shoes - replacement

Refer to illustrations 6.7a, 6.7b, 6.7c, 6.8, 6.9a, 6.9b, 6.10, 6.15a, 6.15b, 6.15c, 6.16, 6.17a, 6.17b and 6.18

Warning: *Brake shoes must be replaced on both rear wheels at the same time - never replace the shoes on only one wheel, as uneven*

braking may result. Also, the dust created by the wear of the shoes may contain asbestos, which is a health hazard. Never blow it out with compressed air, and do not inhale any of it. An approved filtering mask should be worn when working on the brakes. DO NOT use gasoline or petroleum-based solvents to clean brake parts; use brake system cleaner only.

1   Remove the brake drum as described in Section 9.

2   Before beginning work, wash off the brake assembly with brake system cleaner and allow the residue to drain into a drip pan.

3   Measure the thickness of the friction material of each brake shoe at several points; if either shoe is worn at any point to the specified minimum thickness or less, all four shoes must be replaced as a set. The shoes should also be replaced if any are fouled with oil or grease; there is no satisfactory way of degreasing friction material once it has been contaminated.

4   If any of the brake shoes are worn unevenly, or fouled with oil or grease, trace and repair the cause before reassembly.

5   To replace the brake shoes, continue as follows. If all is well, install the brake drum as described in Section 9.

6   Note the position of the brake shoes and springs, and mark the webs of the shoes, if necessary, to aid installation.

7   Remove the hold-down cups and springs by depressing and turning them 90-degrees. This can be accomplished with a special hold-

9

6.7c ... and withdraw the retainer pin from the rear of the backing plate

6.8 Unhook the shoes from the lower pivot point, and remove the lower return spring

6.9a Free the shoes from the wheel cylinder. Note elastic band (arrowed) used to retain pistons ...

6.9b ... then detach the parking brake cable and remove the shoe assembly from the vehicle

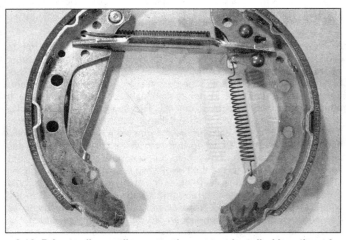

6.10 Prior to dismantling, note the correct installed location of the shoe assembly components

down spring tool or, if you're careful, a pair of pliers. With the cups removed, lift off the springs and withdraw the retainer pins **(see illustrations)**.

8   Ease the shoes out one at a time from the lower pivot point to release the tension of the return spring, then disconnect the lower

return spring from both shoes **(see illustration)**.

9   Ease the upper end of both shoes out from their wheel cylinder locations, taking care not to damage the wheel cylinder seals, and disconnect the parking brake cable from the trailing shoe. The brake shoe assembly can then be maneuvered out of position and away from the

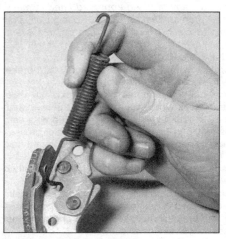

6.15a Hook the tensioning spring into the trailing shoe ...

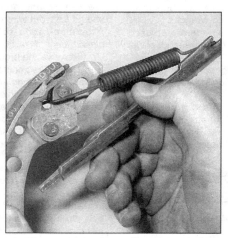

6.15b ... then engage the pushrod with the opposite end of the spring ...

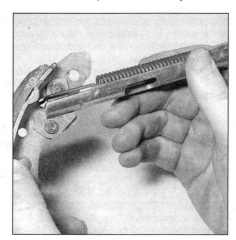

6.15c ... and pivot the strut into position on the shoe

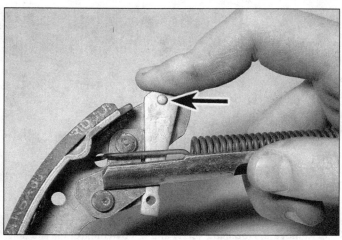

6.16 Slot the wedge key into position, making sure its raised dot (arrowed) is facing away from the shoe

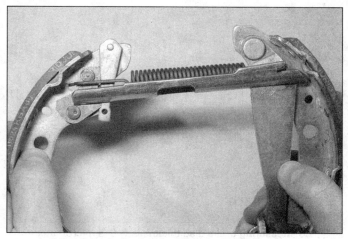

6.17a Locate the leading shoe in the pushrod . . .

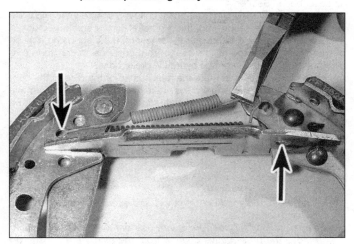

6.17b . . . and hook the upper return spring into position in the leading shoe and pushrod (arrows)

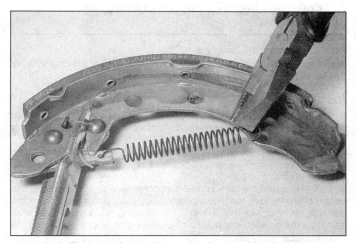

6.18 Fit the spring to the wedge key, and hook it onto the trailing shoe

backing plate. Do not depress the brake pedal until the brakes are reassembled; wrap a strong elastic band around the wheel cylinder pistons to retain them (see illustrations).

10   Make a note of the correct installed positions of all components (see illustration), then unhook the upper return spring, and disengage the wedge key spring.

11   Unhook the tensioning spring, and remove the pushrod from the trailing shoe, together with the wedge key.

12   Examine all components for signs of wear or damage and replace as necessary. Note: All return springs should be replaced, regardless of their apparent condition.

13   Peel back the protective caps, and check the wheel cylinder for fluid leaks or other damage; check that both cylinder pistons are free to move easily. Refer to Section 12, if necessary, for information on wheel cylinder overhaul.

14   Apply a little high-temperature brake grease to the contact areas of the pushrod and parking brake lever.

15   Hook the tensioning spring into the trailing shoe. Engage the pushrod with the opposite end of the spring, and pivot the pushrod into position on the trailing shoe (see illustrations).

16   Install the wedge key between the trailing shoe and pushrod, making sure it is installed correctly (see illustration).

17   Locate the parking brake lever on the leading shoe in the pushrod, and install the upper return spring using a pair of pliers (see illustrations).

18   Install the spring on the wedge key, and hook it onto the trailing

shoe (see illustration).

19   Prior to installation, clean the backing plate and apply a thin smear of high-temperature brake grease or anti-seize compound to the shoe contact areas on the backing plate and to the wheel cylinder pistons and lower pivot point. Do not allow the lubricant to contact the friction material.

20   Remove the elastic band installed on the wheel cylinder, then install the shoe assembly.

21   Connect the parking brake cable to the parking brake lever, and locate the top of the shoes in the wheel cylinder piston slots.

22   Install the lower return spring between the shoes, then lever the bottom of the shoes onto the bottom anchor.

23   Tap the shoes to centralize them with the backing plate, then install the shoe retainer pins and springs, and secure them in position with the spring cups.

24   Install the brake drum as described in Section 9.

25   Repeat the above procedure on the remaining rear brake.

26   Once both sets of rear shoes have been replaced, adjust the lining-to-drum clearance by repeatedly depressing the brake pedal until normal (non-assisted) pedal pressure returns.

27   Check and, if necessary, adjust the parking brake as described in Section 17.

28   On completion, check the brake fluid level as described in Chapter 1. Note: New shoes will not give full braking efficiency until they have bedded-in. Be prepared for this, and avoid hard braking as far as possible for the first hundred miles or so after shoe replacement.

9

7.3 Measuring brake disc thickness with a micrometer

7.8 Remove the retaining screw and remove the front brake disc

7 Brake disc (front) - inspection, removal and installation

Refer to illustrations 7.3 and 7.8
Warning: *Before starting work, refer to the* **Warning** *at the beginning of Section 4 concerning the dangers of asbestos dust.*

Inspection

Note: *If either disc requires replacement, both should be replaced at the same time to ensure even and consistent braking. New brake pads should also be installed.*

1   Loosen the front wheel lug bolts. Apply the parking brake, then raise the front of the car and support it securely on jackstands. Remove the front wheel.
2   Slowly rotate the brake disc so that the full area of both sides can be checked; remove the brake pads if better access is required to the inboard surface (see Section 4). Light scoring is normal in the area swept by the brake pads, but if heavy scoring or cracks are found, the disc must be replaced. If the scoring is light, the disc can be resurfaced by an automotive machine shop.
3   It is normal to find a lip of rust and brake dust around the disc's perimeter; this can be scraped off if required. If, however, a lip has formed due to excessive wear of the brake pad swept area, then the disc's thickness must be measured using a micrometer **(see illustration)**. Take measurements at several places around the disc, at the inside and outside of the pad swept area; if the disc has worn at any point to the specified minimum thickness or less, the disc must be machined or replaced. **Note:** *The minimum thickness dimension is stamped or cast into the disc, either on the edge of the disc, on the center (hat) portion or on the inside, where the disc mates with the hub.* **Warning:** *Never machine the disc to a point less than the minimum thickness found on the disc.* Note that if the disc is to be machined, BOTH discs must be refinished to maintain a consistent thickness on both sides.
4   If the disc is thought to be warped, it can be checked for runout with a dial indicator mounted on any convenient fixed point, while the disc is slowly rotated. If the measurements obtained are at the specified maximum or beyond, the disc is excessively warped and must be machined or replaced; however, it is worth checking first that the hub bearing is in good condition (Chapters 1 and/or 10). If the hub bearing is OK but the runout is still excessive, the disc must be machined or replaced.
5   Check the disc for cracks, especially around the wheel bolt holes and cooling fins (ventilated discs only), and any other wear or damage, and replace if necessary.

Removal

6   On models with VW front brake calipers, remove the brake pads

as described in Section 4.
7   On models with Girling front brake calipers, unscrew the two bolts securing the brake caliper mounting bracket to the steering knuckle, then slide the caliper and bracket assembly off the disc. Using a piece of wire or string, tie the caliper to the front suspension coil spring, to avoid placing any strain on the hydraulic brake hose.
8   Use chalk or paint to mark the relationship of the disc to the hub, then remove the screw securing the brake disc to the hub, and remove the disc **(see illustration)**. If it is tight, lightly tap its rear face with a rubber or plastic mallet.

Installation

9   Installation is the reverse of the removal procedure, noting the following points:
   a) *Ensure that the mating surfaces of the disc and hub are clean and flat.*
   b) *Align the marks made on removal, and securely tighten the disc retaining screw.*
   c) *If a new disc has been installed, use brake system cleaner to remove the preservative coating from the disc, before installing the caliper.*
   d) *On models with Girling brake calipers, slide the caliper into position over the disc, making sure the pads pass either side of the disc. Tighten the caliper bracket mounting bolts to the specified torque setting.*
   e) *On models with VW brake calipers, install the pads as described in Section 4.*
   f) *Install the wheel, then lower the vehicle to the ground and tighten the wheel bolts to the specified torque. On completion, repeatedly depress the brake pedal until normal (non-assisted) pedal pressure returns.*

8 Brake disc (rear) - inspection, removal and installation

Warning: *Before starting work, refer to the* **Warning** *at the beginning of Section 5 concerning the dangers of asbestos dust.*

Inspection

Note: *If either disc requires replacement, BOTH should be replaced at the same time, to ensure even and consistent braking. New brake pads should be installed also.*

1   Loosen the wheel lug bolts, block the front wheels, then jack up the rear of the car and support it securely on jackstands. Remove the rear wheel.
2   Inspect the disc as described in Section 7.

Removal

3   Unscrew the two bolts securing the brake caliper mounting

9.2 Lever out the cap from the center of the brake drum

9.3 Remove the cotter pin and locking cap . . .

9.4a . . . then unscrew the retaining nut and remove the toothed washer

9.4b Withdraw the outer bearing . . .

9.5 . . . and remove the brake drum

bracket in position, then slide the caliper and bracket assembly off the disc. Using a piece of wire, tie the caliper to the rear suspension coil spring to avoid placing any strain on the hydraulic brake hose.

4   Using a hammer and a large flat-bladed screwdriver or chisel, carefully tap and pry the cap out of the center of the brake disc. Replace the cap if it is disfigured during removal.

5   Extract the cotter pin from the hub nut, and remove the locking cap. Discard the cotter pin; a new one must be used on installation.

6   Loosen and remove the rear hub nut, then slide off the toothed washer and remove the outer bearing from the center of the disc.

7   The disc can now be slide off the stub axle.

Installation

8   If a new disc is been installed, use brake system cleaner to wipe any preservative coating from the disc. If necessary, install the bearing races, inner bearing and oil seal as described in Chapter 10, and thoroughly grease the outer bearing.

9   Apply a smear of grease to the hub bearing seal, and carefully slide the assembly onto the stub axle.

10   Install the outer bearing and toothed thrust washer, ensuring its tooth is correctly engaged in the axle slot.

11   Install the hub nut, tightening it to approximately 10 ft-lbs while rotating the disc to seat the hub bearings. Gradually loosen the hub nut until the position is found where it is just possible to move the toothed washer from side-to-side using a screwdriver, without prying or twisting the screwdriver. **Note:** *Only a small amount of force should be needed to move the washer.* When the hub nut is correctly positioned, secure it in position with a new cotter pin.

12   Install the cap to the center of the brake disc, driving it fully into position.

13   Slide the caliper into position over the disc, making sure the pads straddle the disc. Tighten the caliper mounting bolts to the torque listed in this Chapter's Specifications.

14   Install the wheel, then lower the vehicle to the ground and tighten the wheel bolts to the torque listed in this Chapter's Specifications.

9   Brake drum (rear) - removal, inspection and installation

Refer to illustrations 9.2, 9.3, 9.4a, 9.4b, 9.5, 9.7a and 9.7b
Warning: *Before starting work, refer to the **Warning** at the beginning of Section 6 concerning the dangers of asbestos dust.*

Removal

1   Loosen the rear wheel lug bolts. Block the front wheels, then raise the rear of the vehicle and support it securely on jackstands. Remove the appropriate rear wheel.

2   Using a hammer and a large flat-bladed screwdriver or chisel, carefully tap and pry the cap out of the center of the brake drum **(see illustration)**. Discard the cap if it is disfigured during removal.

3   Extract the cotter pin from the hub nut and remove the locking cap **(see illustration)**. Discard the cotter pin; a new one must be used on installation.

4   Loosen and remove the rear hub nut, then slide off the toothed washer and remove the outer bearing from the center of the drum **(see illustrations)**.

5   It should now be possible to withdraw the brake drum assembly from the stub axle by hand **(see illustration)**. It may be difficult to

9

9.7a If the drum is tight, release the brake shoes by inserting a flat-bladed screwdriver in through the brake drum hole . . .

9.7b . . . and levering the wedge key (arrow) upwards

remove the drum due to the tightness of the hub bearing on the stub axle or due to the brake shoes binding on the inner circumference of the drum. If the bearing is tight, tap the periphery of the drum using a rubber or plastic mallet. If the brake shoes are binding, first check that the parking brake is fully released, then continue as follows.

6   Referring to Section 17 for further information, fully loosen the parking brake adjustment, to obtain maximum free play in the cable.

7   Insert a screwdriver through one of the wheel bolt holes in the brake drum, and lever up the wedge key in order to allow the brake shoes to retract fully **(see illustrations)**. The brake drum can now be withdrawn.

Inspection

Note: *If either drum requires replacement, both should be replaced at the same time to ensure even and consistent braking. New brake shoes should also be installed.*

8   Wash off the brake shoe assembly and the drum with brake system cleaner and allow the residue to drain into a drip pan.

9   Clean the outside of the drum, and check it for obvious signs of wear or damage, such as cracks around the wheel bolt holes; replace the drum if necessary.

10   Examine carefully the inside of the drum. Light scoring of the friction surface is normal, but if heavy scoring is found, the drum must be machined or replaced. **Warning:** *If the drum can't be resurfaced without exceeding the maximum allowable diameter (stamped or cast into the drum), then new drums will be required.* It is usual to find a lip on the drum's inboard edge which consists of a mixture of rust and brake dust; this should be scraped away to leave a smooth surface which can be polished with fine (120- to 150-grade) emery paper. If, however, the lip is due to the friction surface being recessed by excessive wear, then the drum must be machined or replaced.

11   If the drum is thought to be excessively worn, or oval, its internal diameter must be measured at several points using an internal micrometer. Take measurements in pairs, the second at right-angles to the first, and compare the two, to check for signs of ovality. Provided that it does not enlarge the drum to beyond the specified maximum diameter, it may be possible to have the drum refinished; if this is not possible, the drums on both sides must be replaced. Note that if the drum is to be machined, BOTH drums must be refinished to maintain a consistent internal diameter on both sides.

Installation

12   If a new brake drum is to be installed, use brake system cleaner to remove any preservative coating that may have been applied to its interior. If necessary, install the bearing races, inner bearing and oil seal as described in Chapter 10, and thoroughly grease the outer bearing.

13   Prior to installation, fully retract the brakes shoes by lifting up the wedge key.

14   Apply a smear of grease to the hub bearing seal and carefully

slide the assembly onto the stub axle.

15   Install the outer bearing and toothed thrust washer, ensuring its tooth is correctly engaged in the axle slot.

16   Install the hub nut, tightening it to approximately 10 ft-lbs while rotating the brake drum to settle the hub bearings in position. Gradually loosen the hub nut until the position is found where it is just possible to move the toothed thrust washer from side-to-side using a screwdriver, without prying or twisting the screwdriver. **Note:** *Only a small amount of force should be needed to move the washer.* When the hub nut is correctly positioned, install the locking cap and secure the nut in position with a new cotter pin.

17   Install the cap to the center of the brake drum, driving it fully into position.

18   Depress the brake pedal several times to operate the self-adjusting mechanism.

19   Repeat the above procedure on the remaining rear brake assembly (where necessary), then check and, if necessary, adjust the parking brake cable as described in Section 17.

20   On completion, install the wheel(s), then lower the vehicle to the ground and tighten the wheel bolts to the torque listed in the Chapter 1 Specifications.

---

10   Brake caliper (front) - removal, overhaul and installation

Warning: *Before starting work, refer to the* **Warning** *at the beginning of Section 2 concerning the dangers of brake fluid, and to the* **Warning** *at the beginning of Section 4 concerning the dangers of asbestos dust.*
Note: *If an overhaul is indicated (usually because of fluid leaks, a stuck piston or broken bleeder screw) explore all options before beginning this procedure. New and factory rebuilt calipers are available on an exchange basis, which makes this job quite easy. If you decide to rebuild the calipers, make sure rebuild kits are available before proceeding. Always rebuild or replace the calipers in pairs - never rebuild just one of them.*

Removal

1   Loosen the wheel lug bolts. Apply the parking brake, then raise the front of the vehicle and support it securely on jackstands. Remove the appropriate wheel. Wash the brake assembly with brake system cleaner and allow the residue to drain into a drip pan.

2   Minimize fluid loss by first removing the master cylinder reservoir cap, then tightening it down onto a piece of cellophane to obtain an airtight seal.

3   Clean the area around the fitting, then loosen the brake hose fitting nut. Don't attempt to unscrew the hose from the caliper. **Note:** *If you are simply removing the caliper for access to other components, don't disconnect the brake hose.*

10.9 To remove the seal from the caliper bore, use a plastic or wooden tool, such as a pencil

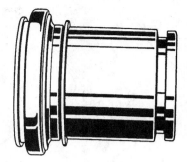

10.19 Position the new dust boot onto the piston like this - this will make it easier to install the flange of the boot in the upper groove in the cylinder bore

4   Remove the brake pads as described in Section 4.
5   On models with VW brake calipers, unscrew the caliper from the end of the brake hose and remove it from the vehicle.
6   On Girling calipers, loosen and remove the caliper upper guide pin bolt, using a slim open-ended wrench to prevent the guide pin itself from rotating, then unscrew the caliper from the brake hose and remove it from the vehicle. Discard the guide pin bolts - a new bolt must be used on installation.

Overhaul

Refer to illustrations 10.9 and 10.19
7   Once again wash the caliper with brake system cleaner.
8   Withdraw the partially ejected piston from the caliper body, and remove the dust seal. If the piston cannot be withdrawn by hand, it can be pushed out by applying compressed air to the brake hose fitting hole. If this is done, first position a wood block or several shop rags in the caliper as a cushion. Use only enough air pressure to ease the piston out of the bore. If the piston is blown out, even with the cushion in place, it may be damaged. **Warning:** *Never place your fingers in front of the piston in an attempt to catch or protect it when applying compressed air, as serious injury could occur.*
9   Using a wood or plastic tool, extract the piston hydraulic seal, taking great care not to damage the caliper bore **(see illustration)**. **Caution:** *Metal tools may cause bore damage.*
10   Thoroughly clean all components, using only isopropyl alcohol, clean brake fluid or brake system cleaner. Allow the components to dry (not necessary if brake fluid was used to clean the parts). Don't use compressed air unless it is dried, filtered and unlubricated.
11   On VW calipers, withdraw the spacers from the caliper body bushings.
12   On Girling calipers, withdraw the guide pins from the caliper mounting bracket, and remove the dust boots.
13   Check all components, and replace any that are worn or damaged. Check particularly the cylinder bore and piston; these should be replaced (note that this means the replacement of the complete body assembly) if they are scratched, worn or corroded in any way. Similarly, check the condition of the spacers/guide pins and their bushings/bores (as applicable); both spacers/pins should be undamaged and (when cleaned) a reasonably tight sliding fit in their bores. If there is any doubt about the condition of any component, replace it.
14   If the assembly is fit for further use, obtain the appropriate repair kit.
15   Replace all rubber seals, dust covers and caps disturbed on disassembly as a matter of course; these should never be re-used.
16   On reassembly, ensure that all components are clean and dry.
17   Soak the piston and the new piston seal in clean brake fluid. Smear clean fluid on the cylinder bore surface.
18   Install the new piston seal in the lower groove in the caliper bore,

using only your fingers (no tools) to manipulate it into the cylinder bore groove. Make sure it seats properly and isn't twisted.
19   Pull the new dust boot over the top of the piston and slide it down to the bottom (closed end) of the piston **(see illustration)**. Hold the piston over the bore and engage the flanged portion of the dust boot with the upper groove in the cylinder. It may be helpful to twist the piston/dust boot assembly to help the flange slip into place. Once the dust boot flange has been seated, push the piston to the bottom of its bore, making sure it enters the bore squarely. Finally seat the lip of the dust boot in the groove in the top of the piston.
20   On VW calipers, apply the grease supplied in the repair kit, or a copper-based high-temperature brake grease or anti-seize compound to the spacers and insert them into their bushings. On Girling calipers, apply the grease supplied in the repair kit, or a copper-based high-temperature brake grease or anti-seize compound to the guide pins, and install the new dust boots. Install the guide pins in the mounting bracket, making sure that the dust boots are correctly located in the grooves on both the sleeve and the bracket.

Installation

21   Screw the caliper fully onto the brake hose fitting.
22   Install the brake pads and caliper as described in Section 4.
23   Securely tighten the brake line fitting nut.
24   Remove the cellophane from the master cylinder reservoir and bleed the hydraulic system as described in Section 2. Note that, providing the precautions described were taken to minimize brake fluid loss, it should only be necessary to bleed the relevant front brake.
25   Install the wheel, then lower the vehicle to the ground and tighten the wheel bolts to the torque listed in the Chapter 1 Specifications.

11   Brake caliper (rear) - removal, overhaul and installation

Warning: *Before starting work, refer to the* **Warning** *at the beginning of Section 2 concerning the dangers of brake fluid, and to the* **Warning** *at the beginning of Section 5 concerning the dangers of asbestos dust.*
Note: *If an overhaul is indicated (usually because of fluid leaks, a stuck piston or broken bleeder screw) explore all options before beginning this procedure. New and factory rebuilt calipers are available on an exchange basis, which makes this job quite easy. If you decide to rebuild the calipers, make sure rebuild kits are available before proceeding. Always rebuild or replace the calipers in pairs - never rebuild just one of them.*

Removal

1   Loosen the wheel lug bolts. Block the front wheels, then jack up the rear of the vehicle and support it securely on jackstands. Remove the relevant rear wheel.
2   Minimize fluid loss by first removing the master cylinder reservoir

9

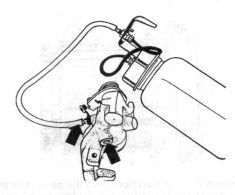

11.16 To pre-bleed the caliper, attach a hose from a bleeder bottle (filled with clean brake fluid) to the bleed screw on the caliper, position the caliper like this and open the bleed screw - when bubble-free fluid flows from the hose fitting hole, the caliper is pre-bled

cap and tightening it down onto a piece of cellophane to obtain an air-tight seal.

3   Clean the area around the fitting, then loosen the brake hose fitting nut but don't attempt to unscrew it from the caliper. **Note:** *If you are simply removing the caliper for access to other components, don't disconnect the brake hose.*

4   Remove the brake pads as described in Section 5.

5   Unscrew the caliper from the end of the brake hose and remove it from the vehicle.

Overhaul

Refer to illustration 11.16

Note: *It is not possible to overhaul the brake caliper parking brake mechanism. If the mechanism is faulty, or fluid is leaking from the parking brake lever seal, the caliper assembly must be replaced.*

6   Once again wash the caliper with brake system cleaner.

7   Using a small screwdriver, carefully pry out the dust seal from the caliper bore, taking care not to damage the piston.

8   Remove the piston from the caliper bore by rotating it in an counterclockwise direction. This can be achieved using a pair of snap-ring pliers or needle-nose pliers engaged in the caliper piston slots. Once the piston turns freely but does not come out any further, the piston can be withdrawn by hand.

9   Using a wood or plastic tool, extract the piston hydraulic seal **(see illustration 10.9)**.

10   Withdraw the guide pins from the caliper mounting bracket, and remove the guide sleeve dust boots.

11   Thoroughly clean all components, using only isopropyl alcohol, clean brake fluid or brake system cleaner. Allow the components to dry (not necessary if brake fluid was used to clean the parts). Don't use compressed air unless it is dried, filtered and unlubricated. Inspect all the caliper components as described in Section 10, Step 13, and replace as necessary, noting that the parking brake mechanism must not be disassembled.

12   Soak the piston and the new piston seal in clean brake fluid. Smear clean fluid on the cylinder bore surface. Install the new piston seal in the lower groove in the caliper bore, using only your fingers (no tools) to manipulate it into the cylinder bore groove. Make sure it seats properly and isn't twisted.

13   Pull the new dust boot over the top of the piston and slide it down to the bottom (closed end) of the piston **(see illustration 10.19)**. Hold the piston over the bore and engage the flanged portion of the dust boot with the upper groove in the cylinder. It may be helpful to twist the piston/dust boot assembly to help the flange slip into place. Once the dust boot flange has been seated, push the piston squarely into the bore until it engages the parking brake actuator screw. Turn the piston in a clockwise direction, using the method employed on disassembly,

until it is fully retracted into the caliper bore.

14   Seat the lip of the dust boot in the groove in the top of the piston.

15   Apply the grease supplied in the repair kit, or a copper-based high-temperature brake grease or anti-seize compound, to the guide pins. Install the new dust boots to the guide pins and install the pins to the caliper mounting bracket, ensuring that the dust boots are correctly located in the grooves on both the pins and caliper bracket.

16   Prior to installation the caliper must be pre-bled. Fill the caliper with fresh brake fluid by loosening the bleed screw and pouring fluid through the caliper until bubble-free fluid is expelled from the fitting hole **(see illustration)**. Tighten the bleed screw and plug the hose fitting hole until the caliper is to be installed.

Installation

17   Screw the caliper fully onto the flexible hose fitting.

18   Install the brake pads and caliper as described in Steps 10 to 12 of Section 5.

19   Securely tighten the brake line fitting nut.

20   Remove the cellophane and bleed the hydraulic system as described in Section 2. Note that, providing the precautions described were taken to minimize brake fluid loss, it should only be necessary to bleed the relevant rear brake.

21   Connect the parking brake cable to the caliper, and adjust the parking brake as described in Section 17.

22   Install the wheel, then lower the vehicle to the ground and tighten the wheel bolts to the torque listed in the Chapter 1 Specifications. Pump the brake pedal several times to bring the pads into contact with the disc. On completion, check the brake fluid level as described in Chapter 1.

12   Wheel cylinder (rear) - removal and installation

Warning: *Before starting work, refer to the* **Warning** *at the beginning of Section 2 concerning the dangers of brake fluid, and to the* **Warning** *at the beginning of Section 6 concerning the dangers of asbestos dust.*

Removal

1   Remove the brake drum as described in Section 9. Clean the brake assembly with brake system cleaner and allow the residue to drain into a drip pan.

2   Using pliers, carefully unhook the upper brake shoe return spring and remove it from both brake shoes. Pull the upper ends of the shoes away from the wheel cylinder to disengage them from the pistons.

3   Minimize fluid loss by first removing the master cylinder reservoir cap, then tightening it down onto a piece of cellophane to obtain an airtight seal.

4   Unscrew the brake line fitting nut at the wheel cylinder. Carefully ease the line out of the wheel cylinder, and plug its end to prevent dirt entry. Wipe off any spilled fluid immediately.

5   Unscrew the two wheel cylinder retaining bolts from the rear of the backing plate and remove the cylinder, taking great care not to allow brake fluid to contaminate the brake shoe linings.

Overhaul

6   Clean the wheel cylinder with brake system cleaner.

7   Pull the rubber dust seals from the ends of the cylinder body.

8   The pistons will normally be ejected by the pressure of the coil spring, but if they are not, tap the end of the cylinder body on a piece of wood to eject the pistons from their bores.

9   Inspect the surfaces of the pistons and their bores in the cylinder body for scoring, or evidence of metal-to-metal contact. If evident, replace the complete wheel cylinder assembly.

10   If the pistons and bores are in good condition, discard the seals and obtain a repair kit, which will contain all the necessary replaceable items.

11   Remove the seals from the pistons noting their correct installed orientation. Lubricate the new piston seals with clean brake fluid and install them onto the pistons with the lips of the seals facing inward.

12   Dip the pistons in clean brake fluid, then install the spring to the cylinder.
13   Insert the pistons into the cylinder bores using a twisting motion.
14   Install the dust seals, and check that the pistons can move freely in their bores.

Installation

15   Ensure that the backing plate and wheel cylinder mating surfaces are clean, then spread the brake shoes and maneuver the wheel cylinder into position.
16   Engage the brake line and screw in the fitting nut two or three turns to ensure that the thread has started.
17   Insert the two wheel cylinder retaining bolts and tighten them to the torque listed in this Chapter's Specifications. Now securely tighten the brake line fitting nut.
18   Remove the cellophane from the master cylinder reservoir.
19   Ensure that the brake shoes are correctly located in the cylinder pistons, then carefully install the brake shoe upper return spring, using a screwdriver to stretch the spring into position.
20   Install the brake drum as described in Section 9.
21   Bleed the brake hydraulic system as described in Section 2. Providing precautions were taken to minimize loss of fluid, it should only be necessary to bleed the relevant rear brake.

13   Master cylinder - removal, overhaul and installation

Warning: *Before starting work, refer to the* **Warning** *at the beginning of Section 2 concerning the dangers of brake fluid.*

Removal

1   Disconnect the battery negative terminal. Where necessary, to improve access to the master cylinder, remove the air inlet duct as described in the relevant Part of Chapter 4.
2   Remove the master cylinder reservoir cap, and siphon the brake fluid from the reservoir. **Note:** *Do not siphon the fluid by mouth, as it is poisonous; use a syringe or an old poultry baster (if a baster is used, never again use it for preparing food).* Alternatively, open any convenient bleed screw in the system, and gently pump the brake pedal to expel the fluid through a plastic tube connected to the screw (see Section 2). Disconnect the electrical connector from the brake fluid level sender unit.
3   Wipe clean the area around the brake line fittings on the side of the master cylinder, and place absorbent rags beneath the line fittings to catch any surplus fluid. Make a note of the correct installed positions of the fittings, then unscrew the fitting nuts and carefully withdraw the lines. Plug or tape over the line ends and master cylinder orifices, to minimize the loss of brake fluid, and to prevent the entry of dirt into the system. Wash off any spilled fluid immediately with cold water.
4   Loosen and remove the two nuts and washers securing the master cylinder to the power brake booster, then withdraw the unit from the engine compartment. Remove the O-ring from the rear of the master cylinder, and discard it.

Overhaul

5   If the master cylinder is faulty, it must be replaced. Repair kits are not available from VW dealers, so the cylinder must be treated as a sealed unit.
6   The only items which can be replaced are the mounting seals for the fluid reservoir; if these show signs of deterioration, pull off the reservoir and remove the old seals. Lubricate the new seals with clean brake fluid, and press them into the master cylinder ports. Ease the fluid reservoir into position, and push it fully home.

Installation

7   Remove all traces of dirt from the master cylinder and booster unit mating surfaces, and install a new O-ring in the groove on the master cylinder body.
8   Install the master cylinder on the booster unit, ensuring that the

booster unit pushrod enters the master cylinder bore centrally. Install the master cylinder mounting nuts and washers and tighten them to the torque listed in this Chapter's Specifications.
9   Connect the brake line fittings to the master cylinder ports and tighten them securely.
10   Refill the master cylinder reservoir with new fluid and bleed the complete hydraulic system as described in Section 2.

14   Brake pedal - removal and installation

Removal

1   Disconnect the battery negative terminal. **Caution:** *If the stereo in your vehicle is equipped with an anti-theft system, be sure you have the correct activation code before disconnecting the battery.*
2   Remove the brake light switch as described in Section 21.
3   It is then necessary to release the brake pedal from the ball on the booster pushrod. To do this, reach up behind the pedal and carefully expand the pedal retaining clip lugs until the pedal can be gently pulled off the booster pushrod ball.
4   Carefully unhook the brake pedal return spring from the pedal bracket.
5   Slide off the pedal pivot shaft right-hand retaining clip, then slide the shaft to the left until the brake pedal is released from its right-hand end. **Note:** *On some models it will be necessary to unclip the plastic wiring bracket from the instrument panel frame to improve access to the retaining clip.*
6   Remove the pedal from the underneath the dash, and recover the return spring.
7   Carefully clean all components, and replace any that are worn or damaged.

Installation

8   Prior to installation, apply a smear of multi-purpose grease to the pivot shaft and pedal bearing surfaces.
9   Install the return spring and maneuver the pedal into position.
10   Slide the pivot shaft into position. Make sure the flats on the end of the pivot shaft are positioned vertically, and slide on the shaft retaining clip, making sure it is securely clipped in position. Where necessary, clip the plastic wiring bracket back into position.
11   Hook the return spring onto the bracket, then retain the booster unit pushrod, and clip the pedal back onto its pushrod ball. Make sure that the pedal is securely retained by its spring clip.
12   Install the brake light switch as described in Section 21 and reconnect the battery.

15   Power brake booster - check, removal and installation

Check

1   To test the operation of the booster, depress the brake pedal several times to exhaust the vacuum, then start the engine while keeping the pedal firmly depressed. As the engine starts, there should be a noticeable "give" in the brake pedal as the vacuum builds up. Allow the engine to run for at least two minutes, then switch it off. If the brake pedal is now depressed, it should feel normal, but further applications should result in the pedal feeling firmer, with the pedal stroke decreasing with each application.
2   If the booster does not operate as described, first inspect the booster unit check valve as described in Section 16. On diesel models, also check the operation of the vacuum pump as described in Section 25.
3   If the booster unit still fails to operate satisfactorily, the fault lies within the unit itself. Repairs to the unit are not possible - if faulty, the booster unit must be replaced.

Removal

Note: *On models equipped with ABS, it is not possible to remove the*

9

power brake booster without first removing the ABS hydraulic unit (see Section 23). Therefore, booster unit removal and installation on models with ABS should be left to a VW dealer service department or other qualified repair shop.

4   Remove the master cylinder as described in Section 13.

5   On models with a manual transaxle, remove the clutch master cylinder as described in Chapter 6.

6   Remove the heat shield (if equipped) from the front of the booster, then carefully ease the vacuum hose out from the booster unit sealing grommet.

7   From inside the vehicle, remove the brake light switch as described in Section 21.

8   Unscrew the four retaining nuts securing the booster unit to the pedal mounting bracket, then return to the engine compartment and maneuver the booster unit out of position, noting the gasket which is installed on the rear of the unit. As the booster is withdrawn, it will be necessary to detach its pushrod ball from the brake pedal spring clip (see Step 3 of Section 14).

Installation

9   Check the booster unit vacuum hose sealing grommet for signs of damage or deterioration and replace it if necessary.

10   Install a new gasket on the rear of the booster unit, then reposition the unit in the engine compartment.

11   From inside the vehicle, ensure that the booster unit pushrod is correctly engaged with the brake pedal, then clip the pedal onto the pushrod ball. Check that the pedal is securely retained, then install the booster unit mounting nuts and tighten them to the torque listed in this Chapter's Specifications.

12   Carefully ease the vacuum hose back into position in the booster, taking great care not to displace the sealing grommet. Where necessary, install the heat shield to the booster.

13   Install the master cylinder as described in Section 13 of this Chapter.

14   Install the brake light switch as described in Section 21.

15   On completion, start the engine and check for air leaks at the vacuum hose-to-booster unit connection; check the operation of the braking system.

16   Power brake booster check valve - removal, check and installation

1   The check valve is located in the vacuum hose running from the intake manifold to the brake booster. If the valve is to be replaced, the complete hose/valve assembly should be replaced.

Removal

2   Carefully ease the vacuum hose out of the booster unit, taking care not to displace the grommet.

3   Note the correct routing of the hose, then loosen the retaining clamp and disconnect the opposite end of the hose assembly from the manifold/pump and remove it from the vehicle.

Check

4   Examine the check valve and vacuum hose for signs of damage, and replace if necessary.

5   The valve may be tested by blowing through it in both directions. Air should flow through the valve in one direction only, when blown through from the booster unit end of the valve. Replace the valve if this is not the case.

6   Examine the booster unit rubber sealing grommet for signs of damage or deterioration, and replace as necessary.

Installation

7   Ensure that the sealing grommet is correctly installed in the booster unit.

8   Ease the hose fitting into position in the booster, taking great care

17.3 Parking brake cable locknuts and adjuster nuts (arrows)

not to displace or damage the grommet.

9   Ensure that the hose is correctly routed, and connect it to the intake manifold/pump, tightening its retaining clamp securely.

10   On completion, start the engine and check the check valve to booster unit connection for signs of air leaks.

17   Parking brake - adjustment

Refer to illustrations 17.3 and 17.8

1   To check the parking brake adjustment, first apply the brake pedal firmly several times to establish correct shoe-to-drum/pad-to-disc clearance, then apply and release the parking brake several times.

2   Applying normal, moderate pressure, pull the parking brake lever to the fully applied position, counting the number of clicks emitted from the parking brake ratchet mechanism. If adjustment is correct, there should be approximately 4 to 7 clicks before the parking brake is fully applied. If this is not the case, adjust as follows.

3   Remove the rear section of the center console as described in Chapter 11 to gain access to the parking brake lever. **Note:** *On some models, the parking brake adjusting nuts can be accessed by simply removing the ashtray from the rear of the center console* (**see illustration**).

4   Block the front wheels, then raise the rear of the vehicle and support it securely on jackstands. Continue as described under the relevant sub-heading.

Rear drum brake models

5   With the parking brake set on the fourth notch of the ratchet mechanism, loosen the locknuts and rotate the adjusting nuts equally until it is difficult to turn both rear wheels/drums. Once this is so, fully release the parking brake lever and check that the wheels/hubs rotate freely. Check the adjustment by applying the parking brake fully, counting the clicks emitted from the parking brake ratchet and, if necessary, re-adjust.

6   Once adjustment is correct hold the adjusting nuts and securely tighten the locknuts. Install the center console section/ashtray (as applicable).

Rear disc brake models

7   With the parking brake fully released, equally loosen the parking brake locknuts and adjusting nuts until both the rear caliper parking brake levers are back against their stops.

8   From this point, equally tighten both adjusting nuts until both parking brake levers just move off the caliper stops. Ensure that the gap between each caliper parking brake lever and its stop is less than 1.5 mm (1/16-inch), and ensure both the right- and left-hand gaps are

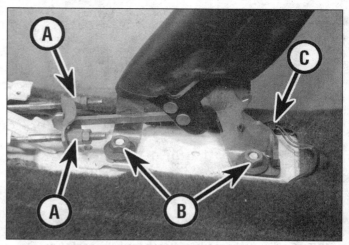

17.8 On rear disc brake models, adjust the parking brake so that the clearance between the parking brake lever and caliper (arrow) is as specified

equal **(see illustration)**. Check that both wheels/discs rotate freely, then check the adjustment by applying the parking brake fully, counting the clicks emitted from the parking brake ratchet. If necessary, re-adjust.

9   Once adjustment is correct, hold the adjusting nuts and securely tighten the locknuts. Install the center console section/ashtray (as applicable).

18   Parking brake lever - removal and installation

Refer to illustration 18.2

Removal

1   Remove the rear section of the center console as described in Chapter 11 to gain access to the parking brake lever.
2   Loosen and remove both the parking brake cable locknuts and adjusting nuts, and detach the cables from the compensator plate **(see illustration)**.
3   Disconnect the electrical connector from the warning light switch, then unscrew the retaining nuts and remove the lever from the vehicle.

Installation

4   Installation is a reversal of the removal. Prior to installing center console, adjust the parking brake as described in Section 17.

18.2 Parking brake cable locknuts and adjuster nuts (A), lever retaining nuts (B) and warning light switch wiring (C)

19   Parking brake cables - removal and installation

Refer to illustrations 19.5, 19.6, 19.7a, 19.7b and 19.7c

Removal

1   Remove the rear section of the center console as described in Chapter 11 to gain access to the parking brake lever. The parking brake cable consists of two sections, a right- and a left-hand section, which are linked to the lever by an equalizer plate. Each section can be removed individually.
2   Loosen the relevant parking brake locknut and adjusting nut to obtain maximum free play in the cable, and disengage the inner cable from the parking brake equalizer.
3   Block the front wheels, then raise the rear of the vehicle and support it securely on jackstands.
4   From the vehicle underbody, free the front end of the outer cable from the body and withdraw the cable from its support guide.
5   Work back along the length of the cable, noting its correct routing, and free it from all of the retaining clips **(see illustration)**.
6   On models with rear drum brakes, remove the rear brake shoes from the relevant side as described in Section 6. Using a hammer and pin punch, carefully tap the outer cable out from the brake backing plate, and remove it from underneath the vehicle **(see illustration)**.

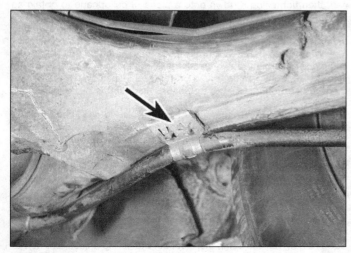

19.5 Release the retaining clips (arrowed) and detach the parking brake cable from the trailing arm

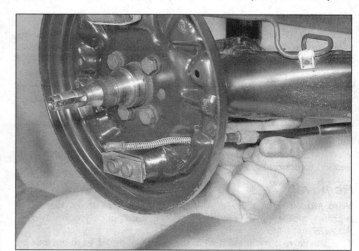

19.6 On drum brake models, remove the brake shoes and detach the cable from the backing plate

9

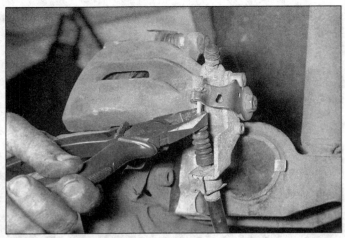

19.7a On disc brake models, detach the inner cable from the caliper lever . . .

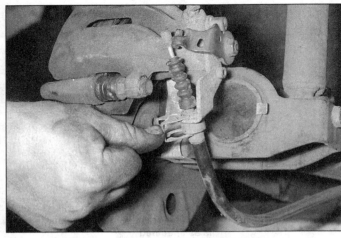

19.7b . . . then remove the retaining clip . . .

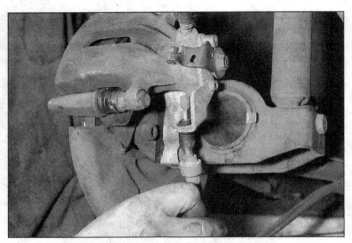

19.7c . . . and free the cable from the caliper bracket

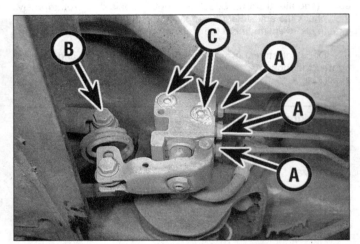

20.1 Load-dependent rear brake pressure-regulating valve brake pipe unions (A), spring pivot bolt (B) and retaining bolts (C)

7   On models with rear disc brakes, disengage the inner cable from the caliper parking brake lever, then remove the outer cable retaining clip and detach the cable from the caliper **(see illustrations)**.

Installation

8   Installation is a reversal of the removal procedure. Prior to installing the center console, adjust the parking brake as described in Section 17.

20 Rear brake pressure-regulating valve - removal and installation

Refer to illustration 20.1
Warning: *Before starting work, refer to the* **Warning** *at the beginning of Section 2 concerning the dangers of brake fluid.*

Removal

Note: *Models equipped with ABS are not equipped with any rear brake pressure-regulating valves; the function is automatically controlled by the ABS unit.*
1   The valve is mounted next to the rear axle, attached to the axle by a spring **(see illustration)**. As the load being carried by the vehicle is altered, the suspension moves in relation to the vehicle body, altering the tension in the spring. The spring then adjusts the pressure-regulat-

ing valve lever so that the correct pressure is applied to the rear brakes to suit the load being carried.
2   Minimize fluid loss by first removing the master cylinder reservoir cap, then tightening it down onto a piece of cellophane to obtain an airtight seal.
3   Loosen and remove the nut and bolt securing the valve spring to the axle.
4   Wipe clean the area around the brake line fittings on the valve, and place absorbent rags beneath the line fittings to catch any surplus fluid. Make identification marks on the brake lines; these marks can then be used on installation to ensure each line is correctly reconnected.
5   Loosen the fitting nuts and disconnect the brake lines from the valve. Plug or tape over the line ends and valve orifices, to minimize the loss of brake fluid and to prevent the entry of dirt into the system. Wash off any spilled fluid immediately with cold water.
6   Unscrew the bolts and remove the pressure-regulating valve and spring from underneath the vehicle.

Installation

7   Installation is the reverse of the removal procedure, noting the following points.
   a) *If a new valve is being installed, set the spring adjustment bolt to the same position as the one on the old valve, and tighten it securely.*

b) Ensure that the brake pipes are correctly connected to the valve, and that their union nuts are securely tightened.

c) Coat the ends of the spring with grease prior to installation.

d) Bleed the complete braking system as described in Section 2.

e) On completion, take the vehicle to a VW dealer to have the valve operation checked and if necessary adjusted.

21 Brake light switch - removal and installation

Removal

1   The brake light switch is located on the pedal bracket behind the dash.

2   Press in the locking buttons, and unclip the fuse box cover from the underside of the driver's side lower dash panel. Carefully pry the trim panel out from the top of the driver's side lower dash panel, then loosen and remove all the panel retaining screws. Carefully move the panel downwards to release it from the dash, then remove it from the vehicle.

3   Reach up under the dash and disconnect the wiring connector from the switch

4   Twist the switch 90-degrees and release it from the mounting bracket.

Installation

5   Prior to installation, fully extend the brake light switch plunger.

6   Fully depress and hold the brake pedal, then maneuver the switch into position. Secure the switch in position it by twisting it 90-degrees and release the brake pedal.

7   Reconnect the wiring connector, and check the operation of the brake lights. The brake lights should illuminate after the brake pedal has traveled approximately 5 mm (13/64-inch). If the switch is not functioning correctly, it is faulty and must be replaced; no adjustment is possible.

8   On completion, install the driver's side lower dash panel.

22 Anti-lock braking system (ABS) - general information

Note: *On models equipped with traction control, the ABS unit is a dual function unit, controlling both the anti-lock braking system (ABS) and the electronic differential locking (EDL) system functions.*

ABS is available as an option on all models covered in this manual. The system comprises a hydraulic block (which contains the hydraulic solenoid valves and accumulators), the electrically driven return pump, and four wheel sensors (one installed on each wheel), the electronic control unit (ECU) and the brake pedal position sensor. The purpose of the system is to prevent the wheel(s) locking during heavy braking. This is achieved by automatic release of the brake on the relevant wheel, followed by re-application of the brake.

The solenoids are controlled by the ECU, which itself receives signals from the four wheel sensors (one installed on each hub), which monitor the speed of rotation of each wheel. By comparing these signals, the ECU can determine the speed at which the vehicle is traveling. It can then use this speed to determine when a wheel is decelerating at an abnormal rate, compared to the speed of the vehicle, and therefore predicts when a wheel is about to lock. During normal operation, the system functions in the same way as a non-ABS braking system. In addition to this, the brake pedal position sensor (which is installed on the power brake booster) also informs the ECU of how hard the brake pedal is being depressed.

If the ECU senses that a wheel is about to lock, it operates the relevant solenoid valve in the modulator block, which then isolates the brake caliper on the wheel which is about to lock from the master cylinder, effectively sealing-in the hydraulic pressure.

If the speed of rotation of the wheel continues to decrease at an abnormal rate, the ECU switches on the electrically driven return pump, which pumps the brake fluid back into the master cylinder, releasing pressure on the brake caliper so that the brake is released. Once the speed of rotation of the wheel returns to an acceptable rate, the pump stops; the solenoid valve opens, allowing the hydraulic master cylinder pressure to return to the caliper, which then re-applies the brake. This cycle can be carried out at up to 10 times a second.

The action of the solenoid valves and return pump creates pulses in the hydraulic circuit. When the ABS system is functioning, these pulses can be felt through the brake pedal.

The operation of the ABS system is entirely dependent on electrical signals. To prevent the system responding to any inaccurate signals, a built-in safety circuit monitors all signals received by the ECU. If an inaccurate signal or low battery voltage is detected, the ABS system is automatically shut down, and the warning light on the instrument panel is illuminated, to inform the driver that the ABS system is not operational. Normal braking should still be available, however.

If a fault does develop in the ABS system, the vehicle must be taken to a VW dealer service department or other qualified repair shop for fault diagnosis and repair.

23 Anti-lock braking system (ABS) components - removal and installation

Hydraulic unit

1   Removal and installation of the hydraulic unit should be entrusted to a VW dealer. Great care has to be taken not to allow any fluid to escape from the unit as the lines are disconnected. If the fluid is allowed to escape, air can enter the unit, causing air locks which cause the hydraulic unit to malfunction.

Electronic control unit (ECU)

Removal

2   The ABS ECU is located underneath the rear seat on the right-hand side of the vehicle. Prior to removal, disconnect the battery negative cable. **Caution:** *If the stereo in your vehicle is equipped with an anti-theft system, be sure you have the correct activation code before disconnecting the battery.*

3   Lift up the right-hand rear seat cushion and unclip the ECU from its mountings. Release the retaining clip and pivot the electrical connector out of position, then remove the ECU from the vehicle.

Installation

4   Installation is a reversal of the removal procedure, ensuring that the ECU wiring connector is correctly and securely reconnected.

Front wheel sensor

Removal

5   Chock the rear wheels, then firmly apply the parking brake, jack up the front of the vehicle and support on axle stands. Remove the appropriate front wheel.

6   Trace the wiring back from the sensor to the connector, freeing it from all the relevant retaining clips, and disconnect it from the main loom.

7   Loosen and remove the bolt securing the sensor to the steering knuckle, and remove the sensor and lead assembly from the vehicle.

Installation

8   Prior to installation, apply a thin coat of multi-purpose grease to the sensor tip (VW recommend the use of lubricating paste G 000 650 - available from your dealer).

9   Ensure that the sensor and steering knuckle sealing faces are clean, then install the sensor on the knuckle. Install the retaining bolt and tighten it to the specified torque.

10   Ensure that the sensor wiring is correctly routed and retained by all the necessary clips, and reconnect it to its wiring connector.

11   Install the wheel, then lower the vehicle to the ground and tighten the wheel bolts to the torque listed in the Chapter 1 Specifications.

9

Rear wheel sensor

Removal

12   Chock the front wheels, then jack up the rear of the vehicle and support it on axle stands. Remove the appropriate wheel.
13   Remove the sensor as described in Steps 7 and 8.

Installation

14   Install the sensor as described above in Steps 9 to 12.

Front reluctor rings

15   The front reluctor rings are fixed onto the rear of wheel hubs. Examine the rings for damage such as chipped or missing teeth. If replacement is necessary, the complete hub assembly must be disassembled and the bearings replaced as described in Chapter 10.

Rear reluctor rings

16   The rear reluctor rings are pressed onto the inside of the rear brake drum disc. Examine the rings for signs of damage such as chipped or missing teeth, and replace as necessary. If replacement is necessary, remove the drum/disc as described in Chapter 9 and take it to a VW dealer service department or other qualified repair shop, who will have access to the necessary tools required to extract the old ring and press on the new one.

Brake pedal position sensor

Removal

17   Release the vacuum from inside the booster unit by depressing the brake pedal several times. Although not absolutely necessary, to improve access to the sensor, remove the master cylinder as described in Section 13.
18   Disconnect the battery negative terminal (see the **Caution** in Step 2). Trace the wiring back from pedal position sensor, and disconnect at the connector.
19   Using a small screwdriver, carefully lever off the sensor retaining clip then withdraw the sensor from the front of the power brake booster. Recover the sealing ring and snap-ring.

24.5 Ensure the vacuum pump slot (arrow) is aligned with the pump drivegear

Installation

20   If a new sensor is being installed, note the color of the spacer installed on the original sensor, and install the relevant color spacer to the new sensor. This is vital to ensure that the correct operation of the anti-lock braking system.
21   Install the new snap-ring in the groove on the front of the power brake booster, positioning its end gap over the booster unit sensor lower locating slot.
22   Install the new sealing ring on the sensor, and lubricate it with a smear of oil to aid installation.
23   Install the sensor on the brake booster, aligning its locating notch with the booster unit upper groove. Push the sensor until it clicks into position, and check that it is securely retained by the snap-ring.
24   Reconnect the sensor wiring, and connect the battery negative terminal.

---

24 Vacuum pump (diesel models) - removal and installation

Refer to illustration 24.5

Removal

1   Release the retaining clip, and disconnect the vacuum hose from the top of the pump.
2   Loosen and remove the retaining bolt, and remove the pump retaining clamp from the cylinder block.
3   Withdraw the vacuum pump from the cylinder block, and recover the O-ring seal. Discard the O-ring - a new one should be used on installation.

Installation

4   Install the new O-ring to the vacuum pump, and apply a smear of oil to the O-ring to aid installation.
5   Maneuver the vacuum pump into position, making sure that the slot in the pump drive gear aligns with the dog on the pump drive gear **(see illustration)**.
6   Install the retaining clamp and securely tighten its retaining bolt.
7   Reconnect the vacuum hose to the pump, and secure it in position with the retaining clip.

---

25 Vacuum pump (diesel models) - testing and overhaul

1   The operation of the braking system vacuum pump can be checked using a vacuum gauge.
2   Disconnect the vacuum line from the pump, and connect the gauge to the pump fitting using a length of hose.
3   Start the engine and allow it to idle, then measure the vacuum created by the pump. As a guide, after one minute, a minimum of approximately 20-inches Hg should be recorded. If the vacuum registered is significantly less than this, it is likely that the pump is faulty. However, seek the advice of a VW dealer before condemning the pump.
4   Overhaul of the vacuum pump is not possible, since no major components are available separately for it; the only spare part readily available is the pump cover sealing ring. If faulty, the complete pump assembly must be replaced.

Chapter 10
Suspension and steering systems

Contents

Specifications

Front suspension
Type .... Independent, with MacPherson struts incorporating coil springs and telescopic shock absorbers. Stabilizer bar fitted to most models.

Rear suspension
Type .... Transverse torsion beam axle with trailing arms. Coil spring/shock absorber assemblies. Stabilizer bar on some models.

Steering
Type .... Rack-and-pinion. Power assistance standard on certain models, optional on others

10

Torque specifications

Ft-lbs (unless otherwise indicated)

Front suspension

Stabilizer bar connecting link retaining nut..	18
Control arm balljoint	
Retaining bolts	26
Retaining nut	33
Control arm pivot bolt	
Stage 1	37
Stage 2	Angle-tighten a further 90°
Control arm rear mounting bolt	
Stage 1	52
Stage 2	Angle-tighten a further 90°
Subframe mounting bolts (long)	
Stage 1	52
Stage 2	Angle-tighten a further 90°
Subframe mounting bolts (short)	48
Suspension strut spring seat retaining nut	
Plus suspension	44
Base suspension	30
Suspension strut-to-steering knuckle bolt nut	70
Suspension strut upper mounting nuts	44

Rear suspension

Rear axle	
Pivot nut	59
Mounting bracket retaining bolts	52
Stub axle/back plate retaining bolts	44
Shock absorber/coil spring assembly	
Upper mounting bottom nut...	132 in-lbs
Upper mounting top nut	18
Lower mounting bolt nut	52
Spring retaining plate nut	132 in-lbs

Steering

Universal joint pinch bolt nut	
Stage 1	22
Stage 2	Angle-tighten a further 45°
Intermediate shaft connecting piece nuts	18
Power steering pump	
Swivel bracket and mounting bracket bolts	18
Mounting bolts	18
Feed line fitting bolt..	22
Pulley retaining bolts	18
Steering gear	
Retaining nuts	22
Steering line fitting nuts..	22
Steering wheel nut	37
Tie-rod end	
Retaining nut	26
Locknut	37

Wheels

Wheel bolts..	See Chapter 1

1　General information

The independent front suspension is of the MacPherson strut type, incorporating coil springs and integral telescopic shock absorbers. The MacPherson struts are located by transverse control arms, which use rubber inner mounting bushings, and incorporate a balljoint at the outer ends. The front steering knuckles, which carry the wheel bearings, brake calipers and the hub/disc assemblies, are bolted to the MacPherson struts, and connected to the control arms through the balljoints. Most models are equipped with a front stabilizer bar. The stabilizer bar is rubber-mounted, and is connected to both control arms.

The rear suspension consists of a torsion axle with shock absorbers/coil spring assemblies. On most models, a stabilizer bar is incorporated into the rear axle assembly; this links both trailing arms, and is situated just to the rear of the axle crossmember.

The steering column incorporates a universal joint, and is connected to the steering gear by a second individual universal joint.

The steering gear is mounted onto the front subframe, and is connected by two tie-rods, with balljoints at their outer ends, to the steering arms projecting rearwards from the steering knuckles.

Power-assisted steering is installed as standard on some models, and is available as an option on all others. The hydraulic steering system is powered by a belt-driven pump, which is driven off the crankshaft pulley.

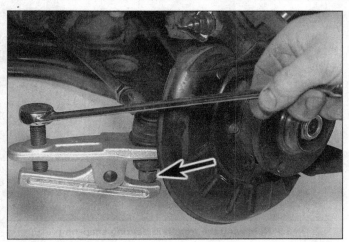

2.6 Using a balljoint separator to free the tie-rod end from the steering knuckle. Leave on the nut (arrow) to protect the balljoint threads

2.7 Removing the retaining plate from the top of the control arm

2 Steering knuckle - removal and installation

Note: *A new driveaxle nut, new suspension strut-to-steering knuckle bolt nuts, and a new tie-rod end nut will be required on installation.*

Removal

Refer to illustrations 2.6, 2.7, 2.8 and 2.9

1   Remove the wheel trim/hub cap (as applicable) and loosen the driveaxle retaining nut with the vehicle resting on its wheels. Also loosen the wheel bolts.

2   Chock the rear wheels of the car, firmly apply the parking brake, then jack up the front of the car and support it on axle stands. Remove the front wheel.

3   Remove the driveaxle retaining nut and (where installed) its washer.

4   On models with ABS, remove the wheel sensor as described in Chapter 9.

5   If the hub bearings are to be disturbed, remove the brake disc as described in Chapter 9. If not, on models with VW calipers, remove the brake pads, or on models with Girling calipers, unscrew the two bolts securing the brake caliper assembly to the hub, and slide the caliper assembly off the disc (see Chapter 9). Using a piece of wire or string, tie the caliper to the front suspension coil spring to avoid placing any

strain on the hydraulic brake hose.

6   Loosen and remove the nut securing the steering gear tie-rod end to the steering knuckle, and release the balljoint tapered shank using a universal balljoint separator **(see illustration)**.

7   Using a suitable marker pen, draw around the end of the suspension control arm, marking the correct installed position of balljoint. Unscrew the balljoint retaining bolts, and remove the retaining plate from the top of the control arm **(see illustration)**. **Note:** *On most models, the balljoint inner retaining bolt hole is slotted; on these models, the inner retaining bolt can be loosened, leaving the retaining plate and bolt in position in the arm, and the balljoint is then disengaged from the bolt.*

8   Using a suitable marker pen, draw around the outline of each suspension strut-to-steering knuckle bolt, marking their positions on the strut. Loosen and remove both nuts and bolts **(see illustration)**.

9   Free the steering knuckle from the strut, then carefully pull the hub assembly outwards and withdraw the driveaxle outer constant velocity joint from the hub assembly **(see illustration)**. The outer joint will be very tight - tap the joint out of the hub using a soft-faced mallet. If this fails to free it from the hub, the joint will have to be pressed out using a puller.

Installation

Refer to illustration 2.12

10   Note that all self-locking nuts disturbed on removal must be replaced as a matter of course. These nuts have threads which are

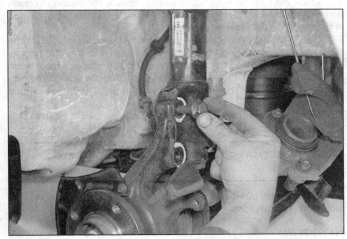

2.8 Mark their positions on the strut, then remove the steering knuckle-to-strut nuts and bolts

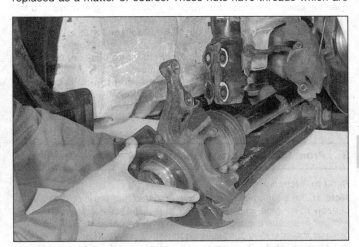

2.9 Free the steering knuckle assembly from the driveaxle splines, and remove it from the vehicle

10

2.12 Install the washer and driveaxle nut, and use the nut to draw the driveaxle joint fully into position

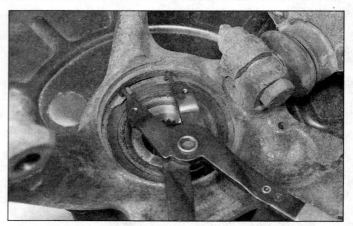

3.3 Removing the front hub bearing snap-ring

pre-coated with locking compound (this is only effective once), and include the driveaxle retaining nut, the tie-rod end nut, and the suspension strut-to-steering knuckle bolt nuts.

11   Ensure that the outer joint and hub splines are clean and dry. On models with Plus suspension, remove all traces of sealant from both sets of splines, and apply a bead of sealant to the outer joint splines (VW recommends the use of part number D 185 400 A2 - available from VW dealers).

12   Maneuver the hub assembly into position, and engage it with the driveaxle outer joint. Ensure that the threads are clean and apply a smear of oil to the contact face of the new driveaxle retaining nut. Install the washer (where installed) and nut, and use the nut to draw the joint fully into position **(see illustration)**.

13   Engage the steering knuckle with the suspension strut while aligning the balljoint with the control arm.

14   Insert the strut-to-steering knuckle; bolts, and install the new retaining nuts. Align the bolts with the marks made prior to removal, and tighten the nuts to the specified torque setting.

15   Install the control arm balljoint retaining bolts. Align the balljoint with the marks made prior to removal, then tighten the retaining bolts to the specified torque.

16   Engage the tie-rod end in the steering knuckle, then install a new retaining nut and tighten it to the specified torque.

17   Where necessary, install the brake disc to the hub, referring to Chapter 9 for further information.

18   On models with VW brake calipers, install the brake pads. On models with Girling calipers, slide the caliper assembly into position over the disc, then install the mounting bolts and tighten them to the specified torque (see Chapter 9).

19   Where necessary, install the ABS wheel sensor as described in Chapter 9.

20   Ensure that the outer joint is drawn fully into position, then install the wheel and lower the vehicle to the ground.

21   Tighten the driveaxle retaining nut to the torque listed in the Chapter 8 Specifications, then tighten the wheel bolts to the torque listed in the Chapter 1 Specifications. **Note:** *On completion, it is advisable to have the camber angle checked and, if necessary, adjusted.*

3   Front hub bearings - replacement

Refer to illustration 3.3

Note 1: *The bearing is a sealed, pre-adjusted and pre-lubricated, double-row roller type, and is intended to last the car's entire service life without maintenance or attention. Never overtighten the driveaxle nut beyond the specified torque wrench setting in an attempt to "adjust" the bearing.* **Note 2:** *A press will be required to disassemble and rebuild the assembly; if such a tool is not available, a large bench vise and spacers (such as large sockets) will serve as an adequate substi-*

tute. *The bearing's inner races are an interference fit on the hub; if the inner race remains on the hub when it is pressed out of the hub carrier, a knife-edged bearing puller will be required to remove it.*

1   Remove the steering knuckle assembly as described in Section 2.

2   Support the steering knuckle securely on blocks or in a vise. Using a tubular spacer which bears only on the inner end of the hub flange, press the hub flange out of the bearing. If the bearing's outboard inner race remains on the hub, remove it using a bearing puller (see note above). If necessary, undo the retaining screws and remove the ABS rotor from the rear of the hub. Install the new rotor and securely tighten the retaining screws.

3   Extract the bearing retaining snap-ring(s) from the steering knuckle assembly **(see illustration)**.

4   Securely support the outer face of the steering knuckle. Using a tubular spacer, press the complete bearing assembly out of the steering knuckle.

5   Thoroughly clean the hub and steering knuckle, removing all traces of dirt and grease, and polish away any burrs or raised edges which might hinder reassembly. Check both for cracks or any other signs of wear or damage, and replace them if necessary. Replace the snap-ring, regardless of its apparent condition.

6   On reassembly, apply a light coating of molybdenum disulfide grease (VW recommends Molycote - available from your dealer) to the bearing outer race and bearing surface of the steering knuckle.

7   Securely support the steering knuckle, and locate the bearing in the hub. Press the bearing fully into position, ensuring that it enters the hub squarely, using a tubular spacer which bears only on the bearing outer race.

8   Once the bearing is correctly seated, secure the bearing in position with the new snap-ring(s), ensuring that they are correctly located in the groove in the steering knuckle.

9   Securely support the outer face of the hub flange, and locate the steering knuckle bearing inner race over the end of the hub flange. Press the bearing onto the hub, using a tubular spacer which bears only on the inner race of the hub bearing, until it seats against the hub shoulder. Check that the hub flange rotates freely, and wipe off any excess oil or grease.

10   Install the steering knuckle assembly as described in Section 2.

4   Front suspension strut - removal, overhaul and installation

Note: *New suspension strut upper and lower retaining nuts and will be required on installation.*

Removal

Refer to illustrations 4.2, 4.3a, 4.3b and 4.3c

1   Loosen the wheel bolts. Chock the rear wheels, apply the parking

brake, then jack up the front of the vehicle and support on axle stands. Remove the appropriate wheel.

2   Using a suitable marker pen, draw around the outline of each suspension strut-to-steering knuckle bolt, marking their positions on the strut. Loosen and remove both nuts and bolts, and unclip the brake hose from the strut **(see illustration)**.

3   Unclip the plastic cover (where installed) from the strut upper mounting, then loosen and remove the upper mounting nut and recover the mounting plate. Note that it may be necessary to retain the strut piston with a suitable Allen key, to prevent it from rotating as the nut is loosened **(see illustrations)**.

4   Free the strut from the steering knuckle and maneuver it out from underneath the wheel arch. Where necessary, recover the mounting bushing from the top of the strut.

Overhaul

Warning: *Before attempting to disassemble the suspension strut, a suitable tool to hold the coil spring in compression must be obtained. Adjustable coil spring compressors are readily available, and are recommended for this operation. Any attempt to disassemble the strut without such a tool is likely to result in damage or personal injury.*

5   With the strut removed from the car, clean away all external dirt, then mount it upright in a vise.

6   Install the spring compressor, and compress the coil spring until all tension is relieved from the upper spring seat. Continue as described under the relevant sub-heading.

Models with Plus suspension

7   Loosen and remove the spring seat retaining nut, while retaining the strut piston with a suitable Allen key, then remove the bearing and upper spring seat.

8   Remove the coil spring, then slide off the damper piston boot and rubber damper stop.

9   With the strut assembly now completely disassembled, examine all the components for wear, damage or deformation, and check the bearing for smoothness of operation. Replace any of the components as necessary.

10   Examine the strut for signs of fluid leakage. Check the strut piston for signs of pitting along its entire length, and check the strut body for signs of damage. While holding it in an upright position, test the operation of the strut by moving the piston through a full stroke, and then through short strokes of 50 to 100 mm. In both cases, the resistance felt should be smooth and continuous. If the resistance is jerky, or uneven, or if there is any visible sign of wear or damage to the strut, replacement is necessary.

11   If any doubt exists about the condition of the coil spring, carefully

4.2 Remove the strut-to-steering knuckle bolts, and free the brake hose (arrow) from the strut

remove the spring compressors, and check the spring for distortion and signs of cracking. Replace the spring if it is damaged or distorted, or if there is any doubt as to its condition.

12   Inspect all other components for signs of damage or deterioration, and replace any that are suspect.

13   Slide the rubber damper and piston boot onto the strut piston rod.

14   Install the coil spring onto the strut, making sure its end is correctly located against the spring seat stop.

15   Install the upper spring seat and bearing, and screw on the retaining nut. Tighten the retaining nut to the specified torque setting while retaining the strut piston.

Models with base suspension

Refer to illustrations 4.16 and 4.19a through 4.19h

16   Using the special slotted socket, loosen and remove the upper spring seat retaining nut and lift off the strut mounting, upper spring seat and washer. **Note:** *A special slotted socket is required to remove and install the upper spring seat retaining nut; alternatives to the VW tool are available from automotive tool manufacturers. In the absence of the special VW tool, a suitable replacement can be fabricated from a long-reach 13 mm socket. Cut the lower end of the socket to leave two*

4.3a Remove the plastic cover . . .

4.3b . . . then unscrew the strut upper mounting nut . . .

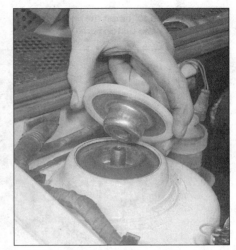

4.3c . . . and lift off the mounting plate

10

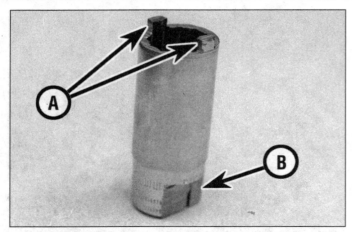

4.16 In the absence of the special VW tool, a replacement can be fabricated from a long-reach 13 mm socket. Cut the lower end of the socket to leave two teeth (A) which will engage with the slots in the strut nut, and file the upper end of the socket (B) so that it can be held with an open-ended spanner

teeth (A) which will engage with the slots in the strut nut, and file the upper end of the socket (B) so that it can be held with an open-ended wrench **(see illustration)**.

17   Lift off the coil spring, and remove the rubber damper and protective sleeve from the strut.

18   Inspect the strut components as described in Steps 9 to 12.

4.19a Slide the rubber damper and protective sleeve onto the strut . . .

19   To reassemble the strut, follow the accompanying photos, beginning with **illustration 4.19a**. Be sure to stay in order, and carefully read the caption underneath each **(see illustrations)**.

Installation

Refer to illustration 4.23

20   Ensure that the mounting bushing (where installed) is in position on the top of the strut, then maneuver the strut into position and engage it with the steering knuckle.

21   Make sure the top of the strut is correctly located, then insert the

4.19b . . . and install the washer to the piston rod

4.19c Install the coil spring to the strut . . .

4.19d . . . and install the upper spring seat to the top of the spring

4.19e Install the strut mounting assembly . . .

4.19f . . . and screw the slotted nut onto the strut piston

4.19g Tighten the slotted nut to the specified torque setting . . .

4.19h . . . then carefully release the spring compressors while ensuring that the spring ends are correctly located against the stops on the upper and lower seats

strut-to-steering knuckle bolts and install the new retaining nuts.

22   Install the mounting plate to the top of the strut, and install the new upper mounting nut. Tighten the nut to the specified torque setting and (where necessary) install the cover.

23   Align the strut-to-steering knuckle bolts with the marks made on removal, and tighten the retaining nuts to the specified torque setting **(see illustration)**. Clip the brake hose back into the strut.

24   Install the wheel, and tighten the wheel bolts to the specified torque. **Note:** *On completion, it is advisable to have the camber angle checked and, if necessary, adjusted.*

5   Front control arm - removal, overhaul and installation

Note: *A new control arm pivot bolt and rear mounting bolt will be required on installation.*

Removal

1   Chock the rear wheels, firmly apply the parking brake, then jack up the front of the vehicle and support on axle stands. Remove the appropriate front wheel.

2   On models with Plus suspension, loosen and remove the nut (or nut and bolt, as applicable) securing the stabilizer bar connecting link to the control arm, and free the link from the arm.

3   On models with base suspension, remove the connecting link as described in Section 8.

4   On all models, using a suitable marker pen, draw around the end of the control arm, marking the correct installed position of balljoint, then loosen and remove the balljoint retaining bolts and lift off the

retaining plate from the top of the control arm.

5   Loosen and remove the control arm pivot bolt and rear mounting bolt.

6   Lower the arm out of position, and remove it from underneath the vehicle.

Overhaul

7   Thoroughly clean the control arm and the area around the arm mountings, removing all traces of dirt and underseal if necessary, then check carefully for cracks, distortion or any other signs of wear or damage, paying particular attention to the pivot and rear mounting bushings. If either bushing requires replacement, the control arm should be taken to a VW dealer or an automotive machine shop. A hydraulic press and suitable spacers are required to press the bushings out of the arm and install the new ones.

Installation

Refer to illustrations 5.11 and 5.14

8   Maneuver the control arm into position, engaging it with the balljoint.

9   Install the new pivot bolt and rear mounting bolt.

10   Position the retaining plate on the top of the arm, then install the control arm balljoint retaining bolts. Align the balljoint with the marks made prior to removal, then tighten the retaining bolts to the specified torque.

11   Tighten the control arm rear mounting bolt to the specified stage 1 torque setting, then angle-tighten it through the specified stage 2 angle

4.23 Align the marks made prior to removal, and tighten the strut-to-steering knuckle bolts to the specified torque

5.11 Tighten the control arm rear mounting bolt to the specified Stage 1 torque setting, and then through the specified Stage 2 angle

10

5.14 With the control arm raised to simulate normal ride height, tighten the control arm pivot bolt to the specified torque setting and then through the specified angle

6.11 Tightening the control arm balljoint retaining bolts to the specified torque

(see illustration). Tighten the pivot bolt lightly only at this stage.
12   On 2.0 litter (GT specification) models, install the connecting link nut (or nut and bolt, as applicable) and tighten it to the specified torque setting.
13   On all other models, install the connecting links as described in Section 8.
14   Using a floor jack, raise the control arm to simulate normal ride height, then tighten the control arm front pivot bolt first to the specified stage 1 torque setting, and then through the specified stage 2 angle **(see illustration).** Install the wheel, then lower the vehicle and tighten the wheel bolts to the torque listed in the Chapter 1 Specifications. **Note:** *On completion, it is advisable to have the camber angle checked and, if necessary, adjusted*

6   Balljoints - removal and installation

Note: *A new balljoint retaining nut/clamp bolt nut (as applicable) will be required on installation*

Removal

1   Loosen the wheel bolts. Chock the rear wheels, firmly apply the parking brake, then jack up the front of the vehicle and support on axle stands. Remove the appropriate front wheel.
2   Loosen and remove the bolts securing the inner driveaxle joint to the transaxle flange. Support the driveaxle by suspending it with wire or string, and do not allow it to hang under its own weight.
3   Using a suitable marker pen, draw around the end of the suspension control arm, marking the correct installed position of balljoint. Unscrew the balljoint retaining bolts and remove the retaining plate from the top of the control arm. **Note:** *On most models, the balljoint inner retaining bolt hole is slotted; on these models the inner retaining bolt can be loosened, leaving the retaining plate and bolt in position in the arm, and the balljoint disengaged from the bolt.*
4   Pull the steering knuckle assembly outwards, and disengage the balljoint from the control arm.
5   On models with Plus suspension, loosen the balljoint retaining nut and unscrew it until it is positioned flush with the end of the balljoint shank threads. Release the balljoint from the steering knuckle, using a universal balljoint separator, then unscrew the nut and remove the balljoint from the vehicle.
6   On models with base suspension, loosen and remove the nut and withdraw the balljoint clamp bolt from the steering knuckle. Free the balljoint shank from the steering knuckle, and remove it from the vehicle.
7   Check that the balljoint moves freely, without any sign of rough-

ness. Check also that the balljoint boot shows no sign of deterioration, and is free from cracks and splits. Replace worn or damaged components as necessary.

Installation

Refer to illustration 6.11
8   On models with Plus suspension,, install the balljoint to the steering knuckle and install the new retaining nut. Tighten the nut to the specified torque setting, noting that the balljoint shank can be retained with an Allen key if necessary to prevent it from rotating.
9   On all other models, slide the balljoint into the steering knuckle and install the clamp bolt. Install a new nut to the clamp bolt, and tighten it to the specified torque.
10   Align the balljoint with the lower suspension arm and slot it into position.
11   Install the control arm balljoint retaining bolts. Align the balljoint with the marks made prior to removal, then tighten the retaining bolts to the specified torque **(see illustration).**
12   Align the driveaxle inner joint with the transmission flange, and tighten its retaining bolts to the specified torque setting (see Chapter 8).
13   Install the wheel, then lower the vehicle to the ground and tighten the wheel bolts to the torque listed in the Chapter 1 Specifications.

7   Front stabilizer bar - removal and installation

Removal

Refer to illustration 7.3
1   Loosen the wheel bolts. Chock the rear wheels, firmly apply the parking brake, then jack up the front of the vehicle and support on axle stands. Remove both front wheels.
2   Remove both connecting links as described in Section 8.
3   Make alignment marks between the mounting bushings and stabilizer bar, then loosen the two stabilizer bar mounting clamp retaining bolts **(see illustration).**
4   Remove both clamps from the subframe, and maneuver the stabilizer bar out from underneath the vehicle. Remove the mounting bushings from the bar.
5   Carefully examine the stabilizer bar components for signs of wear, damage or deterioration, paying particular attention to the mounting bushings. Replace worn components as necessary.

Installation

6   Install the rubber mounting bushings to the stabilizer bar, aligning them with the marks made prior to removal. Rotate each bushing so that its split is positioned at the rear.

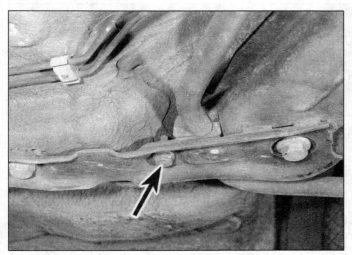

7.3 **The stabilizer bar is secured to the subframe by a mounting bolt (arrow) and clamps**

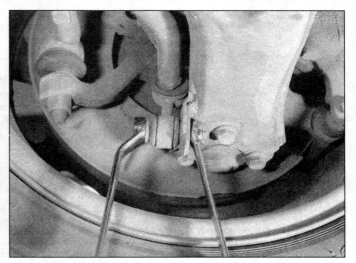

8.2 **On models with Plus suspension, loosen and remove the nut (or nut and bolt) securing the connecting link to the control arm . . .**

7   Install the stabilizer bar, and maneuver it into position. Install the mounting clamps, ensuring that their ends are correctly located in the hooks on the subframe, and install the retaining bolts. Ensure that the bushing markings are still aligned with the marks on the bars, then securely tighten the mounting clamp retaining bolts.
8   Install the connecting links as described in Section 8.
9   Install the wheels, then lower the vehicle to the ground and tighten the wheel bolts to the torque listed in the Chapter 1 Specifications.

8   Front stabilizer bar connecting link - removal and installation

Removal

Models with Plus suspension
Refer to illustrations 8.2 and 8.3
1   Firmly apply the parking brake, then jack up the front of the car and support it on axle stands.
2   Loosen and remove the nut (or nut and bolt, as applicable) securing the stabilizer bar connecting link to the control arm, and free the link from the arm **(see illustration)**.
3   Unscrew the connecting link upper balljoint shank from the end of the stabilizer bar, and remove the connecting link from the vehicle **(see illustration)**.
4   Check that each balljoint moves freely, without any sign of roughness. Check also that the balljoint boots show no sign of deterioration, and are free from cracks and splits. Also check the lower bushing for signs of damage or deterioration. Replace worn or damaged components as necessary.

Models with base suspension
5   Firmly apply the parking brake, then jack up the front of the car and support it on axle stands.
6   Loosen and remove the nut and washer securing the connecting link to the control arm. Recover the lower mounting rubber, noting which way it is installed.
7   Disengage the connecting link from the end of the stabilizer bar, and remove it from the control arm, complete with the upper mounting rubber.
8   Inspect the mounting rubbers for signs of damage or deterioration, and replace as necessary. The connecting link bushing can be pressed out of the link. Coat the new bushing with liquid soap to ease installation, and press it into position.

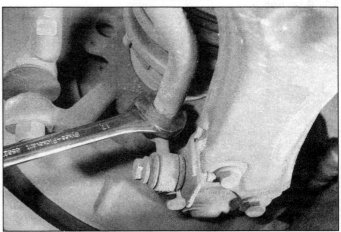

8.3 **. . . then unscrew the link from the end of the stabilizer bar**

Installation
Models with Plus suspension
9   Screw the connecting link upper balljoint into the end of the stabilizer bar, and tighten it securely.
10   Install the lower retaining nut (or nut and bolt) and tighten it to the specified torque setting. Lower the vehicle to the ground.

Models with base suspension
11   Apply liquid soap to the connecting link rubber, to aid installation.
12   Install the upper mounting rubber to the connecting link, making sure its conical side is facing towards the control arm.
13   Maneuver the link into position, and locate it on the end of the arm.
14   Install the lower mounting rubber with its conical surface facing the control arm, then install the washer with its collar facing away from the mounting rubber.
15   Install the connecting link retaining nut, tighten it to the specified torque setting, then lower the vehicle to the ground.

9   Rear hub assembly - removal and installation

The rear hub is an integral part of the brake drum/disc. Refer to Chapter 9 for removal and installation details.

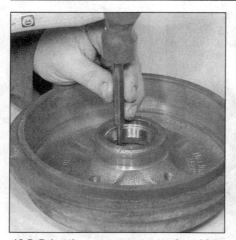

10.5 Drive the outer races out of position using a hammer and punch

10.9 Drive the outer races securely into position using a socket which bears only on the outer edge of the race

10.12 Work grease well into the tapered roller bearings prior to installing them

10 Rear hub bearings - replacement

Refer to illustrations 10.5, 10.9, 10.12 and 10.14

1   Remove the rear brake drum/disc (as applicable) as described in Chapter 9.
2   On disc brake models, lever off the cover ring from the rear of the hub.
3   On all models, using a flat-bladed screwdriver, lever the oil seal out of the rear of the hub, noting which way around it is installed.
4   Remove the inner bearing from the drum/disc.
5   Support the hub and tap the outer bearing outer race out of position **(see illustration)**.
6   Turn the drum/disc over, and tap the inner bearing outer race out of position.
7   Thoroughly clean the hub, removing all traces of dirt and grease, and polish away any burrs or raised edges which might hinder reassembly. Check the hub surface for cracks or any other signs of wear or damage, and replace it if necessary. The bearings and oil seal must be replaced whenever they are disturbed, as removal will almost certainly damage the outer races. Obtain new bearings, an oil seal and a small quantity of the special grease, from your VW dealer or an auto parts store.
8   On reassembly, apply a light film of clean engine oil to each bearing outer race, to aid installation.
9   Securely support the hub, and locate the outer bearing outer race in the hub. Tap the outer race fully into position, ensuring that it enters the hub squarely, using a suitable tubular spacer which bears only on the race outer edge **(see illustration)**.
10   Turn the drum/disc over, and install the inner bearing outer race in the same way.
11   Ensure both outer races are correctly seated in the hub, and wipe them clean.
12   Work grease well into both the tapered roller bearings, and apply a smear of grease to the outer races.
13   Install the tapered roller bearing to the inner bearing outer race **(see illustration)**.
14   Press the oil seal into the rear of the hub, ensuring that its sealing lip is facing inwards **(see illustration)**. Position the seal so that it is flush with the hub face, or until its lip abuts the rear of the hub. If necessary, the seal can be tapped into position using a suitable tubular drift with bears only on the hard outer edge of the seal.
15   On disc brake models, press the new cover ring fully onto the rear of the hub.
16   Turn the drum/disc over, install the tapered roller bearing to the outer race, and install the toothed washer.
17   Install the brake drum/disc as described in Chapter 9.

10.14 Grease the lips of the seal, and press it into the rear of the hub

11 Rear stub axle - removal and installation

Removal

1   Loosen the wheel bolts. Chock the front wheels, then jack up the rear of the vehicle and support it on axle stands. Remove the relevant rear wheel.

Rear drum brake models

2   Remove the brake drum as described in Chapter 9.
3   Minimize fluid loss by first removing the master cylinder reservoir cap, and then tightening it down onto a piece of cellophane, to obtain an airtight seal.
4   Wipe away all traces of dirt around the brake line fitting at the rear of the wheel cylinder, and unscrew the fitting nut. Carefully ease the line out of the wheel cylinder, and plug or tape over its end to prevent dirt entry. Wipe off any spilled fluid immediately.
5   Loosen and remove the bolts and washers securing the brake back plate assembly in position, and remove it along with the stub axle.
6   Inspect the stub axle surface for signs of damage such as scoring, and replace if necessary. Do not attempt to straighten the stub axle.

Rear disc brake models

7   Remove the brake disc as described in Chapter 9.

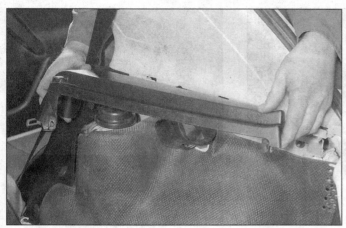

12.2 On Golf models, remove the luggage compartment trim panel to improve access to the strut mounting

12.5 Remove the trim cap from the top of the shock absorber mounting

8 Loosen and remove the bolts securing the disc back plate in position and remove it along with the stub axle.
9 Inspect the stub axle for signs of damage such as scoring and replace if necessary. Do not attempt to straighten the stub axle.

Installation

Rear drum brake models

10 Ensure that the mating surfaces of the axle, stub axle and back plate are clean and dry. Check the back plate for signs of damage, and remove any burrs with a fine file or emery cloth.
11 Install the stub axle and back plate assembly, and install the washers and retaining bolts. Note that the washers are dished, and should be installed with their concave surface facing towards the back plate. Tighten the retaining bolts to the specified torque setting.
12 Unplug the brake line, wipe it clean, and connect it to the rear of the wheel cylinder. Securely tighten the brake line fitting nut.
13 Remove the brake hose clamp or cellophane, as applicable, then install the brake drum as described in Chapter 9.
14 Bleed the hydraulic system as described in Chapter 9, noting that, providing the precautions described were taken to minimize brake fluid loss, it should only be necessary to bleed the relevant rear brake.

Rear disc brake models

15 Install the stub axle and back plate as described in Steps 10 and 11.
16 Install the brake disc and caliper as described in Chapter 9.

12 Rear shock absorber/coil spring assembly - removal, overhaul and installation

Removal

Refer to illustrations 12.2, 12.5, 12.6, 12.7a, 12.7b, 12.7c, 12.8a, 12.8b and 12.8c

1 Loosen the wheel bolts. Chock the front wheels, then jack up the rear of the vehicle and support it on axle stands. Remove the relevant rear wheel.
2 To improve access on Golf models, tilt the seat back forwards, then unclip and remove the load compartment cover panel. Remove the retaining nuts and the trim panel from the side of the luggage compartment **(see illustration)**.
3 To improve access on Jetta models, starting at the bottom of the panel, unclip the left-hand door pillar upper trim panel, and free it from the pillar. Loosen and remove the retaining screw from the top of the rear pillar trim panel. Unclip the rear of the panel from the pillar, then slide the panel towards the front of the vehicle, to disengage its retaining clips.
4 Repeat the procedure on the right-hand side of the vehicle, then carefully unclip the parcel shelf trim panel and remove it from the vehicle.
5 On all models, remove the trim cap from the top of the shock mounting **(see illustration)**.
6 Loosen and remove the upper mounting top nut and remove the dished washer, noting which way it is installed **(see illustration)**.
7 Unscrew the upper mounting bottom nut and lift off the cover plate, upper mounting rubber and shaped washer, noting each compo-

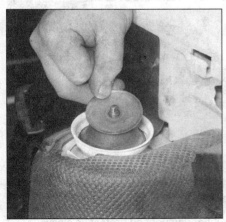

12.6 Unscrew the upper mounting top nut and remove the dished washer . . .

12.7a . . . then unscrew the bottom nut and lift off the cover plate (arrow) . . .

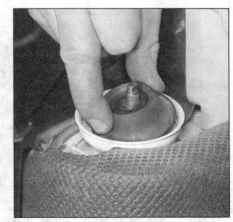

12.7b . . . followed by the upper mounting rubber . . .

10

12.7c . . . and shaped washer

12.8a Free the lower end of the shock
from the trailing arm . . .

12.8b . . . then maneuver the assembly
out from under the wheelwell . . .

12.8c . . . and recover the lower mounting
rubber from the top

12.11 Retain the piston rod with an open-
ended wrench while loosening the spring
plate retaining nut

12.14a Ensure that the cap is securely
clipped onto the shock absorber body . . .

nent's correct installed location **(see illustrations)**.
8   From underneath the vehicle, loosen and remove the shock absorber lower mounting nut and bolt, then maneuver the assembly out of position. Recover the lower mounting rubber from the top of the unit **(see illustrations)**.

Overhaul

Refer to illustrations 12.11 and 12.14a through 12.14h
Warning: *Before attempting to disassemble the shock absorber assembly, a spring suppressor must be obtained. Any attempt to disassemble the strut without such a tool is likely to result in damage or personal injury.*

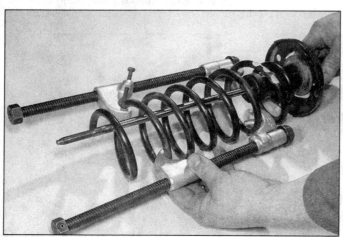

12.14b . . . and install the coil spring, making sure it is installed
the correct way (close-wound coils against the lower seat)

12.14c Slide on the rubber damper and protective sleeve . . .

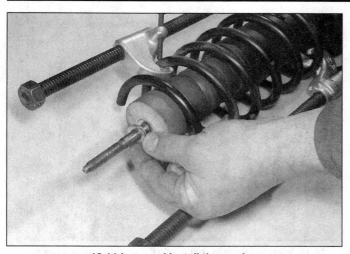

12.14d . . . and install the washer

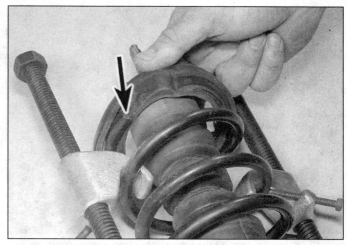

12.14e Install the rubber spring seat, making sure it is correctly located with the spring end (arrow) . . .

12.14f . . . and install the spring retaining plate

12.14g Slide on the spacer and install the retaining nut

9   With the assembly removed from the car, clean away all external dirt, then mount it upright in a vise.

10   Install the spring compressor, and compress the coil spring until

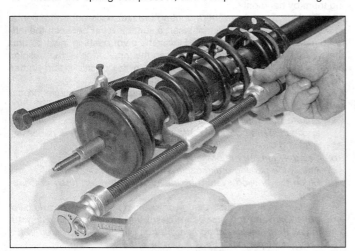

12.14h Tighten the nut to the specified torque, then carefully release the spring compressors, making sure the coil spring ends are correctly located

all tension is relieved from the upper spring seat.

11   Loosen and remove the spring retaining plate nut while retaining the piston rod with an open-ended wrench, then remove the spacer, spring retaining plate, rubber spring seat and washer **(see illustration)**.

12   Remove the coil spring, and recover rubber damper stop and protective sleeve.

13   Inspect the components as described in Steps 9 to 12 of Section 4.

14   To reassemble the strut, follow the accompanying photos, beginning with **illustration 12.14a**. Be sure to stay in order, and carefully read the caption underneath each **(see illustrations)**.

Installation

Refer to illustration 12.18

15   Install the lower mounting rubber to the top of the unit, and maneuver the assembly into position. Make sure the upper end is correctly located, then install the lower mounting bolt, tightening its nut by hand only at this stage.

16   From inside the vehicle, install the washer, upper mounting rubber and cover plate. Note that the rubber should be installed with its tapered serrated face downwards, and the cover with its convex surface towards the rubber.

17   Install the upper mounting bottom nut, and tighten it to the specified torque setting.

18   Install the dished washer with its convex surface downwards, then install the top mounting nut and tighten it to the specified torque

10

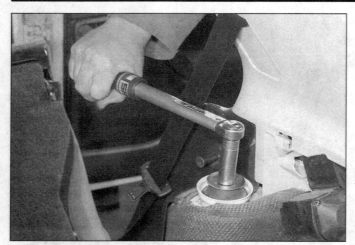

12.18 Ensure that all the upper mounting components are correctly located, and tighten the top mounting nut to the specified torque setting

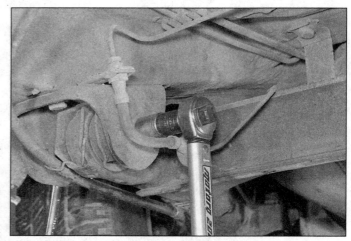

14.15 With the vehicle resting on its wheels, tighten the rear axle pivot bolts

setting **(see illustration)**. Install any trim panels removed to gain access to the upper mounting.
19   Using a floor jack placed under the trailing arm portion of the axle beam, raise the arm to simulate normal ride height, then tighten the lower mounting bolt to the specified torque setting.
20   Install the wheel, then lower the vehicle to the ground and tighten the wheel bolts to the torque listed in the Chapter 1 Specifications.

13   Rear stabilizer bar - removal and installation

The rear suspension stabilizer bar (where installed) runs along the length of the axle beam. It is an integral part of the axle assembly, and cannot be removed. If the stabilizer bar is damaged, which is unlikely, the complete axle assembly must be replaced.

14   Rear axle assembly - removal and installation

Removal
1   Loosen the wheel bolts. Firmly chock the front wheels, then jack up the rear of the vehicle and support it on axle stands. Remove both rear wheels.
2   Referring to Chapter 9, fully loosen the parking brake cable adjuster nut.
3   On models with rear drum brakes, disconnect both cables from the parking brake lever. From underneath the vehicle, work along the length of each cable, and free them from any retaining clips which secure them to the vehicle underbody.
4   On models with rear disc brakes, free the end of the parking brake inner cables from the caliper parking brake levers, then remove the retaining clips and detach the cables from the calipers. Work back along the cables, freeing them from their retaining clips on the axle.
5   On models equipped with ABS, disconnect the ABS wheel sensors at the wiring connectors, and free them from any retaining clips so they are free to be removed with the axle assembly.
6   Referring to Chapter 9, trace the brake lines back from the caliper/back plate to the pressure-regulating valve. Remove all traces of dirt from the valve, and mark the lines for identification purposes. On all models, loosen the fitting nuts and disconnect the lines. Plug the line ends, to minimize fluid loss and prevent the entry of dirt into the hydraulic system. Remove any retaining clips securing the rear section of the line to the vehicle underbody.
7   Make a final check that all necessary components have been disconnected and positioned so that they will not hinder the removal procedure, then position a floor jack beneath the center of the rear axle

assembly. Raise the jack until it is supporting the weight of the axle.
8   Using a suitable marker pen, mark the position of the axle mounting bracket retaining bolts on the bracket.
9   Loosen then remove both the left- and right-hand shock absorber lower mounting nuts and bolts.
10   Loosen and remove the axle mounting bracket retaining bolts, and carefully lower the jack and axle assembly out of position, and remove it from underneath the vehicle. **Note:** *Do not loosen the axle pivot bolts unless absolutely necessary; if the bolts are to be loosened, make alignment marks between the mounting bracket and axle prior to loosening. On installation, ensure that the mounting brackets are correctly positioned in relation to the axle beam, then tighten the pivot bolts to the specified torque.*
11   Inspect the axle mountings for signs or damage or deterioration. If replacement is necessary, the task should be entrusted to a VW dealer, who will have the necessary tools required to press out the old bushings and install the new ones.

Installation
Refer to illustration 14.15
12   Raise the rear axle into position, and insert the mounting bracket retaining bolts.
13   Install the shock absorber lower mounting nuts and bolts, tightening them by hand only.
14   Position the axle so the mounting bracket bolts are in the center of their slots. Have an assistant insert a suitable lever between the left-hand axle mounting bracket, and lever the pivot bushing inwards until there is only a small clearance between the inner edge of the bushing and the mounting bracket. Hold the axle in this position, and tighten the mounting bracket bolts to the specified torque.
15   The remainder of installation is a reversal of the removal procedure, bearing in mind the following points:
  a) *Ensure that the brake lines, parking brake cables and wiring (as applicable) are correctly routed, and retained by all the necessary retaining clips.*
  b) *Securely tighten the brake line fitting nuts.*
  c) *Adjust the parking brake cable as described in Chapter 9.*
  d) *Bleed the braking system hydraulic circuit as described in Chapter 9.*
  e) *Tighten the shock absorber lower mounting bolts and the axle pivot bolts following the procedure described in Section 12, Step 19. Install the wheels and tighten the bolts to the torque listed in the Chapter 1 Specifications.*
  f) *If the pivot bolts were removed, loosen the axle pivot bolts, then rock the vehicle to settle all disturbed components in position. Tighten the pivot bolts to the specified torque* **(see illustrations)**.

15.2 On models without an air bag, pry off the horn pad and disconnect its wiring connectors

15.3 Unscrew the retaining nut . . .

15 Steering wheel - removal and installation

Warning: *On models equipped with airbags, always disable the airbag system before working in the vicinity of the impact sensors, steering column or instrument panel to avoid the possibility of accidental deployment of the airbag, which could cause personal injury (see Chapter 12).*

Removal

1   Set the front wheels in the straight-ahead position, and release the steering lock by inserting the ignition key. Disconnect the cable from the negative terminal of the battery. **Caution:** *If the stereo in your vehicle is equipped with an anti-theft system, make sure you have the correct activation code before disconnecting the battery.*

Models without an airbag

Refer to illustrations 15.2, 15.3 and 15.4

2   Pry the horn pad out from the center of the wheel, and disconnect the horn wiring connectors **(see illustration).**
3   Loosen and remove the steering wheel retaining nut **(see illustration).**
4   Mark the steering wheel and steering column shaft in relation to each other, then lift the steering wheel off the column splines **(see illustration).** If it is tight, tap it up near the center, using the palm of your hand, or twist it from side to side, while pulling upwards to release it from the shaft splines. **Caution:** *Do not hammer on the shaft to remove the wheel.*

Models with an airbag

5   Remove the airbag unit from the center of the steering wheel, as described in Chapter 12.
6   Loosen and remove the retaining screws, and remove the steering column upper and lower shrouds.
7   Trace the wiring back from the airbag contact unit in the steering wheel, and disconnect it at the wiring connector.
8   Remove the steering wheel as described above in Steps 3 and 4.
9   With the steering wheel removed, rotate the contact unit ring slightly so its wiring connector is at the bottom (steering wheel in the straight-ahead position); this will lock the contact unit in the central position, and prevent it from being turned.

Installation

Models without an airbag

10   Installation is a reversal of removal, aligning the marks made on removal. Tighten the steering wheel retaining nut to the specified torque setting. Connect the negative battery cable.

15.4 . . . and pull the steering wheel off the column splines

Models with an airbag

11   Maneuver the wheel into position, making sure the wiring connector is correctly positioned, and engage it with the column splines.
12   Install the steering wheel retaining nut, and tighten it to the specified torque setting.
13   Reconnect the contact unit wiring connector, making sure the wiring is correctly routed.
14   Install the steering column shrouds, and securely tighten the retaining screws.
15   Install the airbag unit as described in Chapter 12..Connect the negative battery cable.

16 Steering column - removal, inspection and installation

Warning: *On models equipped with airbags, always disable the airbag system before working in the vicinity of the impact sensors, steering column or instrument panel to avoid the possibility of accidental deployment of the airbag, which could cause personal injury (see Chapter 12).*
Note: *New steering column shear-bolts and intermediate shaft retaining plate nuts will be required on installation.*

Removal

Refer to illustrations 16.4a, 16.4b, 16.6, 16.7a, 16.7b, 16.7c, 16.8 and 16.9

1   Disconnect the battery negative terminal. **Caution:** *If the stereo in*

10

16.4a   Unclip the trim panel . . .

16.4b   . . . then undo the retaining screws and remove driver's side lower dash panel

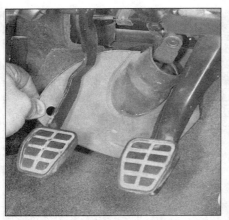

16.6   Pry out the retaining clip and remove the trim cover from the base of the steering column

16.7a   Unscrew the two retaining nuts . . .

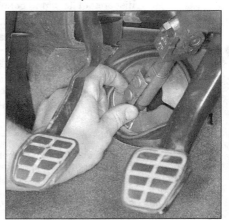

16.7b   . . . then withdraw the retaining plate . . .

your vehicle is equipped with an anti-theft system, make sure you have the correct activation code before disconnecting the battery.

2   Remove the steering wheel as described in Section 15.

3   Remove the steering column combination switches as described in Chapter 12, Section 4.

4   Press in the locking buttons, and unclip the fusebox cover from the underside of the driver's side lower dash panel. Carefully pry the trim panel out from the top of the driver's side lower dash panel, then loosen and remove all the panel retaining screws. Carefully move the panel downwards to release it from the dash, then remove it from the

16.7c   . . . and slide off the connecting piece securing the intermediate shaft halves together

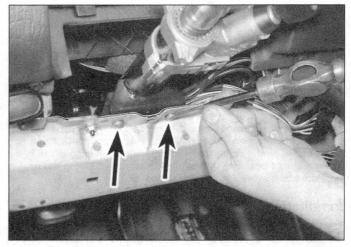

16.8   Tap the shear-bolts (arrows) around with a hammer and chisel until they can be unscrewed by hand . . .

16.9 . . . then lift the steering column out of position and remove it from the vehicle

16.13a Secure the column in position with new shear-bolts . . .

vehicle **(see illustrations)**.

5   Disconnect the wiring connector from the ignition switch, then free the wiring harness from its retaining clips on the column.

6   Pry out the retaining clip and remove the trim cover from the base of the steering column **(see illustration)**.

7   Loosen and remove the nuts from the intermediate shaft connecting piece, and slide out the retaining plate. Slide the connecting piece upwards, then separate the intermediate shaft halves and recover the connecting piece **(see illustrations)**.

8   The steering column is secured in position with shear-bolts. The shear-bolts can be extracted using a hammer and suitable chisel to tap the bolt heads around until they can be unscrewed by hand **(see illustration)**. Alternatively, drill a hole in the center of each bolt head, and extract them using a bolt/stud extractor (sometimes called an "easy-out").

9   Pull the column upwards and away from the firewall, to release its lower retaining clip, and maneuver it out from the vehicle **(see illustration)**.

Inspection

10   The steering column incorporates a telescopic safety feature. In the event of a front-end crash, the shaft collapses and prevents the steering wheel injuring the driver. Before installing the steering column, examine the column and mountings for signs of damage and deformation, and replace as necessary.

11   Check the steering shaft for signs of free play in the column bushings. If any damage or wear is found on the steering column bushings, the column must be replaced as an assembly. Inspect the intermediate shaft universal joint as described in Section 18.

Installation

Refer to illustrations 16.13a and 16.13b

12   Maneuver the steering column into position, and clip the lower retaining clip securely into the firewall.

13   Install the new shear-bolts, and tighten them evenly until both their heads break off **(see illustrations)**.

14   Slide the intermediate shaft connecting piece onto the upper half of the shaft, then align the shaft halves and join them with the connecting piece. Insert the retaining plate and install the new nuts, tightening them to the specified torque setting.

15   Install the trim cover to the base of the column and secure it in position with the retaining clip.

16   Ensure that the wiring harness is correctly routed then secure it in position with the column retaining clips and reconnect the ignition switch wiring.

17   Install the lower dash panel, tighten its retaining screws securely, and clip in the trim cover and fusebox cover.

16.13b . . . and tighten both bolts until their heads break off

18   Install the combination switches as described in Chapter 12.

19   Install the steering wheel as described in Section 15.

17   Ignition switch/steering column lock - removal and installation

Warning: *On models equipped with airbags, always disable the airbag system before working in the vicinity of the impact sensors, steering column or instrument panel to avoid the possibility of accidental deployment of the airbag, which could cause personal injury (see Chapter 12).*

Note: *A new lock assembly shear-bolt will be required on installation.*

Removal

Refer to illustrations 17.5, 17.8a, 17.8b and 17.9

1   Disconnect the battery negative terminal. **Caution:** *If the stereo in your vehicle is equipped with an anti-theft system, make sure you have the correct activation code before disconnecting the battery.* Insert the key into the lock, and turn it to release the steering lock.

2   Remove the steering wheel as described in Section 15.

3   Undo the retaining screws, and remove the steering column upper and lower shrouds.

4   Disconnect the wiring connectors from the steering column com-

10

bination switches. Undo the retaining screws, and remove both switch assemblies.

5   Using a suitable puller, carefully draw the splined collar off from the top of the steering column and recover the spring **(see illustration)**.

6   The lock assembly is secured in position with a shear-bolt. The shear-bolt can be extracted using a hammer and suitable chisel to tap the bolt head around until it can be unscrewed by hand. Alternatively, drill a hole in the center of the bolt head, and extract it using a bolt/stud extractor (sometimes called an "easy-out").

7   Disconnect the wiring connector, then slide the lock assembly upwards and off the steering column.

8   With the lock assembly removed, loosen the retaining screw and remove the ignition switch from the base of the lock assembly **(see illustrations)**.

9   To replace the lock cylinder, carefully drill a 1/8-inch diameter hole in the side of the lock casting at the point shown in **illustration 17.9**. Depress the lock detent plunger, and withdraw the cylinder from the casting. Slide the new lock cylinder into position, and check it is securely retained by the detent plunger. **Note:** *Replacement of the lock cylinder is a tricky operation, and it is recommended that it is entrusted to a VW dealer. If the hole is not accurately drilled, the lock assembly casting will be ruined, and the complete lock assembly will have to be replaced.*

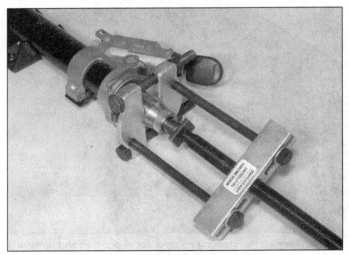

17.5 Draw the splined collar off from the top of the steering column using a suitable puller (shown with the column removed from the vehicle for clarity)

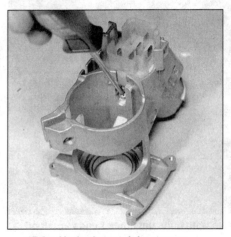

17.8a Undo the retaining screw . . .

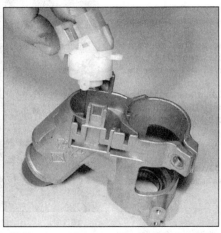

17.8b . . . and remove the ignition switch from the steering column lock assembly

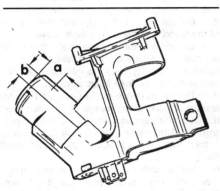

17.9 To replace the lock cylinder, drill a 1/8-inch hole at the point shown to reveal the lock cylinder detent plunger. Depress the plunger, and slide the cylinder out from the housing

A   15/32-inch   B   25/64-inch

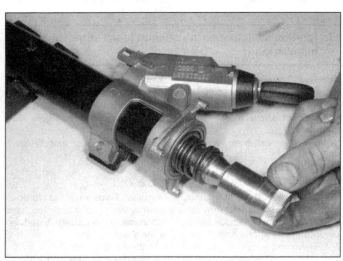

17.12a Install the spring and splined collar to the top of the steering column . . .

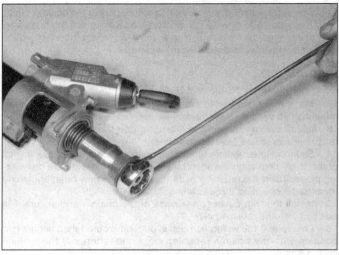

17.12b . . . and press them into position by installing the steering wheel retaining nut and tightening it securely

Installation

Refer to illustrations 17.12a and 17.12b

10   Install the ignition switch (where removed) to the lock assembly, making sure it is correctly engaged with the lock cylinder, and securely tighten its retaining screw.

11   Slide the lock assembly onto the column, aligning it with the column lug, and install the new shear-bolt. Tighten the bolt by hand only at this stage, and reconnect the wiring connector.

12   Install the spring to the top of the column, and install the splined collar to the shaft. Install a washer over the end of the collar, then install the steering wheel retaining nut, and use the nut to press the collar fully onto the steering column shaft **(see illustration)**. Once the collar is securely seated, unscrew the nut and remove the washer.

13   Check the operation of the steering column lock. If all is well, tighten the shear-bolt until its head breaks off.

14   Reconnect the wiring connectors to the combination switches, and securely tighten the switch retaining screws.

15   Install the steering column shrouds, then install the steering wheel as described in Section 15. On completion, reconnect the battery and check the operation of the switches.

18   Steering column intermediate shaft - removal and installation

Warning: *On models equipped with airbags, always disable the airbag system before working in the vicinity of the impact sensors, steering column or instrument panel to avoid the possibility of accidental deployment of the airbag, which could cause personal injury (see Chapter 12). Also, don't allow the steering wheel to turn after the intermediate shaft has been removed. To prevent this, pass the seat belt through the steering wheel and plug it into its latch.*

Note: *New nuts for the intermediate shaft connecting piece retaining plate, and a new clamp bolt nut, will be required on installation.*

Removal

1   Chock the rear wheels, firmly apply the parking brake, then jack up the front of the vehicle and support on axle stands. Set the front wheels in the straight-ahead position.

2   Release the rubber boot from the firewall, then cut the cable-tie and free the boot from the steering gear. Slide the boot downwards to gain access to the intermediate shaft. **Note:** *If necessary, access to the shaft can also be gained from inside the vehicle by prying out the retaining clip and removing the trim cover from the base of the steering column (see Section 16).*

3   Loosen and remove the nuts from the intermediate shaft connecting piece, and slide out the retaining plate. Slide the connecting piece upwards, then separate the intermediate shaft halves and recover the connecting piece.

4   Remove the rubber boot from the steering gear.

5   Using a hammer and punch, white paint or similar, mark the exact relationship between the intermediate shaft universal joint and the steering gear drive pinion. Loosen and remove the pinch bolt securing the joint to the pinion, then free the lower half of the intermediate shaft from the steering gear and remove it from the vehicle.

6   Mark the exact relationship between the intermediate shaft upper universal joint and steering column. Loosen and remove the clamp bolt and nut, then disengage the upper half of the intermediate shaft from the column splines and remove it from the vehicle.

7   Inspect the intermediate shaft universal joints for signs of roughness in its bearings and ease of movement. If either joint is damaged, it should be replaced. Replace the shaft boot if it shows signs of damage or deterioration.

Installation

8   Check that the front wheels are still in the straight-ahead position and the steering wheel is correctly positioned.

9   Aligning the marks made on removal, engage the upper half of the

shaft with the steering column splines. Install the clamp bolt and install the new retaining nut, tightening it to the specified torque setting.

10   Slide the rubber boot into position.

11   Maneuver the lower half of the intermediate shaft into position and, aligning the marks made prior to removal, engage it with the steering gear pinion splines. Install the clamp bolt, and tighten it to the specified torque setting.

12   Slide the intermediate shaft connecting piece onto the upper half of the shaft, then align the shaft halves and join them with the connecting piece. Insert the retaining plate and install the new nuts, tightening them to the specified torque setting.

13   Seat the rubber boot correctly in the firewall, then locate it on the steering gear and secure it in position with a new cable-tie. Lower the vehicle to the ground.

19   Steering gear assembly - removal, overhaul and installation

Warning: *On models equipped with airbags, always disable the airbag system before working in the vicinity of the impact sensors, steering column or instrument panel to avoid the possibility of accidental deployment of the airbag, which could cause personal injury (see Chapter 12). Also, don't allow the steering wheel to turn after the steering gear has been removed. To prevent this, pass the seat belt through the steering wheel and plug it into its latch.*

Note: *New subframe mounting bolts, tie-rod end nuts, steering gear retaining nuts, and intermediate shaft connecting piece retaining plate nuts, will be required on installation*

Removal

1   Loosen the wheel bolts. Chock the rear wheels, firmly apply the parking brake, then jack up the front of the vehicle and support on axle stands. Remove both front wheels.

2   Loosen and remove the nuts securing the steering gear tie-rod ends to the steering knuckles, and release the balljoint tapered shanks using a universal balljoint separator.

3   Release the rubber boot from firewall, then cut the cable-tie and free the boot from the steering gear. Slide the boot downwards to gain access to the intermediate shaft. **Note:** *If necessary, access to the shaft can also be gained from inside the vehicle by prying out the retaining clip and removing the trim cover from the base of the steering column (see Section 16).*

4   Loosen and remove the nuts, and remove the connecting piece retaining plate from the intermediate shaft. Slide the connecting piece upwards, then disengage the shaft halves and recover the connecting piece. Remove the rubber boot.

5   Place a jack with a block of wood beneath the engine, to take the weight of the engine. Alternatively, attach a couple of lifting eyes to the engine, and install a hoist or support bar to take the engine weight.

6   On manual transmission models, where necessary, loosen and remove the bolts securing the gearchange linkage pivot to the top of the steering gear (see Chapter 7A).

7   Loosen and remove all the front subframe mounting bolts, while making sure that the engine/transmission is adequately supported.

Manual steering gear

8   Loosen and remove the steering gear retaining nuts, and remove the mounting clamps.

9   Lower the subframe slightly, and maneuver the steering gear out towards the rear of the subframe. Remove the mounting rubbers from the steering gear, and inspect them for signs of damage or deterioration, replacing them if necessary. **Note:** *If the steering rack is to be removed for some time, lift the engine back into position and install the subframe mounting bolts.*

Power-assisted steering gear

Refer to illustration 19.11

10   Using brake hose clamps, clamp both the supply and return

10

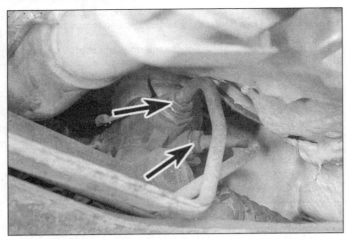

19.11 Power-assisted steering gear fitting nuts (arrows)

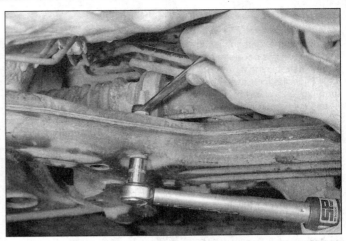

19.15 On installation, tighten the steering gear mounting clamp bolt nuts to the specified torque setting

hoses near the power steering fluid reservoir. This will minimize fluid loss during subsequent operations.

11   Mark the fittings to ensure that they are correctly positioned on reassembly, then unscrew the feed and return line fitting nuts from the steering gear assembly; be prepared for fluid spillage, and position a suitable container beneath the lines while unscrewing the fitting nuts **(see illustration)**. Disconnect both lines, and recover their sealing rings. Plug the line ends and steering gear orifices to prevent fluid leakage and to keep dirt out of the hydraulic system.

12   Remove the steering gear as described in Steps 8 and 9.

Overhaul

13   Examine the steering gear assembly for signs of wear or damage, and check that the rack moves freely throughout the full length of its travel, with no signs of roughness or excessive free play between the steering gear pinion and rack. It is not possible to overhaul the steering gear assembly housing components; if it is faulty, the assembly must be replaced. The only components which can be replaced individually are the steering gear boots, the tie-rod ends and the tie-rods. Tie-rod end and steering gear boot replacement procedures are covered later in this Chapter. Tie-rod replacement should be entrusted to a VW dealer or other repair shop as it is a difficult task, requiring special tools if it is to be carried out correctly and safely.

Installation

Manual steering gear

Refer to illustration 19.15

14   Install the mounting rubbers to the steering gear, and maneuver the assembly into position on the subframe.

15   Install the mounting clamps and install the new retaining nuts. Tighten the retaining nuts to the specified torque setting **(see illustration)**.

16   Carefully raise the subframe into position, and install the new mounting bolts. Tighten the subframe mounting bolts first to the specified stage 1 torque setting, then go around and tighten all the bolts through the specified stage 2 angle.

17   Slide the intermediate shaft connecting piece onto the upper half of the shaft, then align the shaft halves and join them with the connecting piece. Insert the retaining plate and install the new nuts, tightening them to the specified torque setting.

18   Seat the rubber boot correctly in the firewall, then locate it on the steering gear and secure it in position with a new cable-tie.

19   Reconnect the tie-rod ends to the steering knuckles and install the new retaining nuts, tightening them to the specified torque setting.

20   Install the front wheels, and lower the vehicle to the ground. On completion check and, if necessary, adjust the front wheel alignment as described in Section 24.

Power-assisted steering gear

21   Install the steering gear as described in Steps 14 and 15.

22   Wipe clean the feed and return line fittings, then install them to their respective positions on the steering gear, and tighten the fitting nuts securely. Ensure that the lines are correctly routed, and are securely held by all the necessary retaining clips.

23   Carry out the operations described in Steps 16 to 20.

20   Steering gear boots - replacement

1   Remove the tie-rod end as described in Section 23.

2   Mark the correct installed position of the boot on the tie-rod, then loosen the retaining clamp(s) and slide the boot off the steering gear housing and tie-rod end.

3   Thoroughly clean the tie-rod and the steering gear housing, using fine abrasive paper to polish off any corrosion, burrs or sharp edges, which might damage the new boot's sealing lips on installation. Scrape off all the grease from the old boot, and apply it to the tie-rod inner balljoint. (This assumes that grease has not been lost or contaminated as a result of damage to the old boot. Use fresh grease if in doubt).

4   Carefully slide the new boot onto the tie-rod end, and locate it on the steering gear housing. Align the outer edge of the boot with the mark made on the tie-rod prior to removal. Make sure the boot is not twisted, then lift the outer sealing lip of the boot to equalize air pressure within the boot.

5   Secure it in position with a new retaining clamp(s). Where crimped-type clamps are used, pull the clamp as tight as possible, and locate the hooks on the clamp in their slots. Remove any slack in the boot retaining clamp by carefully compressing the raised section of the clip. In the absence of the special tool, a pair of side cutters may be used, taking care not to actually cut the clip.

6   Install the tie-rod end as described in Section 23.

21   Power steering system - bleeding

1   With the engine stopped, fill the fluid reservoir right up to the top with the specified type of fluid.

2   Slowly move the steering from lock-to-lock several times to purge out the trapped air, then top-up the level in the fluid reservoir. Repeat this procedure until the fluid level in the reservoir does not drop any further.

3   Have an assistant start the engine, while you keep watch on the fluid level. Be prepared to add more fluid as the engine starts, as the fluid level is likely to drop quickly. The fluid level must be kept above

the "MIN" mark at all times.

4   With the engine running at idle speed, turn the steering wheel slowly two or three times approximately 45° to the left and right of center, then turn the wheel twice from lock to lock. Do not hold the wheel on either lock, as this imposes strain on the hydraulic system. Repeat this procedure until bubbles cease to appear in fluid reservoir.

5   If, when turning the steering, an abnormal noise is heard from the fluid lines, it indicates that there is still air in the system. Check this by turning the wheels to the straight-ahead position and switching off the engine. If the fluid level in the reservoir rises, then air is present in the system, and further bleeding is necessary.

6   Once all traces of air have been removed from the power steering hydraulic system, turn the engine off and allow the system to cool. Once cool, check that fluid level is up to the maximum mark on the power steering fluid reservoir, topping-up if necessary.

22   Power steering pump - removal and installation

Note: *New feed line fitting bolt sealing washers will be required on installation*

Removal

1   Loosen the steering pump pulley retaining bolts. Working as described in Chapter 1, release the drivebelt tension and unhook the drivebelt from the pump pulley.

2   Using brake hose clamps, clamp both the supply and return hoses near the power steering fluid reservoir. This will minimize fluid loss during subsequent operations. Continue as described under the relevant sub-heading.

3   Unscrew the retaining bolts and remove the pulley from the power steering pump, noting which way it is installed.

4   Loosen the retaining clamp, and disconnect the fluid supply hose from the pump. Where a crimp-type clamp is still installed, cut the clamp and discard it; replace it with a standard worm-drive hose clamp on installation. Loosen the fitting bolt, and disconnect the feed line from the pump, along with its sealing washers; discard the washers - new ones should be used on installation. Be prepared for some fluid spillage as the line and hose are disconnected, and plug the hose/line end and pump unions, to minimize fluid loss and prevent the entry of dirt into the system.

5   Loosen and remove the power steering pump pivot bolt and the adjuster bolt, and remove the pump and swivel bracket assembly from the main mounting bracket. If necessary, loosen and remove the pump mounting bolts, and separate the pump and mounting bracket; the mounting bracket can also be unbolted from the engine.

Installation

6   Where necessary, install the mounting bracket to the engine, and tighten its mounting bolts to the specified torque.

7   Join the pump and swivel bracket, and tighten the mounting bolts to the specified torque setting.

8   Prior to installing, ensure that the pump is primed by injecting hydraulic fluid in through the supply hose fitting and rotating the pump shaft.

9   Move the pump assembly into position and insert the pivot bolt and adjuster bolt, tighten them loosely only at this stage.

10   Position a new sealing washer on each side of the feed line fitting, then install the fitting bolt and tighten it to the specified torque setting. Install the supply line to the pump, and securely tighten its retaining clip. Remove the brake hose clamps used to minimize fluid loss.

11   Install the drive pulley, making sure it is the correct way around, and install its retaining bolts.

12   Install the drivebelt to the pump pulley, and tension it as described in Chapter 1. Once the belt is tensioned, tighten the pulley retaining bolts to the specified torque setting.

13   On completion, bleed the hydraulic system as described in Section 21.

23.9 Tightening the tie-rod end retaining nut to the specified torque setting

23   Tie-rod end - removal and installation

Note: *A new tie-rod retaining nut will be required on installation.*

Removal

1   Apply the parking brake, then jack up the front of the vehicle and support it on axle stands. Remove the appropriate front wheel.

2   Hold the tie-rod and unscrew the tie-rod end locknut by a quarter of a turn. Do not move the locknut from this position, as it will serve as a handy reference mark on installation.

3   Loosen and remove the nut securing the tie-rod end to the steering knuckle, and release the balljoint tapered shank using a universal balljoint separator.

4   Counting the **exact** number of turns necessary to do so, unscrew the balljoint from the tie-rod end.

5   Count the number of exposed threads between the end of the balljoint and the locknut, and record this figure.

6   Carefully clean the balljoint and the threads. Replace the balljoint if its movement is sloppy or too stiff, if excessively worn, or if damaged in any way; carefully check the stud taper and threads.

7   If the balljoint boot is damaged, the complete balljoint assembly must be replaced; it is not possible to obtain the boot separately.

Installation

Refer to illustration 23.9

8   Screw the balljoint into the tie-rod by the number of turns noted on removal. This should bring the balljoint locknut to within a quarter of a turn from the locknut.

9   Connect the tie-rod end to the steering knuckle, then install a new retaining nut and tighten it to the specified torque **(see illustration)**.

10   Install the wheel, then lower the vehicle to the ground and tighten the wheel bolts to the specified torque (see the Chapter 1 Specifications).

11   Check and, if necessary, adjust the front wheel toe setting as described in Section 24, then tighten the tie-rod end locknut securely.

24   Wheel alignment and steering angles - general information

Definitions

1   A car's steering and suspension geometry is defined in four basic settings - all angles are expressed in degrees (toe settings are also expressed as a measurement); the steering axis is defined as an imag-

10

inary line drawn through the axis of the suspension strut, extended where necessary to contact the ground.

2   **Camber** is the angle between each wheel and a vertical line drawn through its center and tire contact patch, when viewed from the front or rear of the car. "Positive" camber is when the wheels are tilted outwards from the vertical at the top; "negative" camber is when they are tilted inwards.

3   Camber angle is adjustable, and can be checked using a camber checking gauge.

4   **Caster** is the angle between the steering axis and a vertical line drawn through each wheel's center and tire contact patch, when viewed from the side of the car. "Positive" caster is when the steering axis is tilted so that it contacts the ground ahead of the vertical; "negative" caster is when it contacts the ground behind the vertical.

5   Caster is not adjustable, and is given for reference only; while it can be checked using a caster checking gauge, if the figure obtained is significantly different from that specified, the vehicle must be taken for careful checking by a professional, as the fault can only be caused by wear or damage to the body or suspension components.

6   **Toe** is the difference, viewed from above, between lines drawn through the wheel centers and the car's center-line. "Toe-in" is when the wheels point inwards, towards each other at the front, while "toe-out" is when they splay outwards from each other at the front.

7   The front wheel toe setting is adjusted by screwing the tie-rods in or out of the tie-rod ends.

8   Rear wheel toe setting is not adjustable, and is given for reference only. While it can be checked, if the figure obtained is significantly different from that specified, the vehicle must be taken for careful checking by a professional, as the fault can only be caused by wear or damage to the body or suspension components.

Checking and adjustment

Front wheel toe setting

9   Due to the special measuring equipment necessary to check the wheel alignment, and the skill required to use it properly, the checking and adjustment of these settings is best left to a VW dealer or similar expert. Note that most tire shops now possess sophisticated alignment equipment.

10   To check the toe setting, a tracking gauge must first be obtained. Two types of gauge are available, and can be obtained from auto parts stores. The first type measures the distance between the front and rear inside edges of the wheels, as previously described, with the vehicle

stationary. The second type, known as a "scuff plate", measures the actual position of the contact surface of the tire, in relation to the road surface, with the vehicle in motion. This is achieved by pushing or driving the front tire over a plate, which then moves slightly according to the scuff of the tire, and shows this movement on a scale. Both types have their advantages and disadvantages, but either can give satisfactory results if used correctly and carefully.

11   Make sure that the steering is in the straight-ahead position when making measurements.

12   If adjustment is necessary, apply the parking brake, then jack up the front of the vehicle and support it securely on axle stands.

13   First clean the tie-rod threads; if they are corroded, apply penetrating fluid before starting adjustment. Release the rubber boot outer clamps, peel back the boots and apply a smear of grease. This will ensure that both boots are free and will not be twisted or strained as their respective tie-rods are rotated.

14   Retain the tie-rod with a suitable wrench, and loosen the locknut fully. Alter the length of the tie-rod end rod, by screwing them into or out of the tie-rod ends. Rotate the tie-rod using an open-ended wrench installed to the tie-rod flats provided; shortening the tie-rods (screwing them onto their tie-rod ends) will reduce toe-in/increase toe-out.

15   When the setting is correct, hold the tie-rod and tighten the balljoint locknut to the specified torque setting. If after adjustment, the steering wheel spokes are no longer horizontal when the wheels are in the straight-ahead position, remove the steering wheel and reposition it (see Section 15).

16   Check that the toe setting has been correctly adjusted by lowering the vehicle to the ground and re-checking the toe setting; re-adjust if necessary. Ensure that the rubber boots are seated correctly and are not twisted or strained, and secure them in position with the retaining clips; where necessary, install a new retaining clamp (see Section 20).

Rear wheel toe setting

17   The procedure for checking the rear toe setting is the same as described for the front setting in Step 10. The setting is not adjustable - see Step 8.

Front wheel camber angle

18   Checking and adjusting the front wheel camber angle should be entrusted to a VW dealer or other suitably equipped specialist. Note that most tire shops now possess sophisticated checking equipment. For reference, adjustments are made by loosening the suspension strut-to-steering knuckle mounting bolts, and repositioning the steering knuckle assembly.

Chapter 11 Body

Contents

Specifications

Torque specifications

	Ft-lbs (unless otherwise indicated)
Hood lock retaining bolts	108 in-lbs
Door hinge retaining bolts	27
Door check link pivot bolt nut	60 in-lbs
Door handle retaining bolt	72 in-lbs
Door lock retaining bolts	72 in-lbs
Door window glass clamp nuts	84 in-lbs
Door window glass regulator retaining bolts	84 in-lbs
Front seat mounting bolt nut	72 in-lbs

1 General information

The body is made of pressed-steel sections, and is available in two-door convertible, three- and five-door Hatchback, and four-door Sedan models. Most components are welded together, but some use is made of structural adhesives; the front wings are bolted on.

The hood, door, and some other vulnerable panels are made of zinc-coated metal, and are further protected by being coated with an anti-chip primer before being sprayed.

Extensive use is made of plastic materials, mainly in the interior, but also in exterior components. The front and rear bumpers and front grille are injection-molded from a synthetic material that is very strong and yet light. Plastic components such as wheelarch liners are installed on the underside of the vehicle, to improve the body's resistance to corrosion.

2 Maintenance - bodywork and underframe

The general condition of a vehicle's bodywork is the one thing that significantly affects its value. Maintenance is easy, but needs to be regular. Neglect, particularly after minor damage, can lead quickly to further deterioration and costly repair bills. It is important also to keep watch on those parts of the vehicle not immediately visible, for instance the underside, inside all the wheelarches, and the lower part of the engine compartment.

The basic maintenance routine for the bodywork is washing - preferably with a lot of water, from a hose. This will remove all the loose solids which may have stuck to the vehicle. It is important to flush these off in such a way as to prevent grit from scratching the finish. The wheelarches and underframe need washing in the same way, to remove any accumulated mud which will retain moisture and tend to

11

These photos illustrate a method of repairing simple dents. They are intended to supplement *Body repair - minor damage* in this Chapter and should not be used as the sole instructions for body repair on these vehicles.

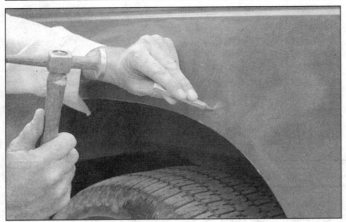

1   If you can't access the backside of the body panel to hammer out the dent, pull it out with a slide-hammer-type dent puller. In the deepest portion of the dent or along the crease line, drill or punch hole(s) at least one inch apart . . .

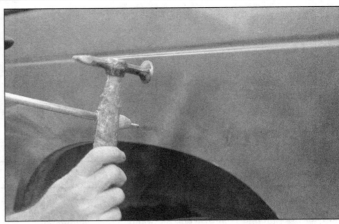

2   . . . then screw the slide-hammer into the hole and operate it. Tap with a hammer near the edge of the dent to help 'pop' the metal back to its original shape. When you're finished, the dent area should be close to its original contour and about 1/8-inch below the surface of the surrounding metal

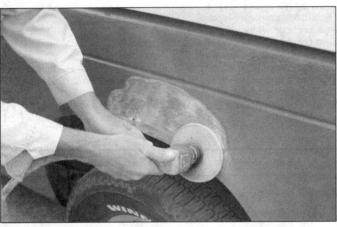

3   Using coarse-grit sandpaper, remove the paint down to the bare metal. Hand sanding works fine, but the disc sander shown here makes the job faster. Use finer (about 320-grit) sandpaper to feather-edge the paint at least one inch around the dent area

4   When the paint is removed, touch will probably be more helpful than sight for telling if the metal is straight. Hammer down the high spots or raise the low spots as necessary. Clean the repair area with wax/silicone remover

5   Following label instructions, mix up a batch of plastic filler and hardener. The ratio of filler to hardener is critical, and, if you mix it incorrectly, it will either not cure properly or cure too quickly (you won't have time to file and sand it into shape)

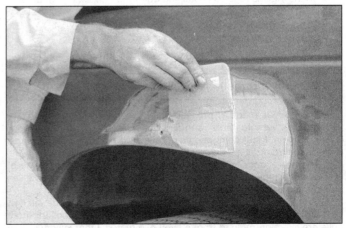

6   Working quickly so the filler doesn't harden, use a plastic applicator to press the body filler firmly into the metal, assuring it bonds completely. Work the filler until it matches the original contour and is slightly above the surrounding metal

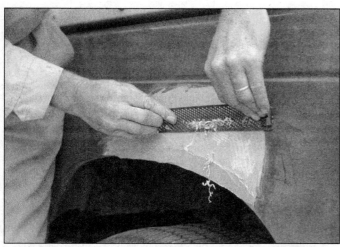

7   Let the filler harden until you can just dent it with your fingernail. Use a body file or Surform tool (shown here) to rough-shape the filler

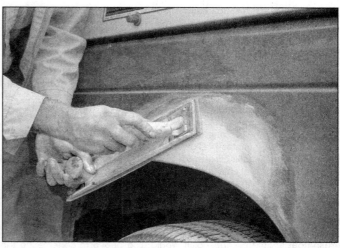

8   Use coarse-grit sandpaper and a sanding board or block to work the filler down until it's smooth and even. Work down to finer grits of sandpaper - always using a board or block - ending up with 360 or 400 grit

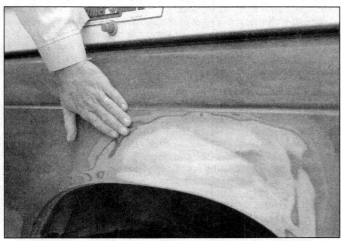

9   You shouldn't be able to feel any ridge at the transition from the filler to the bare metal or from the bare metal to the old paint. As soon as the repair is flat and uniform, remove the dust and mask off the adjacent panels or trim pieces

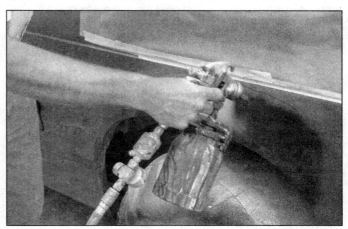

10   Apply several layers of primer to the area. Don't spray the primer on too heavy, so it sags or runs, and make sure each coat is dry before you spray on the next one. A professional-type spray gun is being used here, but aerosol spray primer is available inexpensively from auto parts stores

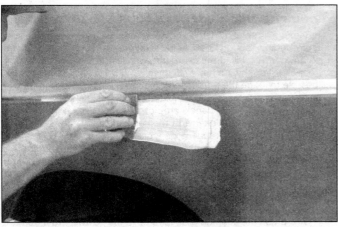

11   The primer will help reveal imperfections or scratches. Fill these with glazing compound. Follow the label instructions and sand it with 360 or 400-grit sandpaper until it's smooth. Repeat the glazing, sanding and respraying until the primer reveals a perfectly smooth surface

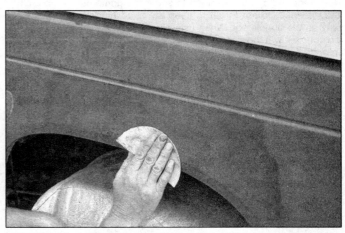

12   Finish sand the primer with very fine sandpaper (400 or 600-grit) to remove the primer overspray. Clean the area with water and allow it to dry. Use a tack rag to remove any dust, then apply the finish coat. Don't attempt to rub out or wax the repair area until the paint has dried completely (at least two weeks)

11

encourage rust. Paradoxically enough, the best time to clean the underframe and wheelarches is in wet weather, when the mud is thoroughly wet and soft. In very wet weather, the underframe is usually cleaned of large accumulations automatically, and this is a good time for inspection.

Periodically, except on vehicles with a wax-based underbody protective coating, it is a good idea to have the whole of the underframe of the vehicle steam-cleaned, engine compartment included, so that a thorough inspection can be carried out to see what minor repairs and renovations are necessary. Steam cleaning is available at many garages, and is necessary for the removal of the accumulation of oily grime, which sometimes is allowed to become thick in certain areas. If steam-cleaning facilities are not available, there are some excellent grease solvents available which can be brush-applied; the dirt can then be simply hosed off. Note that these methods should not be used on vehicles with wax-based underbody protective coating, or the coating will be removed. Such vehicles should be inspected annually, preferably just before Winter, when the underbody should be washed down, and any damage to the wax coating repaired. Ideally, a completely fresh coat should be applied. It would also be worth considering the use of wax-based protection for injection into door panels, sills, box sections, etc., as an additional safeguard against rust damage, where such protection is not provided by the vehicle manufacturer.

After washing the body, wipe off with a chamois leather to give an unspotted clear finish. A coat of clear protective wax polish will give added protection against chemical pollutants in the air. If the paintwork sheen has dulled or oxidized, use a cleaner/polisher combination to restore the brilliance of the shine. This requires a little effort, but such dulling is usually caused because regular washing has been neglected. Care needs to be taken with metallic paintwork, as special non-abrasive cleaner/polisher is required to avoid damage to the finish. Always check that the door and ventilator opening drain holes and lines are completely clear, so that water can be drained out. Glass should be treated in the same way as paintwork. Windscreens and windows can be kept clear of the smeary film which often appears, by proprietary glass cleaner. Never use any form of wax or other body or chromium polish on glass.

3   Maintenance - upholstery and carpets

Mats and carpets should be brushed or vacuum-cleaned regularly, to keep them free of grit. If they are badly stained, remove them from the vehicle for scrubbing or sponging, and make quite sure they are dry before installation. Seats and interior trim panels can be kept clean by wiping with a damp cloth. If they do become stained (which can be more apparent on light-colored upholstery), use a little liquid detergent and a soft nail brush to scour the grime out of the grain of the material. Do not forget to keep the headlining clean in the same way as the upholstery. When using liquid cleaners inside the vehicle, do not over-wet the surfaces being cleaned. Excessive damp could get into the seams and padded interior, causing stains, offensive odorous or even rot. If the inside of the vehicle gets wet accidentally, it is worthwhile taking some trouble to dry it out properly, particularly where carpets are involved. *Do not leave oil or electric heaters inside the vehicle for this purpose.*

4   Minor body damage - repair

See photo sequence

Repairs of minor scratches in bodywork

If the scratch is very superficial, and does not penetrate to the metal of the bodywork, repair is very simple. Lightly rub the area of the scratch with a paintwork renovator or a very fine cutting paste to remove loose paint from the scratch, and to clear the surrounding bodywork of wax polish. Rinse the area with clean water.

Apply touch-up paint to the scratch using a fine paint brush; continue to apply fine layers of paint until the surface of the paint in the

scratch is level with the surrounding paintwork. Allow the new paint at least two weeks to harden, then blend it into the surrounding paintwork by rubbing the scratch area with a paintwork renovator or a very fine cutting paste. Finally, apply wax polish.

Where the scratch has penetrated right through to the metal of the bodywork, causing the metal to rust, a different repair technique is required. Remove any loose rust from the bottom of the scratch with a penknife, then apply rust-inhibiting paint to prevent the formation of rust in the future. Using a rubber or nylon applicator, fill the scratch with bodystopper paste. If required, this paste can be mixed with cellulose thinners to provide a very thin paste which is ideal for filling narrow scratches. Before the stopper-paste in the scratch hardens, wrap a piece of smooth cotton rag around the top of a finger. Dip the finger in cellulose thinners, and quickly sweep it across the surface of the stopper-paste in the scratch; this will ensure that the surface of the stopper-paste is slightly hollowed. The scratch can now be painted over as described earlier in this Section.

Repairs of dents in bodywork

When deep denting of the vehicle's bodywork has taken place, the first task is to pull the dent out, until the affected bodywork almost attains its original shape. There is little point in trying to restore the original shape completely, as the metal in the damaged area will have stretched on impact, and cannot be reshaped fully to its original contour. It is better to bring the level of the dent up to a point which is about 1/8-inch below the level of the surrounding bodywork. In cases where the dent is very shallow anyway, it is not worth trying to pull it out at all. If the underside of the dent is accessible, it can be hammered out gently from behind, using a mallet with a wooden or plastic head. While doing this, hold a suitable block of wood firmly against the outside of the panel, to absorb the impact from the hammer blows and thus prevent a large area of the bodywork from being "belled-out."

Should the dent be in a section of the bodywork which has a double skin, or some other factor making it inaccessible from behind, a different technique is called for. Drill several small holes through the metal inside the area - particularly in the deeper section. Then screw long self-tapping screws into the holes, just sufficiently for them to gain a good purchase in the metal. Now the dent can be pulled out by pulling on the protruding heads of the screws with a pair of pliers.

The next stage of the repair is the removal of the paint from the damaged area, and from an inch or so of the surrounding "sound" bodywork. This is accomplished most easily by using a wire brush or abrasive pad on a power drill, although it can be done just as effectively by hand, using sheets of abrasive paper. To complete the preparation for filling, score the surface of the bare metal with a screwdriver or the tang of a file, or alternatively, drill small holes in the affected area. This will provide a good "key" for the filler paste.

To complete the repair, see the Section on filling and respraying.

Repairs of rust holes or gashes in bodywork

Remove all paint from the affected area, and from an inch or so of the surrounding "sound" bodywork, using an abrasive pad or a wire brush on a power drill. If these are not available, a few sheets of abrasive paper will do the job most effectively. With the paint removed, you will be able to judge the severity of the corrosion, and therefore decide whether to replace the whole panel (if this is possible) or to repair the affected area. New body panels are not as expensive as most people think, and it is often quicker and more satisfactory to install a new panel than to attempt to repair large areas of corrosion.

Remove all fittings from the affected area, except those which will act as a guide to the original shape of the damaged bodywork (e.g. headlamp shells etc.). Then, using tin snips or a hacksaw blade, remove all loose metal and any other metal badly affected by corrosion. Hammer the edges of the hole inwards, to create a slight depression for the filler paste.

Wire-brush the affected area to remove the powdery rust from the surface of the remaining metal. Paint the affected area with rust-inhibiting paint; if the back of the rusted area is accessible, treat this also.

Before filling can take place, it will be necessary to block the hole in some way. This can be achieved with aluminum or plastic mesh, or aluminum tape.

Aluminum or plastic mesh, or glass-fiber matting, is probably the best material to use for a large hole. Cut a piece to the approximate size and shape of the hole to be filled, then position it in the hole so that its edges are below the level of the surrounding bodywork. It can be retained in position by several blobs of filler paste around its periphery.

Aluminum tape should be used for small or very narrow holes. Pull a piece off the roll, trim it to the approximate size and shape required, then pull off the backing paper (if used) and stick the tape over the hole; it can be overlapped if the thickness of one piece is insufficient. Burnish down the edges of the tape with the handle of a screwdriver or similar, to ensure that the tape is securely attached to the metal underneath.

Bodywork repairs - filling and respraying

Before using this Section, see the Sections on dent, deep scratch, rust holes and gash repairs.

Many types of bodyfiller are available, but generally speaking, those proprietary kits which contain a tin of filler paste and a tube of resin hardener are best for this type of repair which can be used directly from the tube. A wide, flexible plastic or nylon applicator will be found invaluable for imparting a smooth and well-contoured finish to the surface of the filler.

Mix up a little filler on a clean piece of cardboard - measure the hardener carefully (follow the maker's instructions on the pack), otherwise the filler will set too rapidly or too slowly. Using the applicator, apply the filler paste to the prepared area; draw the applicator across the surface of the filler to achieve the correct contour and to level the surface. When a contour that approximates to the correct one is achieved, stop working the paste - if you carry on too long, the paste will become sticky and begin to "pick-up" on the applicator. Continue to add thin layers of filler paste at 20-minute intervals, until the level of the filler is just proud of the surrounding bodywork.

Once the filler has hardened, the excess can be removed using a metal plane or file. From then on, progressively-finer grades of abrasive paper should be used, starting with a 40-grade production paper, and finishing with a 400-grade wet-and-dry paper. Always wrap the abrasive paper around a flat rubber, cork, or wooden block - otherwise the surface of the filler will not be completely flat. During the smoothing of the filler surface, the wet-and-dry paper should be periodically rinsed in water. This will ensure that a very smooth finish is imparted to the filler at the final stage.

At this stage, the "dent" should be surrounded by a ring of bare metal, which in turn should be encircled by the finely "feathered" edge of the good paintwork. Rinse the repair area with clean water, until all the dust produced by the rubbing-down operation has gone.

Spray the whole area with a light coat of primer - this will show up any imperfections in the surface of the filler. Repair these imperfections with fresh filler paste or bodystopper, and again smooth the surface with abrasive paper. If bodystopper is used, it can be mixed with cellulose thinners, to form a thin paste which is ideal for filling small holes. Repeat this spray-and-repair procedure until you are satisfied that the surface of the filler, and the feathered edge of the paintwork, are perfect. Clean the repair area with clean water, and allow to dry fully.

The repair area is now ready for final spraying. Paint spraying must be carried out in a warm, dry, windless and dust-free atmosphere. This condition can be created artificially if you have access to a large indoor working area, but if you are forced to work in the open, you will have to pick your day very carefully. If you are working indoors, dousing the floor in the work area with water will help to settle the dust which would otherwise be in the atmosphere. If the repair area is confined to one body panel, mask off the surrounding panels; this will help to minimize the effects of a slight mis-match in paint colors. Bodywork fittings (e.g. chrome strips, door handles etc.) will also need to be masked off. Use genuine masking tape, and several thickness of newspaper, for the masking operations.

Before starting to spray, agitate the aerosol can thoroughly, then spray a test area (an old tin, or similar) until the technique is mastered. Cover the repair area with a thick coat of primer; the thickness should be built up using several thin layers of paint, rather than one thick one. Using 400 grade wet-and-dry paper, rub down the surface of the primer until it is smooth. While doing this, the work area should be thoroughly doused with water, and the wet-and-dry paper periodically rinsed in water. Allow to dry before spraying on more paint.

Spray on the top coat, again building up the thickness by using several thin layers of paint. Start spraying in the center of the repair area, and then, using a circular motion, work outwards until the whole repair area and about 2 inches of the surrounding original paintwork is covered. Remove all masking material 10 to 15 minutes after spraying on the final coat of paint.

Allow the new paint at least two weeks to harden, then, using a paintwork renovator or a very fine cutting paste, blend the edges of the paint into the existing paintwork. Finally, apply wax polish.

Plastic components

With the use of more and more plastic body components by the vehicle manufacturers (e.g. bumpers, spoilers, and in some cases major body panels), rectification of more serious damage to such items has become a matter of either entrusting repair work to a specialist in this field, or renewing complete components. Repair of such damage by the DIY owner is not feasible, owing to the cost of the equipment and materials required for effecting such repairs. The basic technique involves making a groove along the line of the crack in the plastic, using a rotary burr in a power drill. The damaged part is then welded back together, using a hot air gun to heat up and fuse a plastic filler rod into the groove. Any excess plastic is then removed, and the area rubbed down to a smooth finish. It is important that a filler rod of the correct plastic is used, as body components can be made of a variety of different types (e.g. polycarbonate, ABS, polypropylene).

Damage of a less serious nature (abrasions, minor cracks etc.) can be repaired by the DIY owner using a two-part epoxy filler repair material which can be used directly from the tube. Once mixed in equal proportions, this is used in similar fashion to the bodywork filler used on metal panels. The filler is usually cured in twenty to thirty minutes, ready for sanding and painting.

If the owner is renewing a complete component himself, or if he has repaired it with epoxy filler, he will be left with the problem of finding a suitable paint for finishing which is compatible with the type of plastic used. At one time, the use of a universal paint was not possible, owing to the complex range of plastics met with in body component applications. Standard paints, generally speaking, will not bond to plastic or rubber satisfactorily, but professional matched paints, to match any plastic or rubber finish, can be obtained from some dealers. However, it is now possible to obtain a plastic body parts finishing kit which consists of a pre-primer treatment, a primer and colored top coat. Full instructions are normally supplied with a kit, but basically the method of use is to first apply the pre-primer to the component concerned, and allow it to dry for up to 30 minutes. Then the primer is applied, and left to dry for about an hour before finally applying the special-colored top coat. The result is a correctly colored component, where the paint will flex with the plastic or rubber, a property that standard paint does not normally possess.

5 Major body damage - repair

Where serious damage has occurred, or large areas need replacement due to neglect, it means that complete new panels will need welding-in, and this is best left to professionals. If the damage is due to impact, it will also be necessary to check completely the alignment of the bodyshell, and this can only be carried out accurately by a VW dealer using special jigs. If the body is left misaligned, it is primarily dangerous, as the car will not handle properly, and secondly, uneven stresses will be imposed on the steering, suspension and possibly transmission, causing abnormal wear, or complete failure, particularly to such items as the tires.

11

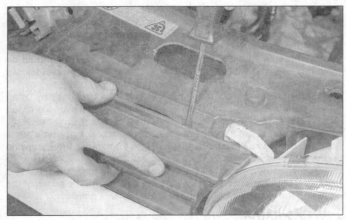

6.2a Using a screwdriver, release the upper . . .

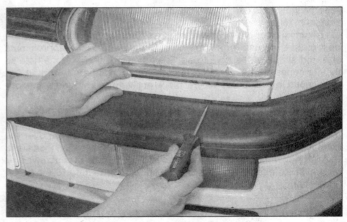

6.2b . . . and lower retaining clips . . .

6.2c . . . then move the radiator grille forwards and
away from the vehicle

6.3 Remove the fasteners (arrows) securing the wheelarch
liner to the bumper

6 Front bumper - removal and installation

Removal

Refer to illustrations 6.2a, 6.2b, 6.2c, 6.3, 6.4a and 6.4b

1   Apply the parking brake, then jack up the front of the vehicle and
support it on axle stands (see "*Jacking and vehicle support*").

2   Using a suitable screwdriver, carefully release the radiator grille
upper and lower retaining lugs then move the grille forwards and away
from the vehicle **(see illustrations)**.

3   Press out their center pins and remove the fasteners securing the
wheelarch liners to the bumper ends. Note that new fasteners will be
required on installation if the center pins are not recovered **(see illustration)**. Also undo the screws securing the liners to the bumper.

4   Working around the bumper, loosen and remove its eight retaining bolts **(see illustrations)**.

5   Disconnect the wiring connectors from the front direction indicators
and (where necessary) foglamps and free the wiring from any relevant
retaining clips so that the lamps are free to be removed with the bumper.

6   On models with headlamp washers, remove the washer jets as
described in Chapter 12.

6.4a Loosen and remove the three upper
retaining screws (arrows) . . .

6.4b . . . and the five lower bumper retaining screws (arrows)

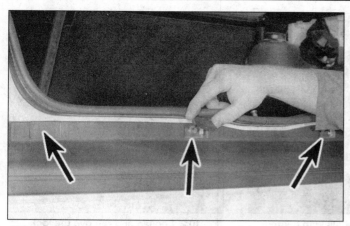

7.3 Unclip the trim caps from the top edge of the rear bumper to gain access to the retaining screws (arrows)

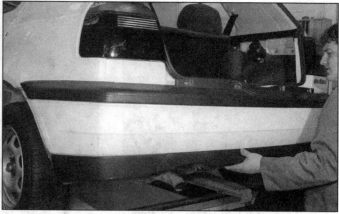

7.4a Remove the rear bumper from the vehicle . . .

7.4b . . . and inspect the bumper shock absorbers for signs of damage

7   Carefully release the bumper left- and right-hand ends and pull the bumper away from the vehicle in a forwards direction.

Installation

8   Installation is a reverse of the removal procedure, ensuring that the bumper ends engage correctly with the slides as the bumper is refitted.

7   Rear bumper - removal and installation

Removal

Refer to illustrations 7.3, 7.4a and 7.4b

1   To improve access, chock the front wheels, then jack up the rear of the vehicle and support it on axle stands (see "*Jacking and vehicle support*").
2   Loosen and remove the bolts securing the bottom of the bumper in position.
3   Carefully pry out the trim caps from the top edge of the bumper to gain access to the upper retaining bolts **(see illustration)**.
4   Loosen and remove the upper retaining bolts then pull the bumper away from the vehicle in a rearwards direction. Inspect the bumper shock absorbers, which are mounted onto the rear of the vehicle, for signs of damage or deformation and replace if necessary **(see illustrations)**.

Installation

5   Installation is a reverse of the removal procedure, ensuring that the bumper ends engage correctly with the slides as the bumper is refitted.

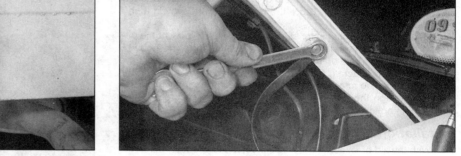

8.2a Unscrew the retaining nut . . .

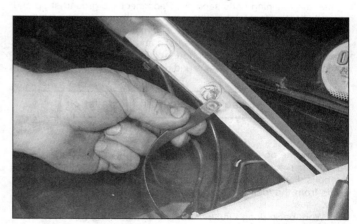

8.2b . . . and free the ground strap from the left-hand side of the hood

8   Hood - removal, installation and adjustment

Removal

Refer to illustrations 8.2a, 8.2b, 8.3a and 8.3b

1   Open the hood and have an assistant support it. Using a pencil or felt tip pen, mark the outline of each hood hinge relative to the hood, to use as a guide on installation.
2   Undo the retaining nut and free the ground strap from the left-hand hood retaining bolt **(see illustrations)**.

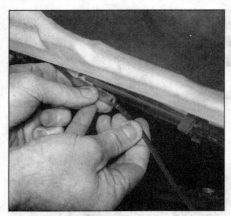

8.3a Disconnect the washer hose from the windshield jets . . .

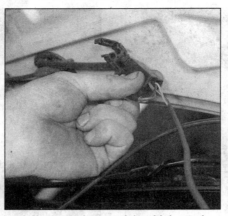

8.3b . . . and on models with heated washer jets also disconnect the wiring connectors

10.2 Loosen and remove the lock retaining bolts (A) and the support strut bolt (B) . . .

3   Disconnect the washer hose from the windshield washer jets and, where necessary, disconnect the wiring from the jet heating elements **(see illustrations)**.

4   Undo the hood retaining bolts and, with the help of an assistant, carefully lift the hood clear. Store the hood out of the way in a safe place.

5   Inspect the hood hinges for signs of wear and free play at the pivots, and if necessary replace. Each hinge is secured to the body by two bolts; note that one of the left-hand hinge retaining bolts will have the ground strap attached to it. Mark the position of the hinge on the body then undo the retaining bolts and remove it from the vehicle. On installation, align the new hinge with the marks and securely tighten the retaining bolts.

Installation and adjustment

6   With the aid of an assistant, install the hood and loosely install the retaining bolts. Align the hinges with the marks made on removal, then tighten the retaining bolts securely. Reconnect the ground strap and securely tighten its retaining nut.

7   Close the hood, and check for alignment with the adjacent panels. If necessary, loosen the hinge bolts and re-align the hood to suit. Once the hood is correctly aligned, securely tighten the hinge bolts and check that the hood fastens and releases satisfactorily.

9   Hood release cable - removal and installation

Removal

1   Mark the position of the hood lock on the crossmember with a suitable marker pen then loosen and remove the two hood lock retaining bolts. Free the lock from the crossmember then release the outer cable from the lock lever and detach the inner cable from the lock body.

2   Work back along the length of the cable, noting its correct routing, and free it from the retaining clips and ties. Tie a length of string to the end of the cable.

3   From inside the vehicle, loosen and remove the screws securing the hood release handle to the vehicle.

4   Release the cable grommet from the firewall and withdraw the lever and cable assembly. Once the cable is free, untie the string and leave it in position in the vehicle; the string can then be used to draw the new cable back into position.

Installation

5   Tie the inner end of the string to the end of the cable, then use the string to draw the hood release cable through into the engine compartment. Once the cable is through, untie the string.

6   Maneuver the hood release lever back into position, and securely

10.3 . . . and remove the support strut from the lock

tighten its retaining screws. Seat the rubber grommet in the firewall.

7   Ensure that the cable is correctly routed, and secured to all the relevant retaining clips.

8   Install the hood lock as described in Section 10.

10   Hood lock - removal and installation

Removal

Refer to illustrations 10.2, 10.3, 10.4a and 10.4b

1   Open up the hood then, using a suitable screwdriver, carefully release the radiator grille upper and lower retaining lugs then move the grille forwards and away from the vehicle.

2   Using a suitable marker pen, mark the outline of the hood lock on the crossmember then loosen and remove the two lock retaining bolts **(see illustration)**.

3   Undo the retaining bolt and slide the lock support strut out from the hood crossmember **(see illustration)**.

4   Free the release outer cable from the lock lever then detach the inner cable from the lock bracket and remove the lock from the vehicle **(see illustrations)**.

Installation

5   Before installation, remove all traces of old locking compound from the hood retaining bolts and their threads in the body.

6   Locate the hood release inner cable in the lock bracket and reconnect the outer cable to the lever. Seat the lock on the crossmember.

10.4a Detach the release cable from the rear of the lock . . .

10.4b . . . and remove the hood lock from the vehicle

11.2a Unscrew the wiring connector . . .

11.2b . . . and disconnect the wiring connector and (where necessary) the central locking vacuum line (arrow) from the pillar

11.3 Unscrew the nut and remove the pivot bolt (arrow) securing the check link to the pillar

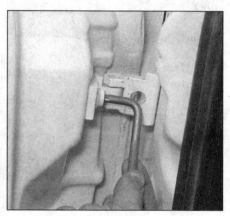

11.4 Unscrew the bolts securing the hinge to the door

7   Apply a suitable locking compound (VW recommends the use of locking fluid D 185 400 A2 - available from your VW dealer) to the threads of the lock retaining bolts.

8   Align the lock with the marks made prior to removal then install the bolts and tighten them to the specified torque setting.

9   Install the support strut and securely tighten its retaining bolts.

10   Clip the radiator grille into position and check that the lock operates smoothly, without any sign of undue resistance. Check that the hood fastens and releases satisfactorily. If adjustment is necessary, loosen the hood lock retaining bolts, and adjust the position of the lock to suit. Once the lock is operating correctly, tighten its retaining bolts to the specified torque.

11   Door - removal, installation and adjustment

Removal

Refer to illustrations 11.2a, 11.2b, 11.3 and 11.4

1   Disconnect the battery negative terminal. **Caution:** *If the stereo in your vehicle is equipped with an anti-theft system, make sure you have the correct activation code before disconnecting the battery.*

2   Open up the door and disengage the wiring boot from the door pillar then rotate the connector counterclockwise and disconnect it from the pillar. On models equipped with central locking also disconnect the vacuum line which passes through the connector **(see illustrations).**

3   Loosen and remove the nut and pivot bolt securing the check link to the pillar **(see illustration).**

4   Draw around the outline of the hinge. Have an assistant support the door then loosen and remove the bolts securing the hinges to the door and remove the door from the vehicle **(see illustration).**

5   Examine the hinges for signs of wear or damage. If replacement is necessary, mark the position of the hinge(s) then undo the retaining bolts and remove them from the vehicle. Install the new hinge(s) and align with the marks made before removal and lightly tighten the retaining bolts.

Installation

6   On models where grub bolts are installed to the hinges, apply a smear of multi-purpose grease to the hinge pins, then, with the aid of an assistant, install the door to the vehicle. Once the door is correctly positioned, tighten the grub bolts to the specified torque.

7   With the aid of an assistant, install the door to the vehicle and install the hinge retaining bolts. Align the hinges with the marks made before removal and tighten the retaining bolts to the specified torque.

8   Align the check link with its bracket and install the pivot bolt and nut, tightening it to the specified torque setting.

9   Reconnect the door wiring connector, making sure it is correctly reconnected, and secure it in position. Where necessary, also reconnect the central locking hose making sure the hose connection is pushed firmly together so that the end of the hose aligns with the colored line.

11

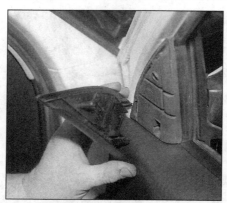

12.2 Unclip the exterior mirror trim panel and remove it from the door . . .

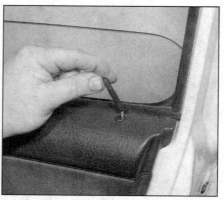

12.3 . . . and unscrew the lock operating knob from its rod

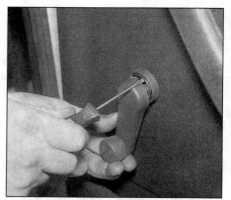

12.4a On models with manual windows, slide the spacer directly away from the handle as shown . . .

12.4b . . . then remove the handle and spacer from the door

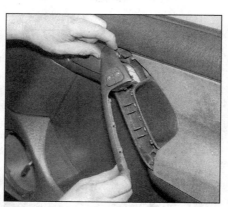

12.5a Unclip the armrest handle cover from the door . . .

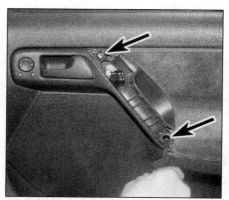

12.5b . . . then loosen and remove the door inner trim panel retaining screws (arrows)

10　Fold the rubber boot back into position making sure it is correctly located on the pillar.

11　Check the door alignment and, if necessary, adjust then reconnect the battery negative terminal. If the paint work around the hinges has been damaged, paint the affected area with a suitable touch-in brush to prevent corrosion.

Adjustment

12　Close the door and check the door alignment with surrounding body panels. If necessary, slight adjustment of the door position can be made by loosening the hinge retaining bolts and repositioning the hinge/door as necessary. Once the door is correctly positioned, tighten the hinge bolts to the specified torque. If the paint work around the hinges has been damaged, paint the affected area with a suitable touch-in brush to prevent corrosion.

12　Door inner trim panel - removal and installation

Removal

Front door

Refer to illustrations 12.2, 12.3, 12.4a, 12.4b, 12.5a, 12.5b, 12.6, 12.7a and 12.7b

1　Disconnect the battery negative terminal then open the door. **Caution:** *If the stereo in your vehicle is equipped with an anti-theft system, make sure you have the correct activation code before disconnecting the battery.*

2　Carefully pry out and remove the exterior mirror inner trim panel **(see illustration)**.

3　Unscrew the door lock inner operating knob from its rod **(see illustration)**.

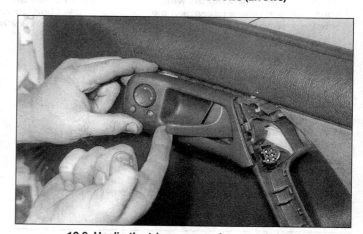

12.6 Unclip the trim cover and remove it from around the door interior handle

4　On models with manual windows, slide the spacer directly away from the regulator handle, to release the retaining clip. Pull the handle off the spindle, and remove the regulator spacer **(see illustrations)**.

5　Carefully unclip the upper trim cover from the door armrest handle and remove it from the vehicle, where necessary, disconnecting the wiring connector as it becomes accessible. Loosen and remove the screws securing the armrest to the door **(see illustrations)**.

6　Unclip the trim cover from around the door interior handle, disconnecting the wiring connector (where necessary), as the cover is removed **(see illustration)**.

7   Release the door trim panel studs, carefully levering between the panel and door with a flat-bladed screwdriver. Work around the outside of the panel, and when all the studs are released, ease the panel away from the door, disconnecting the wiring connector from the speaker as it becomes accessible **(see illustrations)**.

Rear door

8   Disconnect the battery negative terminal then remove the trim panel as described in Steps 3 to 7. **Caution:** *If the stereo in your vehicle is equipped with an anti-theft system, make sure you have the correct activation code before disconnecting the battery.*

Installation

9   Installation of the trim panel is the reverse of removal. Before installation, check whether any of the trim panel retaining studs were broken on removal, and replace them as necessary.

---

13   Door handle and lock components - removal and installation

---

Removal

Interior door handle

Refer to illustration 13.2
1   Remove the door inner trim panel as described in Section 12.
2   Release the handle lower retaining clip with a suitable screwdriver then slide the handle out of the door in a forwards direction and free it from the end of the link rod **(see illustration)**.

Exterior door handle

Refer to illustrations 13.4a, 13.4b, 13.5a and 13.5b
Note: *This task can be performed with the door inner trim panel in*

12.7a Release the inner trim panel from its retaining clips and remove it from the door . . .

position.
3   If work is being carried out on the front door, insert the key into the lock.
4   Loosen and remove the handle retaining bolt from the rear edge of the door then move the handle assembly forwards and pivot it out of position. On the front door, as the handle is being removed, rotate the key through 90° to disengage the handle from the lock operating lever **(see illustrations)**.
5   Recover the handle seals and the lock retaining clip and inspect them for signs of damage or deterioration; renewing them if necessary **(see illustrations)**. **Note:** *Do not drop the clip into the door; if the clip is dropped it will be necessary to remove the inner trim panel to recover it.*

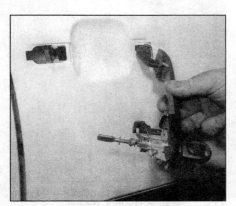

12.7b . . . disconnecting the speaker wiring as it becomes accessible

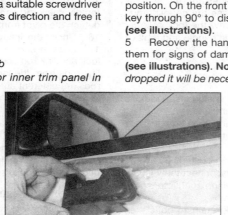

13.2 Release the retaining clip and detach the interior handle from the door and link rod

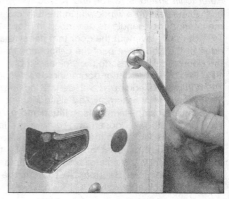

13.4a Loosen and remove the retaining bolt . . .

13.4b . . . then pivot the exterior handle assembly out from the door

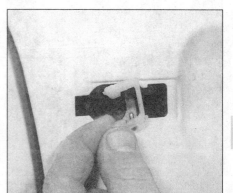

13.5a Recover the rubber seals from the handle . . .

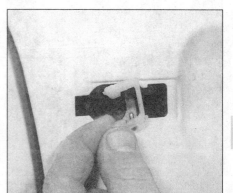

13.5b . . . and remove the handle retaining clip from the door

11

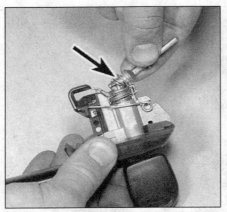

13.7 Unhook the connecting rod from the rear of the lock cylinder and recover its spring (arrow)

13.8a Remove the coupling and spring assembly . . .

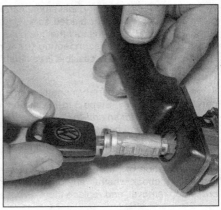

13.8b . . . then withdraw the lock cylinder from the handle . . .

Front door lock cylinder

Refer to illustrations 13.8a, 13.8b and 13.8c

6   Remove the exterior door handle as described in Steps 3 to 5.

7   With the key in the lock, unhook the connecting rod from the rear of the lock cylinder and recover the spring **(see illustration)**.

8   Noting their correct installed position, release the coupling and spring from the rear of the cylinder then withdraw the lock cylinder from the handle. Recover the sealing ring from the handle and replace it if it is damaged **(see illustrations)**.

Front door lock

Refer to illustrations 13.13a and 13.13b

9   Ensure that the window is in the fully closed position then remove the interior door handle as described in Steps 1 and 2.

10   Carefully lever out the door trim panel retaining clips from the rear edge of the door then peel the cellophane insulating panel away from the door to gain access to the lock assembly.

11   Remove the exterior door handle as described in Steps 3 to 5.

12   On models equipped with central locking, disconnect the vacuum line from the lock assembly and disconnect the wiring connectors from the central locking element **(see illustration)**.

13   On all models, loosen and remove the lock retaining bolts then disengage the lock from the link rod and remove it from the door **(see illustrations)**. Note that on some models it may be necessary to loosen the window regulator retaining bolts and release the guide rail from the door (see Section 14) to gain the clearance required to remove the lock assembly.

Rear door lock

Refer to illustrations 13.14, 13.15a, 13.15b, 13.16, 13.17a and 13.17b

14   Carry out the operations described above in Steps 9 to 12 **(see illustration)**.

15   Press out the retaining pin from the center of the interior lock button pivot link rod and free the pivot from the door. Recover the pin and detach the pivot from the link rod **(see illustrations)**.

16   Unclip the link rod guide clips from the door **(see illustration)**.

17   Loosen and remove the lock retaining bolts and maneuver the lock and link rod assembly out from the door. If necessary, detach the link rods from the lock noting their correct installed locations; the link rods are different and must not be interchanged **(see illustrations)**.

Installation

Interior door handle

18   Engage the handle with the link rod and clip it back into position. Make sure the handle operates correctly then install the trim panel as described in Section 12.

Exterior door handle

19   Install the lock retaining clip to the door and fit the seals to the rear of the handle. **Note:** *Do not drop the clip into the door; if the clip is dropped it will be necessary to remove the inner trim panel to recover it.*

20   Hook the lock front pivot into position then clip the rear of the handle into position. On the front door, rotate the key through 90° as the handle clips it into position, this will engage the handle with the

13.8c . . . along with its sealing ring

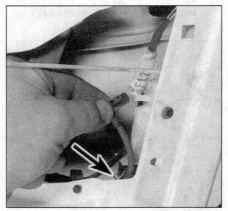

13.12 On models with central locking, disconnect the vacuum line and wiring connector (arrow) from the element

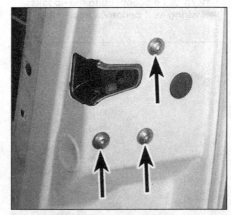

13.13a Loosen and remove the three retaining bolts (arrows) . . .

13.13b . . . and maneuver the lock assembly out of position

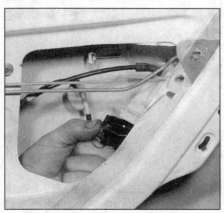
13.14 Disconnecting the rear door lock central locking element wiring connector

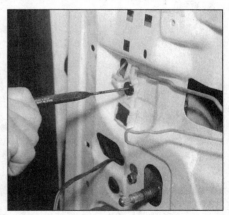
13.15a Press out the retaining pin . . .

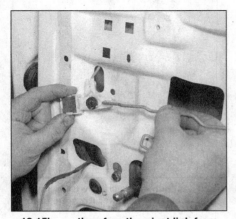
13.15b . . . then free the pivot link from the door and detach it from the link rod

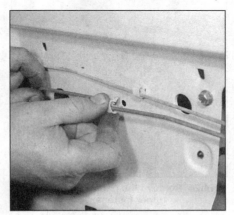
13.16 Remove the guide clips securing both lock link rods to the door

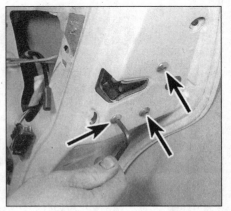
13.17a Undo the retaining bolts (arrows) . . .

lock operating lever.
21   Check the operation of the handle then install the retaining bolt and tighten it to the specified torque setting.

Front door lock cylinder
Refer to illustration 13.24
22   Lubricate the outside of the lock cylinder and the locking plates with a suitable lubricant (VW recommend the use of grease G 000 400

- available from your VW dealer).
23   Install the sealing ring to the handle and insert the cylinder.
24   Install the coupling and spring to the cylinder, making sure they are correctly located, then check the operation of the lock cylinder **(see illustration)**.
25   Install the spring to the connecting rod and hook the rod onto the rear of the lock cylinder.
26   Install the exterior handle as described in Steps 19 to 21.

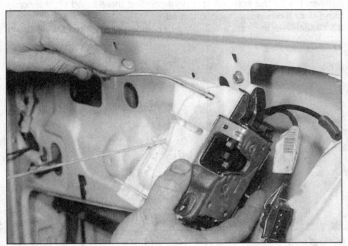
13.17b . . . and remove the lock and link rod assembly from the rear door. The link rods can then be detached from the lock

13.24 On installation, make sure the coupling spring ends are correctly located over the handle tab (arrow)

11

13.31a Remove the rubber plug from the rear edge of the door . . .

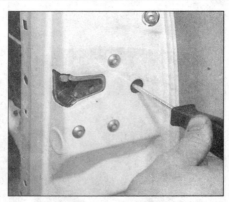

13.31b . . . and adjust the lock as described in text

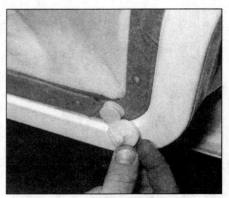

14.2a Slide the door panel clips off from the fasteners . . .

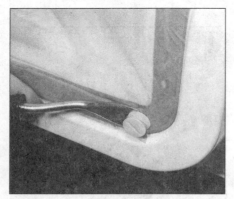

14.2b . . . then carefully pry the fasteners out from the door

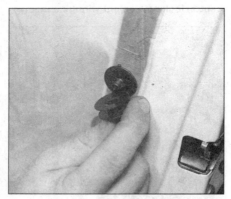

14.2c Remove the panel clip from the upper fastener . . .

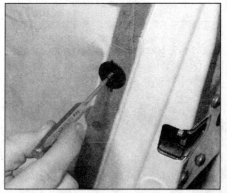

14.2d . . . then press out the retaining pin and unclip the fastener from the door

Front door lock

Refer to illustrations 13.31a and 13.31b

27   Before installation, loosen the lock adjusting (Torx) screw; on the left-hand door the screw has a right-handed thread and on the right-hand door it has a left-handed thread.

28   Maneuver the lock assembly into position and engage it with the link rod.

29   Install the lock retaining bolts and tighten them to the specified torque setting. Where necessary, reconnect the vacuum line and wiring connector(s) (as applicable) to the lock assembly.

30   Install the exterior handle as described in Steps 19 to 21.

31   Remove the rubber plug from the door to gain access to the lock adjustment screw. Tighten the screw to 3 Nm (24 in-lbs) (see Step 27) then install the rubber plug **(see illustrations)**.

32   Where necessary, install the regulator guide rail retaining bolts and adjust as described in Section 14.

33   Check the operation of the lock and handle then press the cellophane insulating panel back onto the door. Press the trim panel retaining clips back into position.

34   Install the interior door handle as described in Step 18.

Rear door lock

35   Install the link rods to the lock making sure they are correctly refitted.

36   Before installation, loosen the lock adjusting (Torx) screw; on the left-hand door the screw has a right-handed thread and on the right-hand door it has a left-handed thread.

37   Maneuver the lock assembly into position and tighten the retaining

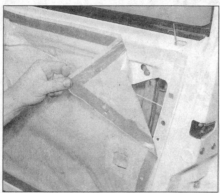

14.2e Once all the clips and fasteners have been removed, peel the cellophane insulating panel away from the door

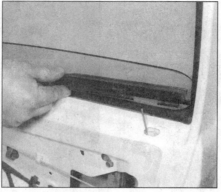

14.4 Unclip the inner sealing strip from the top of the door

14.5a Loosen the window clamp nuts . . .

bolts to the specified torque setting. Where necessary, reconnect the vacuum line and wiring connector(s) (as applicable) to the lock assembly.
38   Attach the link rod to the pivot and clip the pivot into the door. Secure the pivot in position with the retaining pin.
39   Carry out the operations described in Steps 30 to 34 ignoring the remark about the regulator bolts.

14 Door window glass and regulator - removal and installation

Removal

Refer to illustrations 14.2a through 14.2e
1   Remove the interior door handle as described in Section 13.
2   Carefully lever out the door trim panel retaining clips from the door. Carefully peel the cellophane insulating panel away from the door and remove the panel. If the panel is ripped or damaged a new one must be used on installation; any damaged trim clips must also replaced **(see illustrations)**. Continue as described under the relevant sub-heading.

Front door window glass

Refer to illustrations 14.4, 14.5a and 14.5b
3   Position the window glass so the glass clamps on the regulator mechanism are accessible through the door panel cutaways.
4   Carefully ease the window inner sealing strip out from the top edge of the door **(see illustration)**.
5   Loosen the window clamp nuts and release the clamps from the glass then carefully maneuver the window glass out through the top of the door **(see illustrations)**.

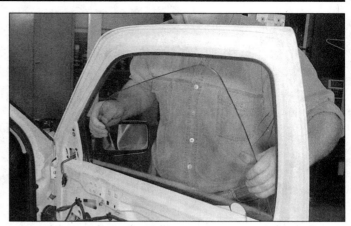
14.5b . . . then free the glass and maneuver it out through the top of the door

Rear door window glass

Refer to illustrations 14.7a, 14.7b, 14.7c, 14.8a and 14.8b
6   Carry out the operations described in Steps 3 and 4.
7   Lower the window then undo the upper and lower retaining screws and remove the window guide rail from the door **(see illustrations)**.
8   Loosen the window clamp nuts then carefully maneuver the glass out through the top of the door **(see illustrations)**.
9   If necessary, the fixed window can then be disengaged from the sealing strip and removed from the door.

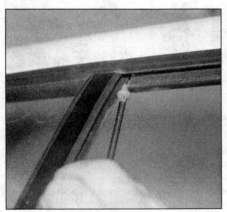
14.7a Undo the upper retaining screw . . .

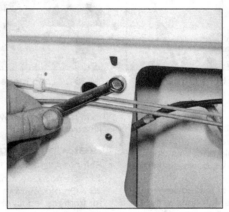
14.7b . . . and the lower retaining bolt . . .

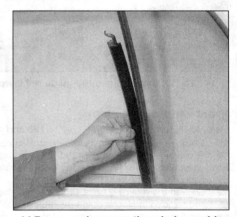
14.7c . . . and remove the window guide rail from the rear door

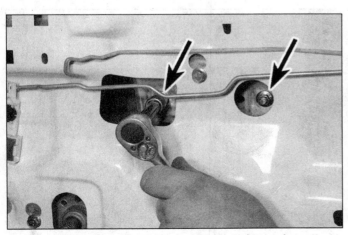
14.8a Loosen the windows clamp nuts (arrows) . . .

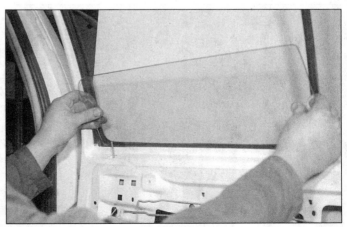
14.8b . . . and maneuver the glass out of the top of the rear door

11

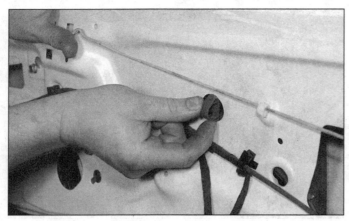

14.11 Release the retaining clip securing the regulator cables to the front door

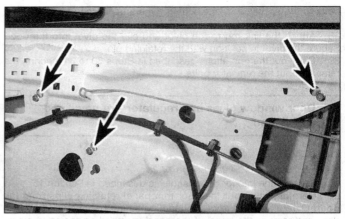

14.12a Loosen the regulator and guide rail upper retaining bolts (arrows) . . .

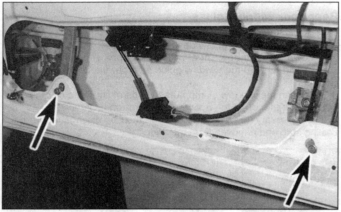

14.12b . . . then remove the guide rail lower bolts (arrows) . . .

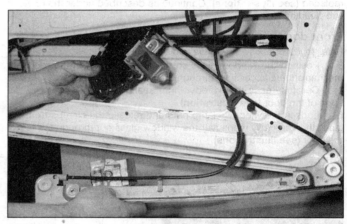

14.13 . . . and maneuver the regulator assembly out from the door

Front window regulator

Refer to illustrations 14.11, 14.12a, 14.12b and 14.13

10   Remove the window glass as described earlier.

11   Release the retaining clip and free the regulator cables from the door **(see illustration)**. On models with electric windows disconnect the wiring connector from the regulator motor.

12   Loosen the retaining bolts situated at the top of the regulator guide rails and remove the lower guide rail retaining bolts. Also loosen the bolt securing the regulator mechanism to the door **(see illustrations)**.

13   Lift the regulator mechanism slightly to disengage it from the door

then maneuver it downwards and out through the door aperture **(see illustration)**.

Rear window regulator

Refer to illustrations 14.16a and 14.16b

14   Remove the window glass as described earlier.

15   On models with electric windows, disconnect the wiring connector from the regulator motor.

16   Loosen the regulator and guide rail retaining bolts then maneuver the regulator out through the door aperture **(see illustrations)**.

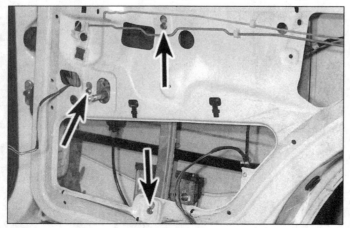

14.16a Loosen the retaining bolts (arrows) . . .

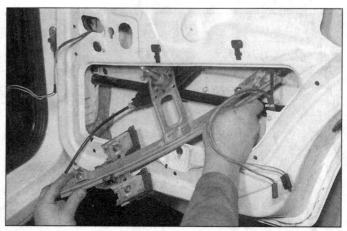

14.16b . . . and remove the regulator assembly from the rear door

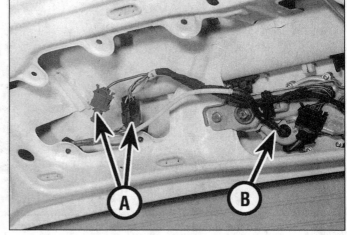

15.2 Remove the trim panel retaining screw

15.4 Disconnect the wiring connectors (A) and detach the washer hose (B) from the wiper motor

Installation

Front door window glass

17   Maneuver the window glass into position and engage it with the regulator clamps. Make sure the glass is correctly seated then lightly tighten the regulator clamp nuts.

18   Install the inner sealing strip to the top of the door.

19   Check that the window glass moves smoothly and easily and closes fully. If necessary, loosen the regulator clamp nuts then reposition the glass as necessary. Once the window operation is correct, tighten the clamp nuts to the specified torque.

20   Once the window is operating correctly, press the cellophane insulating panel back into position, making sure it is correctly seated, and install the trim panel clips. Install the inner trim panel as described in Section 12.

Rear door window glass

21   Where necessary, ease the fixed window into position making sure it is correctly seated in the sealing strip.

22   Maneuver the glass into position, engage it with the regulator clamps and lightly tighten the clamp nuts.

23   Install the window guide rail to the door, engaging it with the glass, and tighten the retaining screws securely.

24   Carry out the operations described in Steps 19 and 20.

Front window regulator

25   Maneuver the regulator into position through the door aperture then install the retaining bolts and tighten all its fixings to the specified torque. Where necessary, reconnect the wiring connector to the regulator motor.

26   Clip the regulator cables into position then install the glass as described above.

Rear window regulator

27   Maneuver the regulator into position through the door aperture then install the retaining bolts and tighten them to the specified torque. Where necessary, reconnect the wiring connector to the regulator motor.

28   Install the glass as described above.

15 Liftgate and support struts - removal and installation

Removal

Liftgate

Refer to illustrations 15.2, 15.4 and 15.7

1   Open up the liftgate then disconnect the battery negative termi-

nal. **Caution:** *If the stereo in your vehicle is equipped with an anti-theft system, make sure you have the correct activation code before disconnecting the battery.*

2   Loosen and remove the liftgate trim panel retaining screw **(see illustration)** then release the trim panel clips, carefully levering between the panel and liftgate with a flat-bladed screwdriver.

3   Work around the outside of the panel, and when all the clips are released, remove the panel.

4   Disconnect the wiring connectors situated behind the trim panel and free the washer hose from the liftgate wiper motor **(see illustration)**. Also disconnect the wiring connectors from the heated rear window terminals and free the wiring grommets from the liftgate.

5   Tie a piece of string to each end of the wiring then, noting the correct routing of the wiring harness, release the harness rubber grommets from the liftgate and withdraw the wiring. When the end of the wiring appears, untie the string and leave it in position in the liftgate; it can then be used on installation to draw the wiring into position.

6   Using a suitable marker pen, draw around the outline of each hinge marking its correct position on the liftgate.

7   Have an assistant support the liftgate, then using a small flat-bladed screwdriver raise the spring clips and pull the support struts off their balljoint mountings on the liftgate. Loosen and remove the bolts securing the hinges to the liftgate and remove the liftgate from the vehicle **(see illustration)**. Where necessary, recover the gaskets which are installed between the hinge and liftgate.

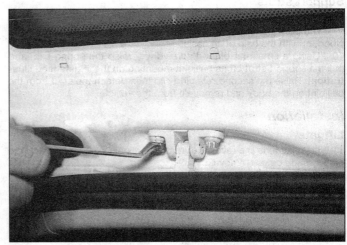

15.7 Unscrew the bolts securing the hinges to the liftgate and remove the liftgate from the vehicle

11

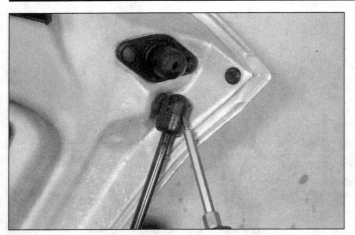

15.10 Carefully lift the spring clip and free the support strut from the liftgate

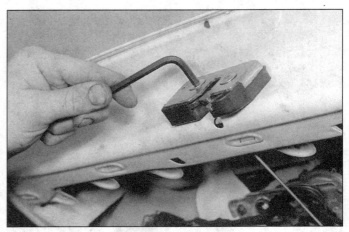

16.1a Undo the retaining screws . . .

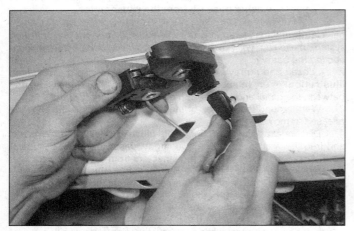

16.1b . . . and remove the lock from the liftgate, disconnecting it from the wiring connector and link rod

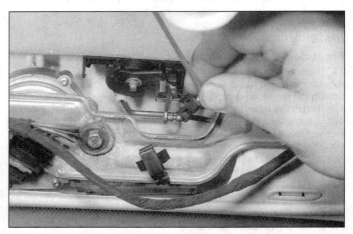

16.3 Release the retaining clips and detach the link rods from the lock button

8   Inspect the hinges for signs of wear or damage and replace if necessary. The hinges are secured to the vehicle by nuts or bolts (depending on model) which can be accessed once the headlining has been freed from the trim strip and peeled back. On installation ensure that the hinge gasket is in good condition and secure the hinge in position.

Support struts
Refer to illustration 15.10
9   Support the liftgate in the open position, using a stout piece of wood, or with the help of an assistant.
10   Using a small flat-bladed screwdriver raise the spring clip, and pull the support strut off its balljoint mounting on the liftgate (**see illustration**). Raise the second retaining clip then detach the strut from the balljoint on the body and remove it from the vehicle.

Installation

Liftgate
11   Installation is the reverse of removal, aligning the hinges with the marks made before removal.
12   On completion, close the liftgate and check its alignment with the surrounding panels. If necessary slight adjustment can be made by loosening the retaining bolts and repositioning the liftgate on its hinges.

Support struts
13   Installation is a reverse of the removal procedure, ensuring that the strut is securely retained by its retaining clips.

16 Liftgate lock components - removal and installation

Removal

Liftgate lock
Refer to illustrations 16.1a and 16.1b
1   Open up the liftgate, then undo the lock retaining screws. Detach the lock link rod from the button then remove the lock, disconnecting its wiring connector as it is withdrawn (**see illustrations**).

Liftgate lock button
Refer to illustrations 16.3, 16.4a, 16.4b and 16.5
2   Remove the liftgate trim panel as described in Step 2 of Section 15.
3   Disconnect the link rod(s) from the handle assembly and, where necessary, disconnect the wiring connector from the switch (**see illustration**).
4   From outside the liftgate, loosen and remove the retaining screws securing the liftgate lock button trim panel in position and remove it from the liftgate (**see illustrations**).
5   Release the retaining clips and remove the lock button from the liftgate (**see illustration**).

Liftgate lock cylinder
Refer to illustrations 16.7a, 16.7b and 16.7c
6   Remove the liftgate lock button as described earlier.
7   Insert the key into the lock then carefully pry off the retaining clip

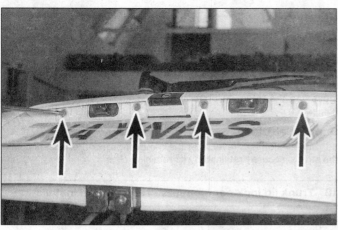

16.4a Undo the retaining screws (arrows) . . .

16.4b . . . and remove the lock button trim panel
from the liftgate . . .

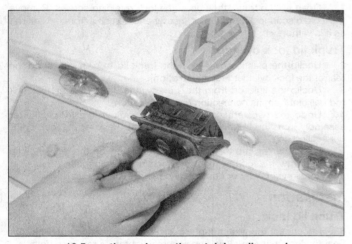

16.5 . . . then release the retaining clips and
withdraw the lock button

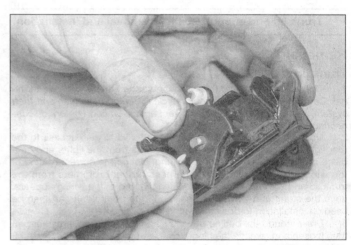

16.7a Remove the retaining clip from the rear of the
lock button . . .

16.7b . . . then lift off the link rod cam . . .

16.7c . . . and withdraw the lock cylinder and sealing ring (arrow)

and remove the link rod cam, noting its correct installed location. Withdraw the lock cylinder from the button and recover its sealing ring.
Inspect the sealing ring for signs of wear or damage and replace if necessary (see illustrations).

Installation

8   Installation is a reversal of the relevant removal procedure. Before
installing the trim panel, check the operation of the lock components
and (where necessary) the central locking system.

11

17.5 Carefully lift the retaining clip and detach the support strut from its upper mounting

17   Trunk lid and support struts - removal and installation

Removal

Trunk lid

1   Open up the trunk lid then disconnect the battery negative terminal. **Caution:** *If the stereo in your vehicle is equipped with an anti-theft system, make sure you have the correct activation code before disconnecting the battery.*

2   Unclip the plastic covers from the trunk lid to gain access to the rear lights. Disconnect the wiring connectors from the lights and tie a piece of string to each end of the wiring. Noting the correct routing of the wiring harness, release the harness rubber grommets from the trunk lid and withdraw the wiring. When the end of the wiring appears, untie the string and leave it in position in the trunk lid; it can then be used on installation to draw the wiring into position.

3   Draw around the outline of each hinge with a suitable marker pen then loosen and remove the hinge retaining bolts and remove the trunk lid from the vehicle.

4   Inspect the hinges for signs of wear or damage and replace if necessary; the hinges are secured to the vehicle by bolts.

Support struts

Refer to illustration 17.5

5   Support the trunk lid in the open position. Using a small flat-bladed screwdriver raise the spring clip, and pull the support strut off its upper mounting **(see illustration)**. Repeat the procedure on the lower strut mounting and remove the strut from the vehicle.

Installation

Trunk lid

6   Installation is the reverse of removal, aligning the hinges with the marks made before removal.

7   On completion, close the trunk lid and check its alignment with the surrounding panels. If necessary slight adjustment can be made by loosening the retaining bolts and repositioning the trunk lid on its hinges.

Support struts

8   Installation is a reverse of the removal procedure, ensuring that the strut is securely retained by its retaining clips.

18   Trunk lid lock components - removal and installation

Removal

Trunk lid lock

Refer to illustrations 18.2, 18.3 and 18.4

1   Open up the trunk, then undo the lock retaining screws. Remove the lock, disconnect its wiring connector and detach it from the link rod as it is withdrawn.

Trunk lid lock cylinder

2   Unclip the plastic cover from the trunk lid to gain access to the rear of the lock cylinder **(see illustration)**.

3   Unclip the link rod from the lock cylinder then free the lock link rod balljoint from its connection **(see illustration)**.

4   Undo the two retaining screws and remove the lock cylinder assembly from the trunk lid **(see illustration)**. Recover the lock cylinder sealing ring.

5   Insert the key into the lock then carefully pry off the retaining clip and withdraw the lock cylinder and sealing ring. Inspect the sealing rings for signs of wear or damage and replace if necessary.

Installation

Trunk lid lock

6   Reconnect the wiring connector and securely attach the link rod. Seat the lock in the trunk lid and securely tighten its retaining bolts.

Trunk lid lock cylinder

7   Install the sealing ring to the lock cylinder and slide it into position in the housing. Secure the cylinder in position with the retaining clip then check the operation of the lock cylinder assembly.

8   Install the sealing ring to the lock cylinder assembly. Insert the assembly into the trunk lid and install the retaining screws tightening them securely.

9   Clip the link rod balljoint back onto the lock cylinder and securely reconnect the link rod. Check the operation of the lock assembly then install the plastic cover to the trunk lid.

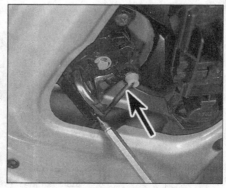

18.2 On Sedan models, unclip the plastic cover from the trunk lid to gain access to the lock assembly

18.3 Pry the lock link rod balljoint off from the lock and disconnect the link rod (arrow) . . .

18.4 . . . then undo the two retaining screws (arrows) and remove the lock cylinder

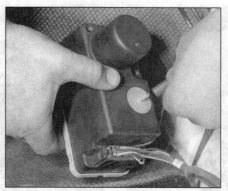

19.3a On Hatchback models, remove the insulation packing from around the central locking pump . . .

19.3b . . . then disconnect the vacuum line and wiring connector and remove the pump from the vehicle

19.5a Remove the retaining screw, then unclip the positioning element from the door lock . . .

19 Central locking components - removal and installation

Removal

Central locking pressure pump

Refer to illustrations 19.3a and 19.3b

1   The central locking operating pump is located in the luggage compartment; on Hatchback models it is located on the right-hand side and on Sedan models it is on the left-hand side. Before removal disconnect the battery negative terminal. **Caution:** *If the stereo in your vehicle is equipped with an anti-theft system, make sure you have the correct activation code before disconnecting the battery.*

2   Unhook the retaining strap then free the pump from the body.

3   Remove the insulation packing from around the pump then disconnect the wiring connector and vacuum line from the pump and remove the pump from the vehicle **(see illustrations)**.

Door lock positioning element

Refer to illustrations 19.5a and 19.5b

4   Remove the door lock as described in Section 13.

5   Turn the lock latch to the "locked" position then loosen and remove the positioning element retaining screw. Release the retaining clips and remove the positioning element from the lock, noting how the element plunger is engaged with the lock lever **(see illustrations)**.

Front door lock microswitch

Refer to illustration 19.7

6   Remove the door lock positioning element as described in Steps 4 and 5.

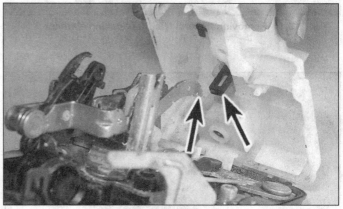

19.5b . . . noting how the element plunger is engaged with the lock lever (arrows)

7   Release the retaining clips and slide out the microswitch **(see illustration)**.

Liftgate lock positioning element - Hatchback models

Refer to illustration 19.10

8   Remove the liftgate trim panel as described in Steps 1 to 3 of Section 15.

9   Disconnect the vacuum line from the positioning element and free its link rod from the lock/linkage (as applicable).

10   Undo the retaining screws and remove the positioning element and link rod from the liftgate **(see illustration)**.

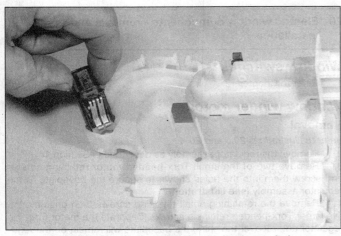

19.7 Removing the front door lock microswitch

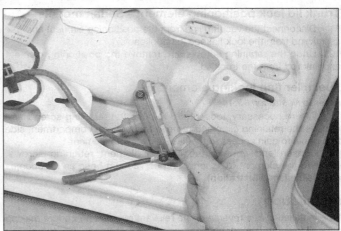

19.10 Removing the liftgate lock positioning element

11

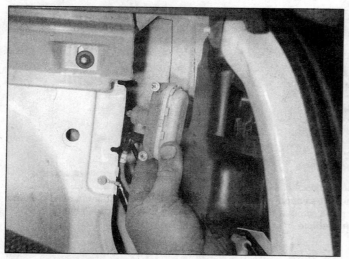

19.14   Removing the fuel filler flap locking element

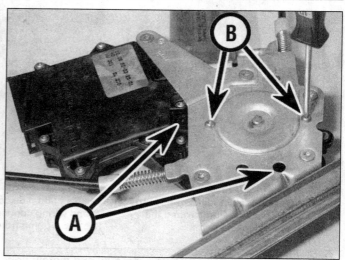

20.3   Unscrew two of the small motor retaining screws (A) and screw them into the locations (B) to secure the baseplate to the regulator

20.4   Loosen and remove the remaining retaining screws and carefully lift the window winder motor off from the regulator assembly

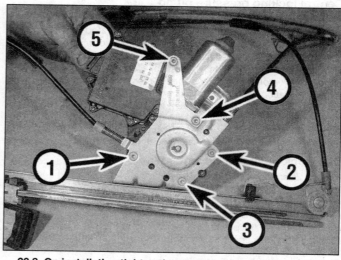

20.8   On installation tighten the motor large retaining screws in the order shown

Trunk lid lock positioning element - Sedan models

11   Disconnect the vacuum line from the positioning element and free its link rod from the lock/linkage (as applicable).

12   Undo the retaining screws and remove the positioning element and link rod from the trunk lid.

Fuel filler flap locking element

Refer to illustration 19.14

13   Where necessary, loosen and remove the retaining screws and remove the retaining clips and remove the luggage compartment side trim panel to gain access to the filler flap locking element.

14   Loosen and remove the locking element retaining screws. Remove the element, disconnecting its vacuum hose as it becomes accessible **(see illustration)**.

Installation

15   Installation is a reverse of the relevant removal procedure making sure all vacuum line connections are securely remade. On completion check the operation of all central locking system components.

20   Electric window components - removal and installation

Window switches

1   Refer to Chapter 12.

Window winder motors

Removal

Refer to illustrations 20.3 and 20.4

2   Remove the regulator assembly as described in Section 14.

3   Remove two of the small Torx-headed motor retaining screws and screw them into the holes shown to secure the baseplate to the regulator assembly **(see illustration)**.

4   Remove the remaining small retaining screws then unscrew the five larger Torx-headed retaining screws. Separate the motor from the regulator and recover the shim; the baseplate will remain on the regulator **(see illustration)**. Ensure that the motor assembly is kept clean.

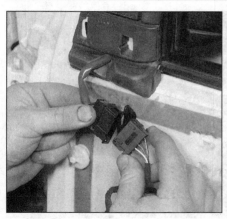

21.4 Disconnect the exterior mirror wiring connector . . .

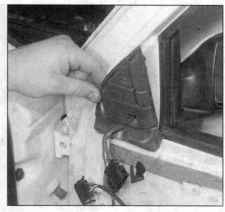

21.5a . . . and remove the insulation panel from the door

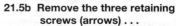

21.5b Remove the three retaining screws (arrows) . . .

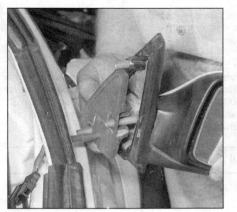

21.5c . . . and lift the mirror and rubber seal assembly away from the door

21.7 Unclip the glass from the mirror and (where necessary) disconnect its wiring connectors (arrows)

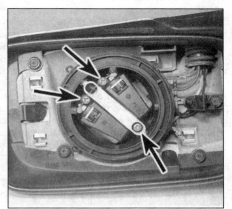

21.10 Exterior mirror motor retaining screws (arrows)

Installation

Refer to illustration 20.8

5   If a new motor is being installed remove the fitting cover.
6   Make sure that the motor drive gear components are sufficiently lubricated (VW recommend the use of grease G 000 450 02 - available from your VW dealer) and free from dust and dirt.
7   Ensure that the regulator and baseplate is free from dirt and install the shim to the motor shaft. Carefully align the motor and engage it with the regulator.
8   Install the motor retaining bolts, tightening them loosely only, then unscrew the two bolts used to secure the baseplate in position and install them to the motor. With all bolts loosely in position, go around and securely tighten the larger bolts in the sequence shown (see illustration). Then securely tighten all the smaller retaining bolts.
9   Install the regulator assembly as described in Section 14.

21 Exterior mirrors and associated components - removal and installation

Removal

Manually operated mirror

1   Remove the door inner trim panel as described in Section 12.
2   Remove the insulation from the door frame then undo the retaining screw and free the mirror adjustment mechanism.
3   Loosen and remove the mirror retaining screws and remove the mirror assembly from the door.

Electrically operated mirror

Refer to illustrations 21.4, 21.5a and 21.5b

4   Remove the door inner trim panel as described in Section 12 and disconnect the mirror wiring connector (see illustration).
5   Remove the mirror insulation from the door frame then undo the retaining screws and remove the mirror assembly from the door. Recover the rubber seal from the mirror; the seal must be replaced if it shows signs of damage or deterioration (see illustrations).

Mirror glass

Refer to illustration 21.7

Note: *The mirror glass is clipped onto the motor. Removal of the glass without the VW special tool (number 800-200) is likely to result in breakage of the glass.*

6   Insert a wide plastic or wooden wedge between the mirror glass and mirror housing and carefully pry the glass from the motor. Take great care when removing the glass; do not use excessive force as the glass is easily broken.
7   Remove the glass from the mirror, where necessary, disconnect the wiring connectors from the mirror heating element (see illustration).

Mirror switch (electrically operated mirror)

8   Refer to Chapter 12.

Electrically operated mirror motor

Refer to illustration 21.10

9   Remove the mirror glass as described above.
10 Undo the retaining screws and remove the motor, disconnecting its wiring connector as it becomes accessible (see illustration).

11

Installation

11   Installation is the reverse of the relevant removal procedure.

22 Windshield, liftgate and fixed rear quarter window glass - general information

These areas of glass are secured by the tight fit of the weather-strip in the body aperture, and are bonded in position with a special adhesive. Replacement of such fixed glass is a difficult, messy and time-consuming task, which is beyond the scope of the home mechanic. It is difficult, unless one has plenty of practice, to obtain a secure, waterproof fit. Furthermore, the task carries a high risk of breakage; this applies especially to the laminated glass windshield. In view of this, owners are strongly advised to have this sort of work carried out by one of the many specialist windshield fitters.

23 Sunroof - general information

Due to the complexity of the sunroof mechanism, considerable expertise is needed to repair, replace or adjust the sunroof components successfully. Removal of the roof first requires the headlining to be removed, which is a complex and tedious operation, and not a task to be undertaken lightly. Therefore, any problems with the sunroof should be referred to a VW dealer.

On models with an electric sunroof, if the sunroof motor fails to operate, first check the relevant fuse. If the fault cannot be traced and rectified, the sunroof can be opened and closed manually using an Allen key to turn the motor spindle (a suitable key is supplied with the vehicle, and should be clipped onto the underside of the sunroof motor). To gain access to the motor, unclip the rear of the access cover then slide the cover to the rear to free it from the headlining. Unclip the Allen key then pivot the spindle cover out of the way and insert the Allen key. Rotate the key to move the sunroof to the required position.

24 Body exterior fittings - removal and installation

Wheelarch liners and body under-panels

1   The various plastic covers installed on the underside of the vehicle are secured in position by a mixture of screws, nuts and retaining clips and removal will be fairly obvious on inspection. Work methodically around the panel removing its retaining screws and releasing its retaining clips until the panel is free and can be removed from the underside of the vehicle. Most clips used on the vehicle, except for the fasteners which are used to secure the wheelarch liners in position, are simply pried out of position. The wheelarch liner clips are released by pressing out their center pins and then removing the outer section of the clip; new clips will be required on installation if the center pins are not recovered.

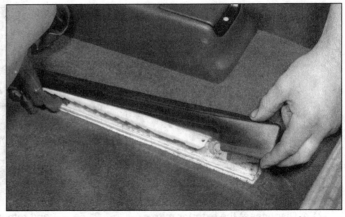

25.1 Removing the seat inner guide rail trim panel

2   On installation, replace any retaining clips that may have been broken on removal, and ensure that the panel is securely retained by all the relevant clips and screws.

Body trim strips and badges

3   The various body trim strips and badges are held in position with a special adhesive tape. Removal requires the trim/badge to be heated, to soften the adhesive, and then cut away from the surface. Due to the high risk of damage to the vehicle's paintwork during this operation, it is recommended that this task should be entrusted to a VW dealer.

25 Seats - removal and installation

Removal

Front seat

Refer to illustrations 25.1, 25.2, 25.3 and 25.4

1   Slide the seat forwards and unclip the trim cover from the seat inner guide rail **(see illustration)**. On some models the trim cover is secured in position by a clip; the clip is released by pressing out its center pin then prying out the outer section; if the center pin is not recovered a new clip will be required on installation.

2   Release the outer seat rail end plug by pulling out its securing wedge and remove the plug from the end of the rail **(see illustration)**.

3   Slide the seat backwards and remove the spring clip from the front of the seat center guide rail **(see illustration)**.

4   Slide the seat fully backwards, disengaging it from the outer guide rails and remove it from the vehicle. Recover the plastic guide pieces from each of the seat guides and replace them if they show signs of damage **(see illustration)**.

25.2 Pull out the retaining wedge and remove the end plug from the rear of the seat outer guide rail

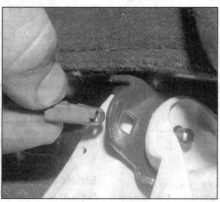

25.3 Remove the spring clip from the front of the seat center guide rail . . .

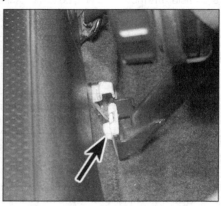

25.4 . . . then slide the seat backwards out of position and recover the guide piece (arrow) from the seat

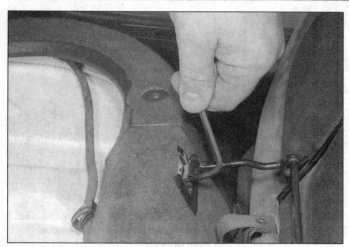

25.5 Unscrew the hinge retaining bolts and remove the seat cushion from the vehicle

Rear seat assembly

Refer to illustrations 25.5, 25.7 and 25.8

5   Lift up the rear seat cushion(s) then loosen and remove the hinge retaining bolts and remove the seat cushion(s) from the vehicle **(see illustration)**.

6   Fold down the rear seat backs.

7   Carefully pry out the retaining clip out from the top of the center hinge pivot **(see illustration)**.

8   Using a small flat-bladed screwdriver, release the outer hinge pivot retaining clip then move the seat cushion upwards to release its pivot pin **(see illustration)**. Disengage the seat back from the center hinge and remove it from the vehicle. Remove the opposite seat back in the same way.

Installation

Front seats

Refer to illustration 25.9

9   Before installation examine the front and rear seat guide pieces for signs of wear or damage and replace if necessary. Installation is a reverse of the removal procedure, ensuring that the seat adjustment lever engages correctly with the center guide locking plunger as the seat is refitted **(see illustration)**.

Rear seat assembly

10   Installation is the reverse of removal making sure the seat backs are clipped securely in position.

26 Front seat belt tensioning mechanism - general information

Most models covered in this manual are equipped with a front seat belt tensioner system. The system is designed to instantaneously take up any slack in the seat belt in the case of a sudden frontal impact, therefore reducing the possibility of injury to the front seat occupants. Each front seat is equipped with its system, the tensioner being situated behind the sill trim panel.

The seat belt tensioner is triggered by a frontal impact above a pre-determined force. Lesser impacts, including impacts from behind, will not trigger the system.

When the system is triggered, the explosive gas in the tensioner mechanism retracts and locks the seat belt through a cable which acts on the inertia reel. This prevents the seat belt moving and keeps the occupant firmly in position in the seat. Once the tensioner has been triggered, the seat belt will be permanently locked and the assembly must be replaced.

There is a risk of injury if the system is triggered inadvertently when working on the vehicle, and it is therefore strongly recommended that any work involving the seat belt tensioner system is entrusted to a VW dealer. Note the following warnings before contemplating any work on the front seat belts.

Warning: *Do not expose the tensioner mechanism to temperatures in excess of 100° C (212° F).*

Warning: *If the tensioner mechanism is dropped, it must be replaced, even it has suffered no apparent damage.*

Warning: *Do not allow any solvents to come into contact with the tensioner mechanism.*

Warning: *Do not attempt to open the tensioner mechanism as it contains explosive gas.*

Warning: *Tensioners must be discharged before they are disposed of, but this task should be entrusted to a VW dealer.*

27 Seat belt components - removal and installation

Removal

Warning: *On models equipped with seat belt tensioners refer to Section 26 before proceeding; under no circumstances should you attempt to separate the tensioner assembly from the inertia reel.*

Front seat belt - four- and five-door models

Refer to illustrations 27.1a, 27.1b, 27.1c, 27.1d, 27.2, 27.3, 27.4, 27.5a, 27.5b, 27.6, 27.8a and 27.8b

1   Unscrew the center screw from the sill trim panel retaining clip then remove the retaining clip. Press down on the top of the front trim panel, to release its lower edge from the sill, then pull the panel

25.7 Pry out the retaining clip from the seat back center hinge pivot . . .

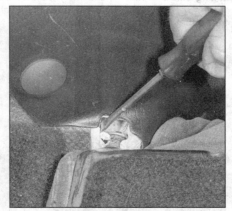

25.8 . . . then release the outer hinge pivot clip and remove the seat back from the vehicle

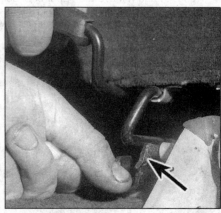

25.9 On installation ensure that the seat adjustment lever engages correctly with the center guide locking plunger (arrow)

11

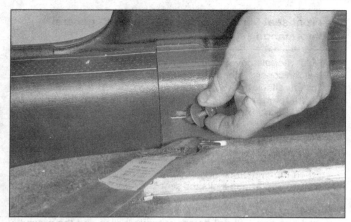

27.1a Unscrew the center screw and unclip the fastener
from the sill trim panel . . .

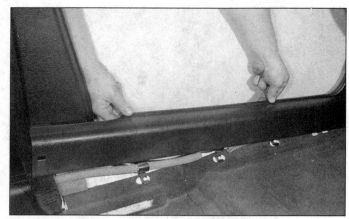

27.1b . . . then unclip the front trim panel and remove
it from the vehicle

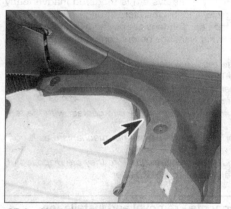

27.1c Unscrew the fasteners and release
the retaining clip (arrow) . . .

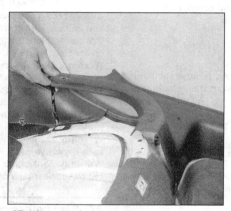

27.1d . . . and remove the rear trim panel
from the sill

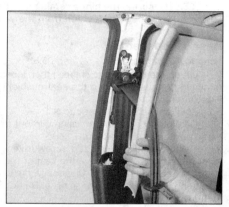

27.2 Unclip the upper trim panel from
the door pillar

upwards and remove it from the vehicle. Fold the rear seat cushion up then unscrew the fasteners securing the rear section of the sill trim panel in position. Working through the aperture in the seat frame, release the retaining clip by forcing it rearwards, and remove the rear section of the sill trim panel (see illustrations).

2   Starting at the bottom of the panel, unclip the door pillar upper trim panel and free it from the pillar (see illustration).

3   Loosen and remove the seat belt lower mounting bolt and free the seat belt from its lower anchorage (see illustration). The upper trim panel can then be removed.

4   Loosen and remove the upper seat belt mounting bolt and free the belt from the door pillar (see illustration).

5   Loosen and remove the door pillar lower trim panel retaining screw then unclip the panel from the pillar (see illustrations).

6   Undo the retaining screws and remove the belt guide from the door pillar (see illustration). The lower trim panel can then be removed.

7   On models with seat belt tensioners, loosen and remove the tensioner assembly retaining nut. This will disable the tensioner making it safe to remove the seat belt assembly.

8   Loosen and remove the inertia reel mounting bolt and remove the seat belt assembly from the vehicle. On models with seat belt tensioners, as the seat belt assembly is removed, release the tensioner assembly from its retaining clip. If necessary, undo the retaining bolt and remove the height adjuster mechanism from the door pillar (see illustrations).

27.3 Unscrew the retaining bolt securing
the seat belt to the floor . . .

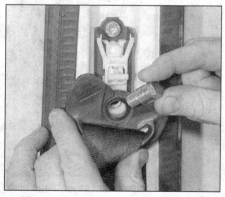

27.4 . . . and the upper retaining bolt
securing the seat belt to the pillar

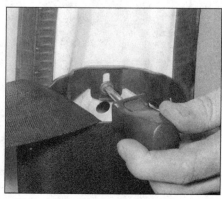

27.5a Unscrew the retaining screw . . .

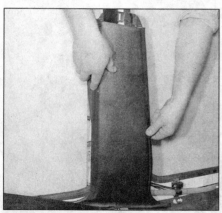

27.5b ... and unclip the lower trim panel from the door pillar

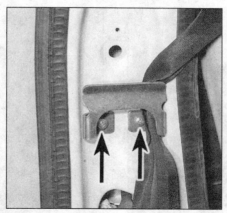

27.6 Undo the retaining screws (arrows) and remove the seat belt guide from the door pillar

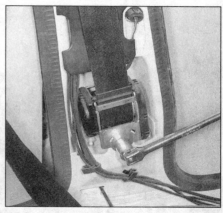

27.8a Loosen and remove the inertia reel retaining bolt and remove the seat belt assembly from the vehicle

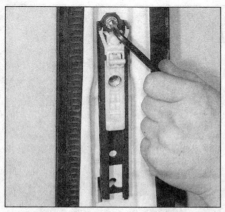

27.8b The height adjuster mechanism is secured to the pillar by one bolt

27.22a On five-door models, unscrew the seat back catch pin ...

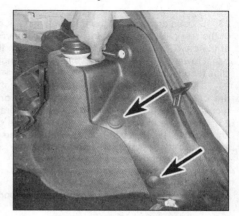

27.22b ... then unscrew the fasteners (arrows) ...

Front seat belt - three-door models

9   Remove the relevant rear seat back as described in Section 25.

10   Remove the sill trim panel and the door pillar upper trim panel as described in Steps 1 to 3.

11   Loosen and remove the rear seat back catch pin from the rear of the rear seat side trim panel.

12   Loosen and remove the retaining screws from the front edge of the rear seat side trim panel then pry out the panel retaining clips. Check that all the fasteners have been removed, then unclip the panel

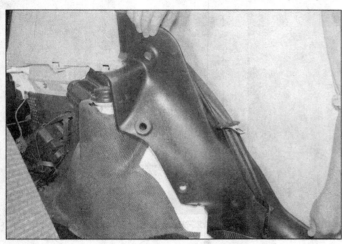

27.22c ... and remove the rear seat side trim panel

and remove it from the vehicle.

13   On models with seat belt tensioners, loosen and remove the tensioner assembly retaining nut. This will disable the tensioner making it safe to remove the seat belt assembly.

14   Loosen and remove the retaining bolt(s) and free the seat belt lower retaining rail from the floor. Disengage the rail from the belt and remove it from the vehicle.

15   Remove the seat belt as described in Steps 7 and 8.

Front seat belt stalk - all models

16   Remove the seat as described in Section 25.

17   Loosen and remove the bolt securing the stalk to the seat, and remove the stalk.

Rear seat side belt - Hatchback models

Refer to illustrations 27.22a, 27.22b, 27.22c, 27.23a, 27.23b, 27.23c, 27.24 and 27.25

18   On three-door models remove the rear seat side trim panel as described in Steps 9 to 12.

19   On five-door models remove the sill trim panel and free the door pillar upper trim panel as described in Steps 1 and 2.

20   From within the luggage compartment, fold down the rear seat then loosen and remove the retaining nuts securing the relevant side trim panel in position. Unclip the panel from the rear pillar trim panel and remove it from the vehicle; where necessary free the luggage compartment light from the panel as it is removed.

21   Loosen and remove the seat belt lower mounting bolt.

22   On five-door models, loosen and remove the rear seat back catch pin from the body then unscrew the retaining fasteners and remove the trim panel from the side of the seat (see illustrations).

11

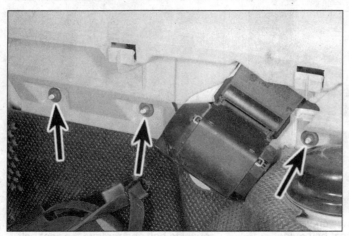

27.23a Undo the lower retaining nuts (arrows) . . .

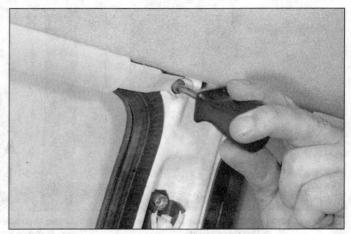

27.23b . . . and the upper retaining screw . . .

23   Loosen and remove the three retaining nuts from the lower edge of the rear pillar trim panel then undo the retaining screw from the top of the panel. Unclip the rear of the panel from the pillar then slide the panel towards the front of the vehicle, to disengage its upper retaining clips **(see illustrations)**.

24   Loosen and remove the seat belt upper mounting bolt and free the belt from the rear pillar **(see illustration)**. Where necessary, recover the spacer from behind the belt anchorage.

25   Undo the inertia reel mounting then free the reel from the pillar and remove the seat belt from the vehicle **(see illustration)**. If necessary, undo the retaining bolt and remove the height adjuster mechanism from the pillar (where installed).

Rear seat side belt - Sedan models

26   Remove the sill trim panel and free the door pillar upper trim panel as described in Steps 1 and 2.

27   Loosen and remove the retaining screw from the top of the rear pillar trim panel. Unclip the rear of the panel from the pillar then slide the panel towards the front of the vehicle, to disengage its retaining clips.

28   Remove the seat belt as described in Steps 21 to 25.

Rear seat center belt and buckles

Refer to illustration 27.29

29   Fold the rear seat cushion forwards then loosen and remove the bolt and washers securing the center belt and/or buckle assembly to the floor, and remove it from the vehicle **(see illustration)**.

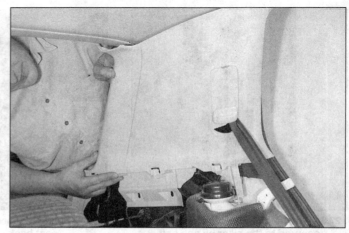

27.23c . . . then unclip the rear pillar trim panel and remove it from the vehicle

Installation

30   Installation is a reversal of the removal procedure, ensuring that all the seat belt mounting bolts are securely tightened, and all disturbed trim panels are securely retained by all the relevant retaining clips. When installing the upper trim panels, ensure that the height adjustment levers engage correctly with the seat belt upper mounting bolt head.

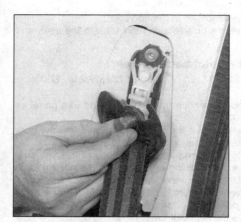

27.24 Unscrew the upper mounting bolt and free the belt from the pillar . . .

27.25 . . . then unscrew the inertia reel retaining bolt and remove the belt assembly from the vehicle

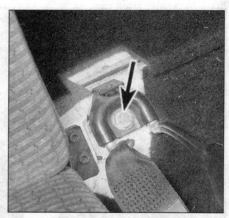

27.29 Rear seat center belt and buckles are secured to the floor by a single bolt (arrow)

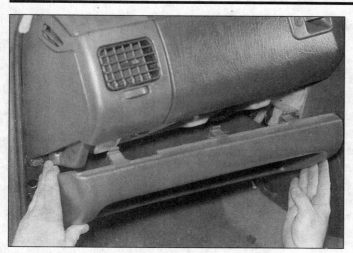

28.6 Undo the retaining screws and remove the passenger side shelf from the dash

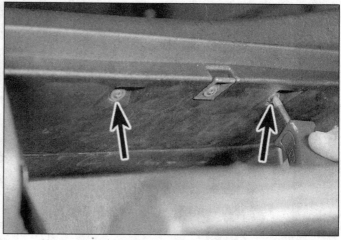

28.7a Loosen and remove the screws from inside the glovebox (arrows) . . .

28 Interior trim - removal and installation

Warning: *On models equipped with airbags, always disable the airbag system before working in the vicinity of the impact sensors, steering column or instrument panel to avoid the possibility of accidental deployment of the airbag, which could cause personal injury (see Chapter 12).*

Interior trim panels

Note: *Specific details for most interior panels are contained within Section 27.*

1   The interior trim panels are secured using either screws or various types of trim fasteners, usually studs or clips.

2   Check that there are no other panels overlapping the one to be removed; usually there is a sequence that has to be followed that will become obvious on close inspection.

3   Remove all obvious fasteners, such as screws. If the panel will not come free, it is held by hidden clips or fasteners. These are usually situated around the edge of the panel and can be pried up to release them; note, however that they can break quite easily so replacements should be available. The best way of releasing such clips without the correct type of tool, is to use a large flat-bladed screwdriver. Note in many cases that the adjacent sealing strip must be pried back to release a panel.

4   When removing a panel, never use excessive force or the panel may be damaged; always check carefully that all fasteners have been

removed or released before attempting to withdraw a panel.

5   Installation is the reverse of the removal procedure; secure the fasteners by pressing them firmly into place and ensure that all disturbed components are correctly secured to prevent rattles.

Glovebox

Refer to illustrations 28.6, 28.7a, 28.7b and 28.7c

6   Loosen and remove the passenger side dash shelf retaining screws. Move the shelf downwards, to release its upper retaining clips and remove it from the dash **(see illustration)**.

7   Open up the glovebox lid then loosen and remove the five retaining screws (two inside the glovebox and three along its lower edge). Slide the glovebox out of position, disconnecting the wiring connector from the glovebox illumination light (where installed) as it becomes accessible **(see illustrations)**.

8   Installation is the reverse of removal.

Carpets

9   The passenger compartment floor carpet is in one piece and is secured at its edges by screws or clips, usually the same fasteners used to secure the various adjoining trim panels.

10   Carpet removal and installation is reasonably straightforward but very time-consuming because all adjoining trim panels must be removed first, as must components such as the seats, the center console and seat belt lower anchors.

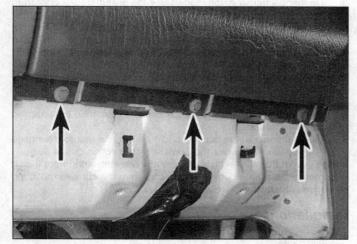

28.7b . . . and those located along the lower edge (arrows) . . .

28.7c . . . then withdraw the glovebox from the dash

11

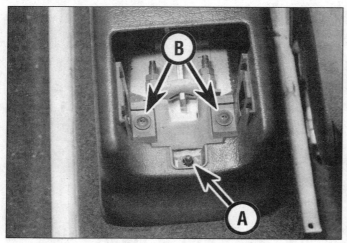

29.1a Unclip the illumination light (A) then undo
the retaining screws (B) . . .

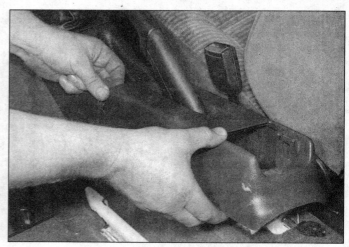

29.1b . . . and remove the rear section of the center console

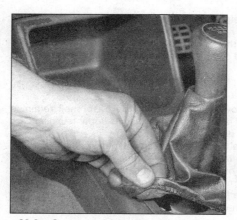

29.3a On manual transmission models,
free the gearshift lever boot
from the trim panel . . .

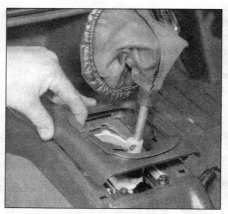

29.3b . . . then unclip the boot trim panel
from the front section of the
center console

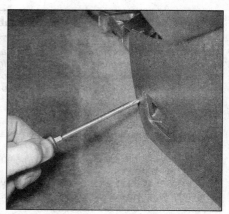

29.5a Loosen and remove the retaining
screws from the front edge of the
center console . . .

Headlining

11 The headlining is clipped to the roof and can be withdrawn only once all fittings such as the grab handles, sun visors, sunroof (if equipped), windshield and rear quarter windows and related trim panels have been removed and the door, liftgate and sunroof aperture sealing strips have been pried clear.

12 Note that headlining removal requires considerable skill and experience if it is to be carried out without damage and is therefore best entrusted to an expert.

29 Center console - removal and installation

Warning: *On models equipped with airbags, always disable the airbag system before working in the vicinity of the impact sensors, steering column or instrument panel to avoid the possibility of accidental deployment of the airbag, which could cause personal injury (see Chapter 12).*

Removal

Refer to illustrations 29.1a, 29.1b, 29.3a, 29.3b, 29.5a, 29.5b and 29.6

1 On low trim level models, remove the ashtray from the rear of the center console and free the ashtray illumination light (where equipped) from the console. Loosen and remove the two retaining screws located behind the ashtray then lift the rear section of the center console

upwards and off the parking brake lever **(see illustrations)**.

2 On high trim level models, loosen and remove the retaining screws and fasteners, located on the side and at the base of the storage compartment, then free the boot from the parking brake lever and remove the rear section of the center console from the vehicle.

3 On manual transmission models free the gearshift lever boot from the console then carefully unclip the boot trim panel and lift it off over the gearshift lever **(see illustrations)**.

4 On automatic transmission models, undo the retaining screw and remove the handle from the top of the selector lever (see Chapter 7B). Carefully unclip the selector lever position display panel from the center console and slide it off the lever, disconnect the wiring connector from the program switch (where equipped) as it becomes accessible.

5 On all models, loosen and remove the retaining screws from the left- and right-hand front edges of the console then undo the retaining nuts **(see illustrations)**.

6 Lift up the rear of the console and slide it to the rear to disengage it from the dash. Maneuver the console front section out of position and remove it from the vehicle, freeing any relevant wiring from it as it becomes accessible. Recover the spacers from the console mounting studs **(see illustration)**.

Installation

7 Installation is the reverse of removal making sure all fasteners are securely tightened.

29.5b ... then undo the retaining nuts from the rear
of the section (arrows)

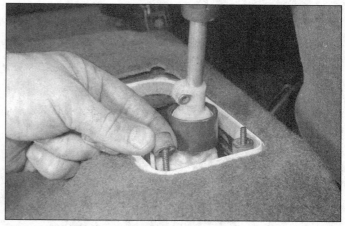

29.6 Remove the front section of the center console and recover
the spacers from the mounting studs

30.10a Loosen and remove the driver's side screws (arrows) ...

30.10b ... the center retaining screws (arrows) ...

30 Dash panel assembly - removal and installation

Warning: *On models equipped with airbags, always disable the airbag system before working in the vicinity of the impact sensors, steering column or instrument panel to avoid the possibility of accidental deployment of the airbag, which could cause personal injury (see Chapter 12).*

Note: *Label each wiring connector as it is disconnected from its relevant component. The labels will prove useful on installation, when routing the wiring and feeding the wiring through the dash apertures.*

Removal

Refer to illustrations 30.10a, 30.10b, 30.10c and 30.14

1   Disconnect the battery negative terminal. **Caution:** *If the stereo in your vehicle is equipped with an anti-theft system, make sure you have the correct activation code before disconnecting the battery.*

2   Remove the center console as described in Section 29.

3   Remove the steering column as described in Chapter 10.

4   Remove the instrument panel assembly, cigarette lighter and radio/cassette unit as described in Chapter 12. Also remove the front (upper) treble loudspeakers from the dash.

5   On models equipped with a passenger side airbag, remove the airbag unit as described in Chapter 12.

6   On models not equipped with a passenger side airbag, remove the glovebox as described in Section 28.

7   On diesel models remove the cold start accelerator cable as described in Chapter 4C.

8   Press in the locking buttons and unclip the fusebox cover from

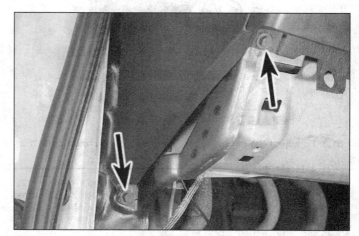

30.10c ... and the passenger side dash retaining screws (arrows)

the underside of the driver's side lower dash panel.

9   Carefully pry the trim panel from the top of the driver's side lower dash panel then loosen and remove all the panel retaining screws. Carefully move the trim panel downwards to release it from the dash then remove it from the vehicle.

10 Working along the base of the dash panel loosen and remove the retaining bolts securing the dash to its mounting frame (**see illustrations**).

11

30.14 Removing the dash assembly

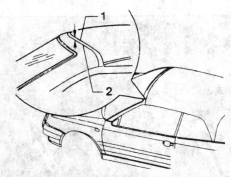

31.2 The convertible top (1) should be seated 3/16-inch below the windshield frame (2)

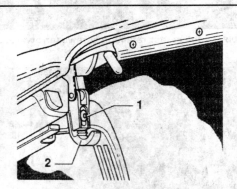

31.3 To adjust the latch, loosen the set screw (1) and adjust the hook (2) in or out as required

11   Remove both windshield wiper arms as described in Chapter 12.
12   Unscrew the windshield wiper motor trim cover retaining fastener screws and pull out the fasteners. Peel off the rubber seal the top of the firewall then release the two halves of the trim cover from the windshield and remove them from the vehicle.
13   Loosen and remove the two dash retaining nuts which are located beneath the center of the windshield. If necessary, remove the wiper motor (see Chapter 12) to improve access to the nuts.
14   From inside the vehicle carefully ease the dash assembly away from the firewall. As it is withdrawn, release the wiring harness from its retaining clips on the rear of the dash, while noting its correct routing (see Note at the start of this Section). Remove the dash assembly from the vehicle (see illustration). Recover the sealing grommets which are attached to the dash mounting studs; replace them if they show signs of wear or damage.

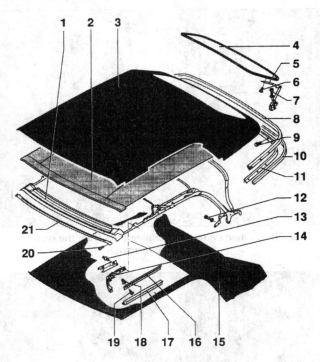

31.1 Convertible top components

1	Convertible top frame	12	Fastener
2	Padding	13	Catch pin
3	Convertible top cover	14	Catch
4	Rear window frame	15	Roof lining
5	Hinge	16	Retaining rail
6	Fastener	17	Roof frame seal
7	Fastener	18	Spreader nut
8	Frame seal	19	Fastener
9	Fastener	20	Fastener
10	Sealing band	21	Clamping strip
11	Fabric mount		

Installation

15   Installation is a reversal of the removal procedure, noting the following points:

 a) *Install the sealing grommets to the dash studs and maneuver the dash into position. Using the labels stuck on during removal, ensure that the wiring is correctly routed and securely retained by its dash clips.*
 b) *Clip the dash back into position, making sure all the wiring connectors are fed through their respective apertures, then install all the dash fasteners, and tighten them securely.*
 c) *On completion, reconnect the battery and check that all the electrical components and switches function correctly.*

31   Convertible top - adjustment

Refer to illustrations 31.1, 31.2 and 31.3

1   The convertible top must be adjusted properly to prevent water leaks and wind noises **(see illustration)**.
2   To check the adjustment, put the top up and latch it securely. The leading edge of the top should be seated 3/16-inch below the windshield frame **(see illustration)**.
3   If the leading edge of the top is less than 3/16-inch below the windshield frame, open the latches, loosen the set screw and adjust the latch hook as necessary **(see illustration)**.

Chapter 12
Chassis electrical systems

Contents

Specifications

System type	12-volt negative ground
Fuses	See Wiring diagrams
Torque specifications	**Ft-lbs** (unless otherwise indicated)
Air bag unit	60 in-lbs

1 General information and precautions

Warning: *Before carrying out any work on the electrical system, read through the precautions given in Safety First! at the beginning of this manual and Chapter 5.*

The electrical system is of 12-volt negative ground type. Power for the lights and all electrical accessories is supplied by a lead/acid type battery which is charged by the alternator.

This Chapter covers repair and service procedures for the various electrical components not associated with the engine. Information on the battery, alternator and starter motor can be found in Chapter 5.

It should be noted that prior to working on any component in the electrical system, the battery negative terminal should first be disconnected to prevent the possibility of electrical short circuits and/or fires. **Caution:** *If the stereo in your vehicle is equipped with an anti-theft system, make sure you have the correct activation code before disconnecting the battery.*

2 Electrical fault-finding - general information

Note: *Refer to the precautions given in 'Safety first!' and in Section 1 of this Chapter before starting work. The following tests relate to testing of the main electrical circuits, and should not be used to test delicate electronic circuits (such as anti-lock braking systems), particularly where an electronic control module is used.*

General

1 A typical electrical circuit consists of an electrical component, any switches, relays, motors, fuses, fusible links or circuit breakers related to that component, and the wiring and connectors which link the component to both the battery and the chassis. To help to pinpoint a problem in an electrical circuit, wiring diagrams are included at the end of this Manual.

2 Before attempting to diagnose an electrical fault, first study the appropriate wiring diagram to obtain a complete understanding of the components included in the particular circuit concerned. The possible sources of a fault can be narrowed down by noting if other components related to the circuit are operating properly. If several components or circuits fail at one time, the problem is likely to be related to a shared fuse or ground connection.

3 Electrical problems usually stem from simple causes, such as loose or corroded connections, a faulty ground connection, a blown fuse, a melted fusible link, or a faulty relay (refer to Section 3 for details of testing relays). Visually inspect the condition of all fuses, wires and connections in a problem circuit before testing the components. Use the wiring diagrams to determine which terminal connections will need to be checked in order to pinpoint the trouble spot.

4 The basic tools required for electrical fault-finding include a circuit tester or voltmeter (a 12-volt bulb with a set of test leads can also be used for certain tests); a self-powered test light (sometimes known as a continuity tester); an ohmmeter (to measure resistance); a battery and set of test leads; and a jumper wire, preferably with a circuit breaker or fuse incorporated, which can be used to bypass suspect

wires or electrical components. Before attempting to locate a problem with test instruments, use the wiring diagram to determine where to make the connections.

5   To find the source of an intermittent wiring fault (usually due to a poor or dirty connection, or damaged wiring insulation), a "wiggle" test can be performed on the wiring. This involves wiggling the wiring by hand to see if the fault occurs as the wiring is moved. It should be possible to narrow down the source of the fault to a particular section of wiring. This method of testing can be used in conjunction with any of the tests described in the following sub-Sections.

6   Apart from problems due to poor connections, two basic types of fault can occur in an electrical circuit - open-circuit, or short-circuit.

7   Open-circuit faults are caused by a break somewhere in the circuit, which prevents current from flowing. An open-circuit fault will prevent a component from working, but will not cause the relevant circuit fuse to blow.

8   Short-circuit faults are caused by a "short" somewhere in the circuit, which allows the current flowing in the circuit to "escape" along an alternative route, usually to ground. Short-circuit faults are normally caused by a breakdown in wiring insulation, which allows a feed wire to touch either another wire, or an grounded component such as the bodyshell. A short circuit fault will normally cause the relevant circuit fuse to blow.

Finding an open-circuit

9   To check for an open-circuit, connect one lead of a circuit tester or voltmeter to either the negative battery terminal or a known good ground.

10   Connect the other lead to a connector in the circuit being tested, preferably nearest to the battery or fuse.

11   Switch on the circuit, bearing in mind that some circuits are live only when the ignition switch is moved to a particular position.

12   If voltage is present (indicated either by the tester bulb lighting or a voltmeter reading, as applicable), this means that the section of the circuit between the relevant connector and the battery is problem-free.

13   Continue to check the remainder of the circuit in the same fashion.

14   When a point is reached at which no voltage is present, the problem must lie between that point and the previous test point with voltage. Most problems can be traced to a broken, corroded or loose connection.

Finding a short-circuit

15   To check for a short-circuit, first disconnect the load(s) from the circuit (loads are the components which draw current from a circuit, such as bulbs, motors, heating elements, etc.).

16   Remove the relevant fuse from the circuit, and connect a circuit tester or voltmeter to the fuse connections.

17   Switch on the circuit, bearing in mind that some circuits are live only when the ignition switch is moved to a particular position.

18   If voltage is present (indicated either by the tester bulb lighting or a voltmeter reading, as applicable), this means that there is a short circuit.

19   If no voltage is present, but the fuse still blows with the load(s) connected, this indicates an internal fault in the load(s).

Finding an ground fault

20   The battery negative terminal is connected to "ground:"- the metal of the engine/transaxle and the car body - and most systems are wired so that they only receive a positive feed, the current returning through the metal of the car body. This means that the component mounting and the body form part of that circuit. Loose or corroded mountings can therefore cause a range of electrical faults, ranging from total failure of a circuit, to a puzzling partial fault. In particular, lights may shine dimly (especially when another circuit sharing the same ground point is in operation), motors (e.g. wiper motors or the radiator cooling fan motor) may run slowly, and the operation of one circuit may have an apparently unrelated effect on another. Note that on many vehicles, ground straps are used between certain components, such as the engine/transaxle and the body, usually where there is no metal-to-metal contact between components due to flexible rubber mountings, etc.

21   To check whether a component is properly grounded, disconnect

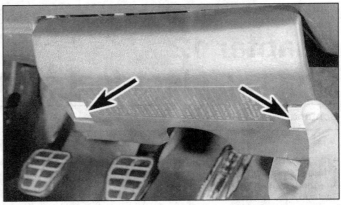

3.2 Depress the locking buttons (arrows) and remove the fusebox cover from the underside of the driver's lower dash panel

the battery and connect one lead of an ohmmeter to a known good ground point. **Caution:** *If the stereo in your vehicle is equipped with an anti-theft system, make sure you have the correct activation code before disconnecting the battery.* Connect the other lead to the wire or ground connection being tested. The resistance reading should be zero; if not, check the connection as follows.

22   If an ground connection is thought to be faulty, disassemble the connection and clean back to bare metal both the bodyshell and the wire terminal or the component ground connection mating surface. Be careful to remove all traces of dirt and corrosion, then use a knife to trim away any paint, so that a clean metal-to-metal joint is made. On reassembly, tighten the joint fasteners securely; if a wire terminal is being refitted, use serrated washers between the terminal and the bodyshell to ensure a clean and secure connection. When the connection is remade, prevent the onset of corrosion in the future by applying a coat of petroleum jelly or silicone-based grease or by spraying on (at regular intervals) a proprietary ignition sealer or a water dispersant lubricant.

3   Fuses and relays - general information

Main fuses

Refer to illustrations 3.2 and 3.8

1   The fuses are located behind the fusebox cover in the driver's side lower dash panel.

2   To remove the fusebox cover, press in both the cover buttons then unclip the cover from the dash **(see illustration).**

3   The main fuses are located in a row below the relays. A list of the circuits each fuse protects is stamped on the fusebox cover (a list is also given in the wiring diagrams at the end of this Chapter). On some models (depending on specification), some additional fuses are located in separate holders which can be found either above the relays or in the engine compartment.

4   To remove a fuse, first switch off the circuit concerned (or the ignition), then pull the fuse out of its terminals **(see illustration).** The wire within the fuse should be visible; if the fuse is blown it will be broken or melted.

5   Always replace a fuse with one of an identical rating; never use a fuse with a different rating from the original or substitute anything else. Never replace a fuse more than once without tracing the source of the trouble. The fuse rating is stamped on top of the fuse; note that the fuses are also color-coded for easy recognition.

6   If a new fuse blows immediately, find the cause before renewing it again; a short to ground as a result of faulty insulation is most likely. Where a fuse protects more than one circuit, try to isolate the defect by switching on each circuit in turn (if possible) until the fuse blows again. Always carry a supply of spare fuses of each relevant rating on the vehicle, a spare of each rating should be clipped into the base of the fusebox.

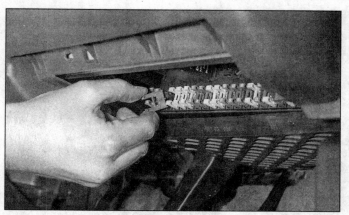

3.4 Removing a fuse

3.8 Diesel glow plug supply fusible link

Fusible links

7   On Diesel models the glow plug electrical supply circuit is protected by a fusible link. On both gasoline and diesel models, the radiator cooling fan run-on supply is also protected by a fusible link.

8   Prior to renewing the link, first ensure that the ignition is turned off. In the case of the glow plug fusible link also ensure that the driver's door is securely shut - the door switch is used to operate the glow plug system. Unclip the cover to gain access to the metal link; if the link has blown it will be broken or melted. Loosen the retaining screws then slide the link out of position (see illustration).

9   Install the new link (noting the information given in Steps 5 and 6) then tighten securely its retaining screws and clip the lid into position.

Relays

Refer to illustration 3.10

10   The relays are located behind the fusebox cover on the driver's side lower dash panel (see illustration).

11   To gain access to the relays, press in the locking buttons and unclip the fusebox cover from the underside of the driver's side lower dash panel. Carefully pry the trim panel from the top of the driver's side lower dash panel then loosen and remove all the panel retaining screws. Carefully move the panel downwards to release it from the dash then remove it from the vehicle. Release the retaining clips and lower the fusebox assembly out from underneath the dash.

12   If a circuit or system controlled by a relay develops a fault and the relay is suspect, operate the system; if the relay is functioning it should be possible to hear it click as it is energized. If this is the case the fault lies with the components or wiring of the system. If the relay is not being energized then either the relay is not receiving a main supply or a switching voltage or the relay itself is faulty. Testing is by the substitution of a known good unit but be careful; while some relays are identical in appearance and in operation, others look similar but perform different functions.

13   To replace a relay, first ensure that the ignition switch is off. The relay can then simply be pulled out from the socket and the new relay pressed in.

14   On installation ensure that the fusebox is securely retained by the clips then install the lower dash panel.

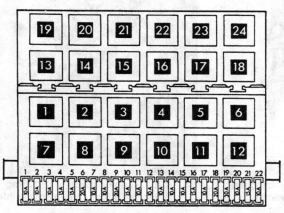

3.10 Fusebox relay and additional fuse locations

1   Air conditioning system relay
2   Liftgate wiper/washer relay
3   Fuel injection/ignition system relay
4   Load reduction relay
5   Not used
6   Emergency lighting relay for trailer electrical supply
7   Headlight washer system relay
8   Windshield wiper/washer relay
9   Seat belt warning system relay/control unit
10   Wiring jumper for foglights
11   Horn relay (models with dual tone horns) or horn wiring jumper
12   Fuel pump relay (gasoline models) or preheating system relay (Diesel models)
13   Intake manifold preheating system relay or starter inhibitor relay
14   ABS relay
15   ABS hydraulic pump relay
16   ABS relay
17   ABS fuse
18   Air conditioning/electric seat adjustment fuse
19   Not used
20   Starter inhibitor/back-up light relay
21   Oxygen sensor heating cut-off relay
22   Not used
23   Not used
24   Not used

4   Switches - removal and installation

Warning: On models equipped with airbags, always disable the airbag system before working in the vicinity of the impact sensors, steering column or instrument panel to avoid the possibility of accidental deployment of the airbag, which could cause personal injury (see Section 24).

Note: Disconnect the battery negative cable before removing any switch, and reconnect the cable after installing the switch.

Caution: If the stereo in your vehicle is equipped with an anti-theft system, make sure you have the correct activation code before disconnecting the battery.

Ignition switch/steering column lock

1   Refer to Chapter 10.

12

4.3 Undo the retaining screws and remove the steering column upper and lower shrouds

4.4a Undo the retaining screws (arrows) . . .

4.4b . . . then remove the relevant combination switch assembly

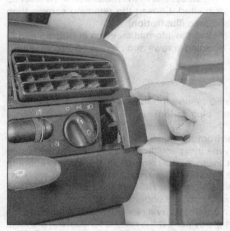

4.6 Remove the blanking plate from the side of the lighting switch . . .

4.7a . . . then depress the retaining lug and withdraw the switch assembly . . .

4.7b . . . disconnecting its wiring connector as it becomes accessible

Steering column combination switches

Refer to illustrations 4.3, 4.4a and 4.4b

2   Remove the steering wheel as described in Chapter 10.

3   Undo the retaining screws and remove the steering column upper and lower shrouds **(see illustration)**.

4   Loosen and remove the three retaining screws then disconnect the wiring connectors and remove both switch assemblies from the steering column **(see illustrations)**.

5   Installation is a reversal of the removal procedure.

Lighting switch (incorporating instrument panel dimmer and headlight dimmer switch)

Refer to illustrations 4.7a and 4.7b

6   Using a suitable flat-bladed screwdriver, carefully lever out the blanking plate from the side of the lighting switch and remove it from the dash **(see illustration)**. Note: *On models equipped with ABS or an airbag the blanking plate contains warning lights; it will be necessary to disconnect the wiring connectors from the lights as the plate is removed.*

7   Reaching in through the dash aperture, depress the lighting switch retaining lug and withdraw the switch assembly. Disconnect the wiring connector(s) and remove the switch from the dash **(see illustrations)**.

8   If necessary, release the retaining clips and slide out the instrument panel dimmer/headlight dimmer switch(es) from the lighting switch.

9   Installation is the reverse of removal.

Heated rear window switch

Refer to illustration 4.11

10   Using a suitable flat-bladed screwdriver, carefully pry out the blanking plug from next to the switch (the blanking plug covers a diagnostic wiring connector).

11   Carefully pry the heated rear window switch out of position, disconnecting its wiring connector as it becomes accessible **(see illustration)**.

12   On installation securely connect the wiring connector then clip the switch back into the dash. Check the operation of the switch then install the blanking plug.

Driver's electric rear window switch assembly

13   Remove the heated rear window switch as described above.

14   Carefully pry the rear window switch assembly out of position, disconnecting its wiring connectors as they become accessible.

15   Installation is the reverse of removal.

Air conditioning system switches (models with standard heating/ventilation system controls)

Note: *On models with the Climatronic automatic air conditioning system do not attempt to remove the switches from the control unit; the control unit should be treated as a sealed unit. If any switch fails to work, seek the advice of your VW dealer.*

16   Refer to Steps 10 to 12.

4.11 Removing the heated rear window switch

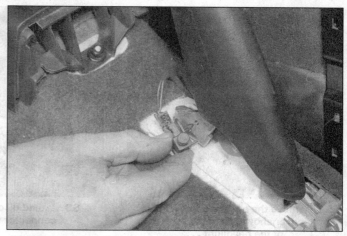

4.29 Unclip the parking brake warning light switch from the lever and disconnect its wiring connector

Heater blower motor switch (models with standard heating/ventilation system controls)

17   The switch is an integral part of the heater control panel assembly and cannot be renewed separately.

Heated front seat switch assembly

18   Using a suitable flat-bladed screwdriver, carefully pry out the small blanking plate from the side of the switch assembly. **Note:** *On models with the Climatronic automatic air conditioning system the plate will contain a temperature sensor, disconnect the wiring connector from the sensor as the plate is removed.*

19   Carefully pry the switch assembly out of position and disconnect its wiring connector.

20   Installation is the reverse of removal.

Door electric window switches

21   Carefully unclip the upper trim cover from the door armrest handle and disconnect its wiring connector.

22   Depress the retaining clips and remove the switch from the trim cover.

23   Installation is the reverse of removal.

Electric mirror switch

24   Carefully unclip the upper trim cover from the door armrest handle and, where necessary, disconnect the wiring connector from the window switch.

25   Carefully unclip the trim cover from around the interior handle and disconnect the wiring connector from the mirror switch.

26   Release the retaining clips and press the switch out from the trim cover.

27   Installation is the reverse of removal.

Parking brake warning light switch

Refer to illustration 4.29

28   Remove the rear section of the center console as described in Chapter 11 to gain access to the parking brake lever.

29   Disconnect the wiring connector from the warning light switch then unclip the switch and remove it from the parking brake lever bracket **(see illustration)**.

30   Installation is the reverse of removal. Check the operation of the switch before installation the center console.

Brake light switch

31   Refer to Chapter 9.

Courtesy light switches

32   Open up the door and remove the rubber cover from the switch.

33   Carefully pry the switch out of position and withdraw it, disconnecting its wiring connector as it becomes accessible. Tie a piece of string to the wiring to prevent it falling back into the door pillar.

34   Installation is a reverse of the removal procedure.

Luggage compartment light switch

35   The luggage compartment switch is built into the liftgate/trunk lock (as applicable).

36   Remove the lock assembly as described in Chapter 11.

37   Release the switch retaining clips and remove it from the lock assembly.

38   Installation is the reverse of removal.

Electric sunroof switch

39   Carefully lever the courtesy light assembly out from the overhead console with a suitable screwdriver. Disconnect the wiring connectors and remove the light assembly.

40   Release the retaining clips and remove the sunroof switch from the light assembly.

41   Installation is the reverse of removal.

---

5   Bulbs (exterior lights) - replacement

General

1   Whenever a bulb is renewed, note the following points:

a) *Disconnect the battery negative cable before starting work.* **Caution:** *If the stereo in your vehicle is equipped with an anti-theft system, make sure you have the correct activation code before disconnecting the battery.*

b) *Remember that if the light has just been in use the bulb may be extremely hot.*

c) *Always check the bulb contacts and holder, ensuring that there is clean metal-to-metal contact between the bulb, its terminals and ground. Clean off any corrosion or dirt before installing a new bulb.*

d) *Wherever bayonet-type bulbs are installed ensure that the terminal contact(s) bear firmly against the bulb contact.*

e) *Always ensure that the new bulb is of the correct rating and that it is completely clean before installing it; this applies particularly to headlight/foglight bulbs (see below).*

Headlight

Refer to illustrations 5.2, 5.3, 5.4a and 5.4b

2   Working in the engine compartment, remove the access cover from the rear of the headlight unit and recover its seal. On models with

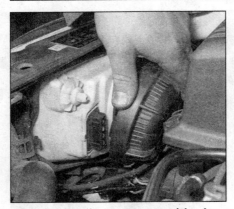

5.2 Remove the access cover (circular cover shown) from the rear of the headlight . . .

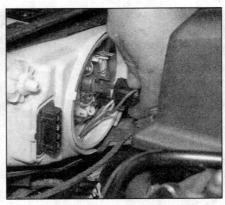

5.3 . . . and disconnect the wiring connector from the bulb

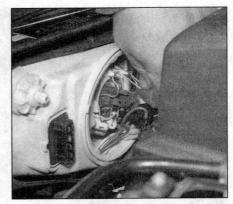

5.4a Rotate the bulb holder 1/4-turn counterclockwise . . .

5.4b . . . and withdraw the bulb from the headlight

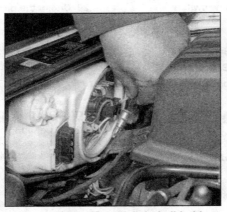

5.9a Withdraw the sidelight bulbholder from the rear of the headlight . . .

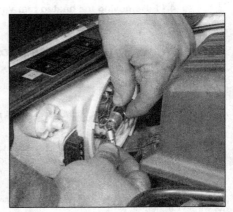

5.9b . . . and remove the bulb by pushing it in and twisting it counterclockwise

a circular cover turn the cover counterclockwise to release it, and on models with an elongated cover depress the retaining clips to release it **(see illustration)**.

3   Disconnect the wiring connector from the rear of the bulb **(see illustration)**.

4   Rotate the bulb holder 1/4-turn counterclockwise to release it from the rear of the light unit. Withdraw the bulb **(see illustrations)**.

5   When handling the new bulb, use a tissue or clean cloth to avoid touching the glass with the fingers; moisture and grease from the skin can cause blackening and rapid failure of this type of bulb. If the glass is accidentally touched, wipe it clean using rubbing alcohol.

6   Install the new bulb, ensuring that its locating tabs are correctly located in the light cut-outs, and secure it in position.

7   Reconnect the wiring connector and install the access cover, making sure it is securely refitted.

Front sidelight

Refer to illustrations 5.9a and 5.9b

8   Remove the access cover and seal from the rear of the headlight unit (see Step 2).

9   Withdraw the sidelight bulbholder from the headlight unit. The bulb is a bayonet installed in the holder and can be removed by pressing it and twisting it in an counterclockwise direction **(see illustrations)**.

10   Installation is the reverse of the removal procedure making sure the access cover is securely refitted.

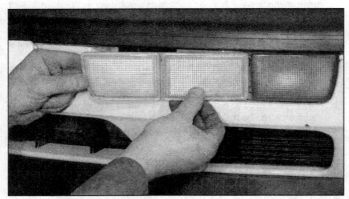

5.11 Removing the front reflector from the bumper (model without foglights shown)

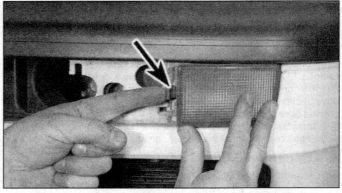

5.12 Depress the retaining clip (arrow) and withdraw the direction indicator light

5.13 Twist the bulbholder counterclockwise and free it from the light unit. The bulb can then be removed by pushing it in and turning it counterclockwise

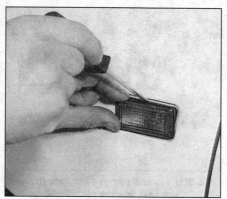

5.15 Carefully ease the side repeater light out of the fender using a screwdriver on its upper edge

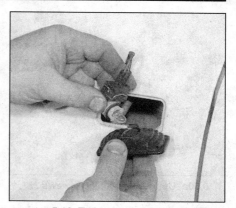

5.16 Twist the bulbholder counterclockwise to free it from the light unit and pull the bulb out of the holder

Front direction indicator

Refer to illustrations 5.11, 5.12 and 5.13

11   Insert a flat-bladed screwdriver, between the direction indicator light and the reflector and carefully pry the reflector out from the bumper **(see illustration)**.

12   Release the direction indicator light retaining clip and withdraw the light unit from the bumper **(see illustration)**.

13   Twist the bulbholder counterclockwise and remove it from the rear of the light unit **(see illustration)**. The bulb is a bayonet installed in the holder and can be removed by pressing it and twisting it in an counterclockwise direction.

14   Installation is a reverse of the removal procedure making sure the light unit and reflector are securely retained by their clips.

Front direction indicator side repeater

Refer to illustrations 5.15 and 5.16

15   Carefully pry the upper edge of the indicator side repeater light out from the fender, taking great care not to damage the painted finish of the fender **(see illustration)**.

16   Withdraw the light unit from the fender then twist the bulbholder counterclockwise and remove it from the light **(see illustration)**. The bulb is of the capless (push-fit) type and can be removed by simply pulling it out of the bulbholder.

17   Installation is a reverse of the removal procedure.

Front foglight

18   Insert a flat-bladed screwdriver between the direction indicator

light and the reflector and carefully pry the reflector out from the bumper.

19   Loosen and remove the foglight retaining screws and withdraw the light from the bumper.

20   Rotate the foglight cover counterclockwise and release it from the rear of the light unit.

21   Disconnect the bulb wiring from the cover terminal then release the spring clip and withdraw the foglight bulb.

22   When handling the new bulb, use a tissue or clean cloth to avoid touching the glass with the fingers; moisture and grease from the skin can cause blackening and rapid failure of this type of bulb. If the glass is accidentally touched, wipe it clean using rubbing alcohol.

23   Insert the new bulb, making sure it is correctly located, and secure it in position with the spring clip.

24   Connect the bulb wire to the cover terminal then install the cover to the rear of the unit.

25   Install the foglight to the bumper, securely tightening its retaining screws. Prior to installing the reflector cover, check the aim of the foglight beam. If necessary the foglight aim can be adjusted by rotating the adjustment screw situated next to the lower retaining screw. Once the beam aim is correct, clip the reflector back into position.

Rear light cluster

Refer to illustrations 5.26 and 5.27

26   From inside the vehicle luggage compartment, unclip the plastic cover (where equipped) to gain access to the rear of the light cluster **(see illustration)**.

27   Release the retaining catches and free the bulbholder assembly from the rear of the light unit **(see illustration)**.

5.26 Unclip the plastic cover from the rear light cluster . . .

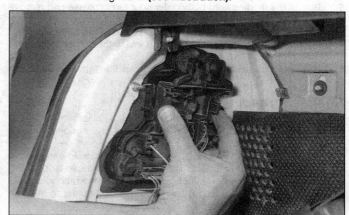

5.27 . . . then depress the retaining catches and withdraw the bulbholder assembly (Hatchback model shown)

12

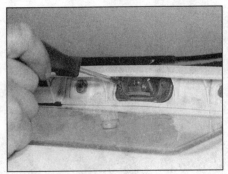

5.30a Undo the retaining screws . . .

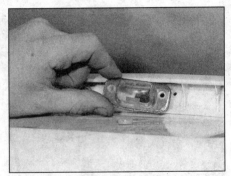

5.30b . . . and remove the lens from the rear license plate light (Hatchback shown)

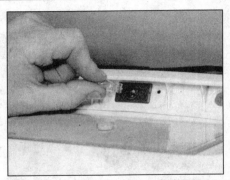

5.31 The license plate light bulb is a push-fit in the light unit

28   The relevant bulb can then be renewed, all bulbs have a bayonet fitting. Note that the stop/tail light bulb has offset locating pins to prevent it being installed incorrectly.

29   Installation is the reverse of the removal sequence ensuring that the bulbholder is securely clipped into position.

License plate light

Refer to illustrations 5.30a, 5.30b and 5.31

30   Loosen and remove the retaining screws and withdraw the lens from the liftgate/trunk lid. Recover the lens seal and examine it for signs of damage or deterioration, renewing it if necessary **(see illustrations)**.

31   The bulb is of the capless (push-fit) type and can be removed by simply pulling it out of the bulbholder **(see illustration)**.

32   Press the new bulb into position and install the seal and lens. Do not overtighten the lens retaining screws as the plastic is easily cracked.

6   Bulbs (interior lights) - replacement

General

1   Refer to Section 5, Step 1.

Courtesy light

2   Using a small, flat-bladed screwdriver, carefully pry the light lens out of position and release the bulb from the light unit contacts.

3   Install the new bulb, ensuring it is securely held in position by the contacts, and clip the lens back into position.

Front seat reading light

4   Carefully lever the courtesy light assembly out from the overhead console with a suitable screwdriver. Disconnect the wiring connectors and remove the light assembly.

5   Rotate the reading light bulbholder counterclockwise and remove it from the rear of the light unit. The bulb is of the capless (push-fit) type and can be removed by simply pulling it out of the bulbholder.

6   Push the new bulb into position and install the holder to the light unit.

7   Reconnect the wiring connector and clip the light unit back into position.

Rear seat reading light

8   Carefully lever the light assembly out from the pillar with a suitable screwdriver. Disconnect the wiring connectors and remove the light assembly.

9   Rotate the reading light bulbholder counterclockwise and remove it from the rear of the light unit. The bulb is of the capless (push-fit) type and can be removed by simply pulling it out of the bulbholder.

10   Push the new bulb into position and install the holder to the light unit.

11   Reconnect the wiring connector and clip the light unit back into position.

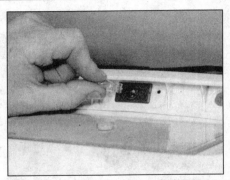

6.14 Removing an instrument panel illumination bulbholder

Luggage compartment light

12   Refer to the information given above in Steps 2 and 3.

Instrument panel illumination/warning lights

Refer to illustration 6.14

13   Remove the instrument panel as described in Section 9.

14   Twist the relevant bulbholder counterclockwise and withdraw it from the rear of the panel **(see illustration)**.

15   All bulbs are integral with their holders. Be very careful to ensure that the new bulbs are of the correct rating, the same as those removed; this is especially important in the case of the ignition/battery charging warning light.

16   Install the bulbholder to the rear of the instrument panel then install the instrument panel as described in Section 9.

Glovebox illumination light bulb

17   Open up the glovebox. Using a small flat-bladed screwdriver carefully pry the top of the light assembly and withdraw it. Release the bulb from its contacts.

18   Install the new bulb, ensuring it is securely held in position by the contacts, and clip the light unit back into position.

Cassette storage box illumination bulb

19   Carefully slide the storage box out from the dash and disconnect its wiring connector.

20   Using a small flat-bladed screwdriver, carefully release the retaining clips and remove the rear cover from the storage box. Remove the bulb from its holder.

21   Installation is the reverse of removal.

Cigarette lighter/ashtray illumination bulb

22   On models equipped with standard manual heating/ventilation system controls, carefully pry out the surround from around the control

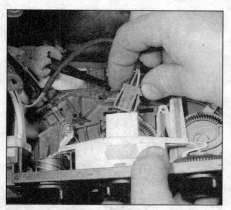

6.28a Disconnect the wiring connector . . .

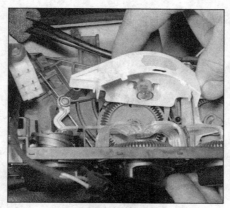

6.28b . . . and unclip the bulbholder from the rear of the heater control panel

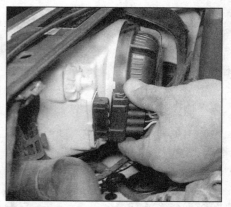

7.2 Disconnecting the headlight wiring connector

7.3a Loosen and remove the retaining screws (arrows) . . .

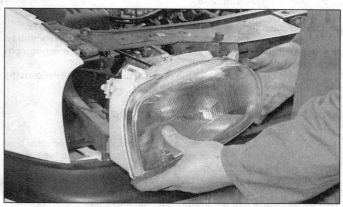

7.3b . . . and remove the headlight from the vehicle

knobs. Loosen and remove the four retaining screws and free the control unit from the rear of the switch panel.

23 On models with the automatic "Climatronic" heating system, carefully insert a flat-bladed screwdriver between the bottom of the display panel and the switches and gently ease the panel out of position. Loosen and remove the four retaining screws and free the electronic control unit from the rear of the switch panel.

24 Remove the ashtray then undo the two dash switch panel retaining screws.

25 Withdraw the switch panel from the dash until access can be gained to the rear of the cigarette lighter. Unclip the bulbholder from the lighter and remove the bulb.

26 Installation is the reverse of removal.

Heater control panel illumination bulb

Refer to illustrations 6.28a and 6.28b

27 Withdraw the heater control panel as described in Section 9 of Chapter 3 so that access to the rear of the panel can be gained. Note that there is no need to remove the panel completely, the control cables can be left attached.

28 Disconnect the wiring connector then unclip the bulbholder assembly from the rear of the control panel **(see illustrations)**.

29 Unclip the surround from around the bulb then carefully pull the bulb out of its holder.

30 Installation is the reverse of removal.

Switch illumination bulbs

31 All of the switches are fitted with illuminating bulbs; some are also fitted with a bulb to show when the circuit concerned is operating. These bulbs are an integral part of the switch assembly and cannot be obtained separately. Bulb replacement will therefore require the replacement of the complete switch assembly.

7 Exterior light units - removal and installation

Note: *Disconnect the battery negative cable before removing any light unit, and reconnect the cable after installation the light unit.*
Caution: *If the stereo in your vehicle is equipped with an anti-theft system, make sure you have the correct activation code before disconnecting the battery.*

Headlight

Refer to illustrations 7.2, 7.3a and 7.3b

1 Using a suitable screwdriver, carefully release the radiator grille upper and lower retaining lugs then move the grille forwards and away from the vehicle.

2 Disconnect the wiring connector from the rear of the headlight unit **(see illustration)**.

3 Loosen and remove the headlight retaining screws and withdraw the headlight from the vehicle **(see illustrations)**. On models with a headlight beam adjustment system it will be necessary to disconnect the wiring connector from the adjustment motor as the headlight is removed.

4 On models equipped with a headlight beam adjustment system, if necessary, rotate the adjustment motor counterclockwise to free the motor from the rear of the headlight unit and pull the motor squarely away to disconnect its balljoint. On installation, align the motor balljoint with the light unit socket and clip it into position. Engage the motor assembly with the light and twist it clockwise to secure it in position.

5 Installation is a direct reversal of the removal procedure. On completion check the headlight beam alignment using the information given in Section 8.

Front direction indicator light

Refer to illustration 7.7

6 Insert a flat-bladed screwdriver between the direction indicator light

12

7.7 Removing the front direction indicator light

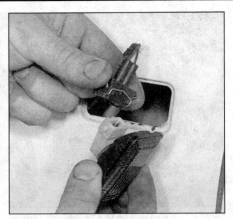

7.9 Removing the front direction indicator side repeater light

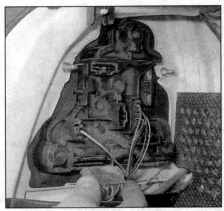

7.15 Disconnect the wiring connectors from the rear of the light unit . . .

and the reflector and carefully pry the reflector out from the bumper.

7 Release the direction indicator light retaining clip then withdraw the light unit from the bumper, disconnecting it from the wiring connector **(see illustration)**.

8 Installation is a reverse of the removal procedure making sure the light unit and reflector are securely retained by their clips.

Front direction indicator side repeater

Refer to illustration 7.9

9 Carefully pry the upper edge of the indicator side repeater light out from the fender, if necessary using a suitable plastic wedge and taking great care not damage the painted finish of the fender. Disconnect it from the wiring connector **(see illustration)**. Tie a piece of string to the wiring to prevent it falling back into the fender.

10 Installation is a reverse of the removal procedure.

Front foglight

11 Insert a flat-bladed screwdriver between the direction indicator light and the reflector and carefully pry the reflector out from the bumper.

12 Loosen and remove the foglight retaining screws then withdraw the light from the bumper and disconnect it from the wiring connector.

13 Install the foglight to the bumper, securely tightening its retaining screws. Prior to installation the reflector cover, check the aim of the foglight beam. If necessary the foglight aim can be adjusted by rotating the adjustment screw situated next to the lower retaining screw. Once the beam aim is correct, clip the reflector back into position.

Rear light cluster

Refer to illustrations 7.15, 7.16a and 7.16b

14 From inside the vehicle luggage compartment, unclip the plastic cover (where equipped) to gain access to the rear of the light cluster.

15 Disconnect the wiring connectors from the rear of the bulbholder **(see illustration)**.

16 Loosen and remove the rear light unit retaining nuts and withdraw the light unit from the rear of the vehicle. Recover the rubber seal from the rear of the light unit; if the seal shows signs of damage or deterioration, replace it **(see illustrations)**.

17 Installation is a reverse of the removal procedure tightening the retaining nuts securely.

License plate light

18 Loosen and remove the retaining screws and withdraw the lens from the liftgate/trunk lid. Recover the lens seal and examine it for signs of damage or deterioration, renewing it if necessary.

19 Withdraw the light unit and disconnect it from the wiring connector.

20 Installation is the reverse of removal. Do not overtighten the lens retaining screws as the plastic is easily cracked.

---

8   Headlight beam alignment - general information

Accurate adjustment of the headlight beam is only possible using optical beam setting equipment and this work should therefore be carried out by a VW dealer or suitably equipped workshop.

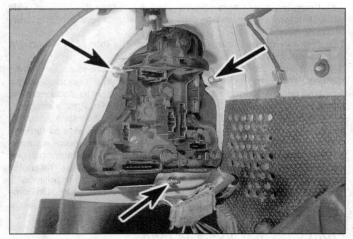

7.16a . . . then undo the retaining nuts (arrows) . . .

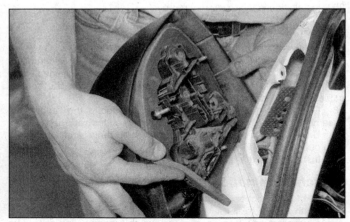

7.16b . . . and withdraw the rear light unit and rubber seal from the vehicle (Hatchback model shown)

9.2 On models not equipped with heated seats, remove the blanking plate from the side of the instrument panel

9.4 Undo the two screws (arrows) and remove the instrument panel shroud

For reference the headlights can be adjusted using the adjuster assemblies installed on the top of each light unit. The outer adjuster alters the horizontal position of the beam while the inner adjuster alters the vertical aim of the beam.

Some models are equipped with an electrically operated headlight beam adjustment system which is controlled through the switch in the dash. On these models ensure that the switch is set to the off position before adjusting the headlight aim.

9 Instrument panel - removal and installation

Warning: *On models equipped with airbags, always disable the airbag system before working in the vicinity of the impact sensors, steering column or instrument panel to avoid the possibility of accidental deployment of the airbag, which could cause personal injury (see Section 24).*

Removal

Refer to illustrations 932, 9.4, 9.5a, 9.5b and 9.5c

1   Disconnect the battery negative terminal. **Caution:** *If the stereo in your vehicle is equipped with an anti-theft system, make sure you have the correct activation code before disconnecting the battery.*
2   Remove the lighting switch and heated front seat switch as described in Section 4. On models not equipped with heated seats, carefully pry out the blanking plate (which is installed in place of the heated seat switch) from the side of the instrument panel **(see illustration).**
3   Loosen the retaining screws and remove the steering column shrouds.
4   Undo the two retaining screws and remove the instrument panel shroud from the dash **(see illustration).**

5   Loosen and remove the two retaining screws from either side of the instrument panel then carefully withdraw the instrument panel from the dash, disconnecting the wiring connector(s) from the rear of the panel **(see illustrations).**

Installation

6   Installation is the reverse of removal making sure that the instrument panel wiring is securely reconnected. On completion reconnect the battery and check the operation of the panel warning lights to ensure that they are functioning correctly.

10 Instrument panel components - removal and installation

1   At the time of writing, no individual components are available for the instrument panel and therefore the panel must be treated as a sealed unit. If there is a fault with one of the instruments, remove the panel as described in Section 9 and take it to your VW dealer for testing. They have access to a special diagnostic tester which will be able to locate the fault and will then be able to advise you on the best course of action.

11 Cigarette lighter - removal and installation

Warning: *On models equipped with airbags, always disable the airbag system before working in the vicinity of the impact sensors, steering column or instrument panel to avoid the possibility of accidental deployment of the airbag, which could cause personal injury (see Section 24).*

9.5a Undo the two retaining screws (arrows) . . .

9.5b . . . then withdraw the instrument panel from the dash . . .

9.5c . . . and disconnect its wiring connector

11.2a   Remove the surround from around the heater controls . . .

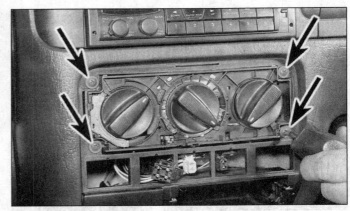

11.2b   . . . then undo the four control unit retaining screws (arrows)

11.5   Undo the two screws (arrows) . . .

11.6a   . . . then withdraw the switch panel from the dash

11.6b   Undo the diagnostic wiring connector retaining screws (arrows) . . .

Removal

Refer to illustrations 11.2a, 11.2b, 11.5, 11.6a, 11.6b, 11.6c and 11.7

1   Disconnect the battery negative terminal. **Caution:** *If the stereo in your vehicle is equipped with an anti-theft system, make sure you have the correct activation code before disconnecting the battery.*

2   On models equipped with standard manual heating/ventilation system controls, carefully pry out the surround from around the control knobs. Loosen and remove the four retaining screws and free the control unit from the rear of the switch panel **(see illustrations)**.

3   On models with the automatic "Climatronic" heating system,

carefully insert a flat-bladed screwdriver between the bottom of the display panel and the switches and gently ease the panel out of position. Loosen and remove the four retaining screws and free the electronic control unit from the rear of the switch panel.

4   Pry out the blanking plugs and remove the switch(es) from the dash panel as described in Section 4.

5   Remove the ashtray and cigarette lighter insert then undo the two dash switch panel retaining screws **(see illustration)**.

6   Withdraw the switch panel from the dash and unscrew the retaining screws securing the diagnostic wiring connectors to the panel. Disconnect the wiring connector from the cigarette lighter then release the

11.6c   . . . then disconnect the cigarette lighter wiring connector and remove the switch panel

11.7   Depress the retaining tangs and push the cigarette lighter out of the switch panel

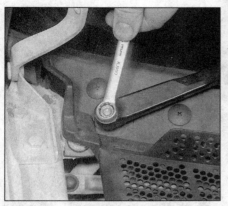

13.3 Unscrew the retaining nut and remove the wiper arm from its spindle

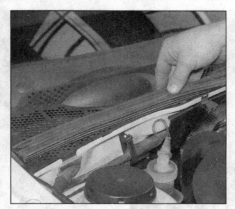

14.3 Unclip the rubber seal and remove it from the engine compartment firewall

14.4a Unscrew the trim cover fastener screws and pull out the fasteners . . .

wiring harness from the panel and remove the switch panel from the dash **(see illustrations)**.

7   Unclip the bulbholder from the lighter then depress the retaining tangs and push out the lighter out of the panel **(see illustration)**.

Installation

8   Installation is a reversal of the removal procedure, ensuring that all the wiring connectors are securely reconnected.

12   Horn(s) - removal and installation

Removal

1   The horn(s) is/are located behind the front bumper. To improve access, apply the parking brake then jack up the front of the vehicle and support it on axle stands (see *"Jacking and vehicle support"*).

2   Undo the retaining bolt and remove the horn, disconnecting its wiring connectors as they become accessible.

Installation

3   Installation is the reverse of removal.

13   Wiper arm - removal and installation

Removal

Refer to illustration 13.3

1   Operate the wiper motor then switch it off so that the wiper arm returns to the at-rest position.

2   Stick a piece of masking tape along the edge of the wiper blade

to use as an alignment aid on installation.

3   Pry off the wiper arm spindle nut cover then loosen and remove the spindle nut. Lift the blade off the glass and pull the wiper arm off its spindle. If necessary the arm can be levered off the spindle using a suitable flat-bladed screwdriver **(see illustration)**.

Note: *If both windshield wiper arms are to be removed at the same time mark them for identification; the arms are not interchangeable.*

Installation

4   Ensure that the wiper arm and spindle splines are clean and dry, then install the arm to the spindle, aligning the wiper blade with the tape installed on removal. Install the spindle nut, tightening it securely, and clip the nut cover back in position.

14   Windshield wiper motor and linkage - removal and installation

Removal

Refer to illustrations 14.3, 14.4a, 14.4b, 14.4c, 14.5, 14.6a and 14.6b

1   Disconnect the battery negative terminal. **Caution:** *If the stereo in your vehicle is equipped with an anti-theft system, make sure you have the correct activation code before disconnecting the battery.*

2   Remove the wiper arms as described in the previous Section.

3   Unclip the rubber seal from the top of the engine compartment firewall **(see illustration)**.

4   Unscrew the windshield wiper motor trim cover fastener screws and pull out the fasteners. Release the two halves of the trim cover from the windshield and remove them from the vehicle **(see illustrations)**.

5   Disconnect the wiring connector from the wiper motor and free the wiring from its retaining clips **(see illustration)**.

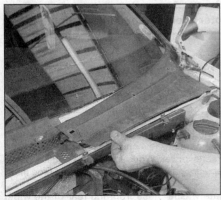

14.4b . . . then remove the passenger's side . . .

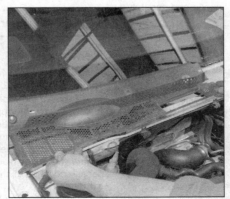

14.4c . . . and driver's side of the windshield wiper motor trim cover

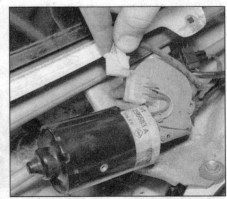

14.5 Disconnect the wiper motor wiring connector and free the wiring from its retaining clips

12

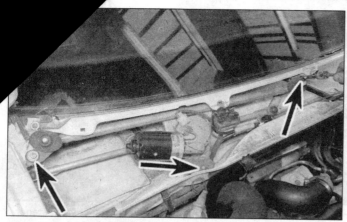

14.6a Loosen and remove the retaining nuts and bolts (arrows) . . .

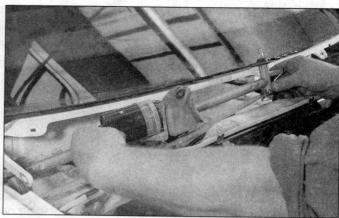

14.6b . . . and maneuver the wiper motor assembly out of position

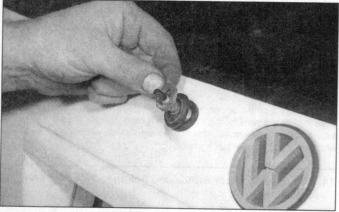

15.2 Unscrew the nut and remove the washer from the tailgate wiper motor spindle

15.3a On Hatchback models, undo the retaining screw . . .

6   Loosen and remove the three wiper motor retaining nuts/bolts (as applicable) and maneuver the motor and linkage assembly out of position **(see illustrations)**. Recover the washers and spacers from the motor mounting rubbers and inspect the rubbers for signs of damage or deterioration, and replace if necessary.

7   If necessary, mark the relative positions of the motor shaft and linkage arm then unscrew the retaining nut from the motor spindle. Free the wiper linkage from the spindle then remove the three motor retaining bolts and separate the motor and linkage.

Installation

8   Where necessary, assemble the motor and linkage and securely

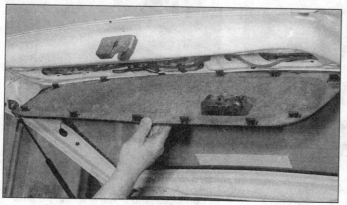

15.3b . . . then unclip the trim panel from the tailgate

tighten the motor retaining bolts. Locate the linkage arm on the motor spindle, aligning the marks made prior to removal, and securely tighten its retaining nut.

9   Ensure that the mounting rubbers are in position then maneuver the motor assembly back into position in the vehicle. Install the spacers and washers and tighten the motor mounting nuts/bolts securely.

10   Reconnect the wiring connector and clip it into the retaining clips.

11   Install the wiper motor trim covers to the vehicle and secure them in position with their retaining clips.

12   Install the rubber seal to the firewall and install the wiper arms.

---

15   Liftgate wiper motor - removal and installation

Removal

Refer to illustrations 15.2, 15.3a, 15.3b, 15.4a, 15.4b, 15.5a, 15.5b, 15.6, 15.7a and 15.7b

1   Remove the wiper arm as described in Section 13.

2   Unscrew the nut from the wiper motor spindle and remove the washer **(see illustration)**.

3   Loosen and remove the liftgate trim panel retaining screw then release the trim panel clips, carefully levering between the panel and liftgate with a flat-bladed screwdriver. Work around the outside of the panel, and when all the clips are released, remove the panel **(see illustrations)**.

4   Disconnect the liftgate washer hose from the rear of the wiper motor then disconnect its wiring connector. Free the wiring from its retaining clips **(see illustrations)**.

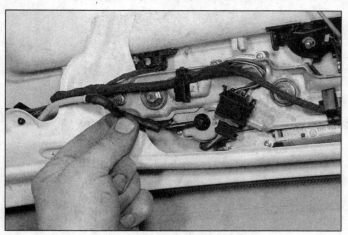

15.4a Disconnect the washer hose . . .

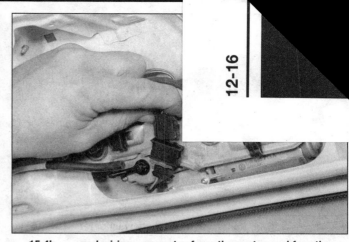

15.4b . . . and wiring connector from the motor and free the wiring from its retaining clips

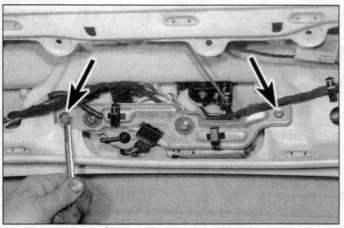

15.5a Undo the two retaining bolts (arrows) . . .

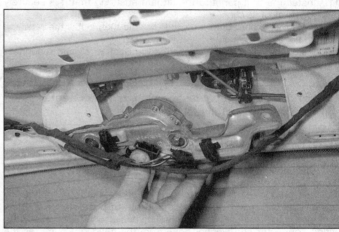

15.5b . . . then remove the motor assembly

5   Loosen and remove the wiper motor retaining bolts and maneuver the assembly out from the liftgate (see illustrations).
6   Recover the rubber grommet from the liftgate; the grommet should be replaced if it is damaged (see illustration).
7   If necessary, loosen and remove the retaining bolts then separate the motor from its mounting bracket and recover the spacers and mounting rubbers (see illustrations). Inspect the rubbers for signs of damage or deterioration and replace if necessary.

Installation

8   Where necessary, install the motor to its mounting, making sure the rubbers and spacers are correctly positioned, and tighten the retaining bolts securely.
9   Install the rubber grommet to the liftgate and maneuver the assembly into position in the liftgate. Install the motor mounting bolts and tighten them securely.

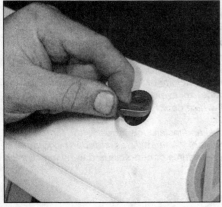

15.6 Recover the rubber sealing grommet from the tailgate

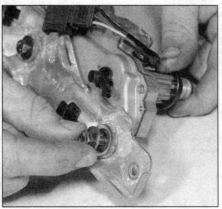

15.7a Unscrew the two retaining bolts then separate the motor and mounting bracket . . .

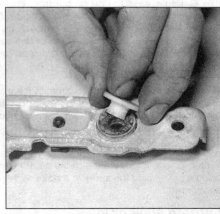

15.7b . . . and recover the spacers from the mounting bushings

12

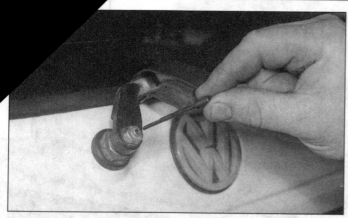

16.12 Removing the tailgate washer jet

10   Reconnect the wiper motor wiring connector and the washer hose.
11   Install the trim panel to the liftgate ensuring that it is securely retained by all of its clips.
12   Slide the washer onto the wiper spindle then install the retaining nut and tighten it securely.
13   Install the wiper arm as described in Section 13 and reconnect the battery.

16   Windshield/liftgate washer system components - removal and installation

Washer system reservoir

1   Remove the battery as described in Chapter 5.
2   Disconnect the wiring connector and the hose(s) from the washer pump.
3   Loosen and remove the retaining nuts from the top of the reservoir and lift the reservoir upwards and out of position. On models equipped with headlight washers it will be necessary to disconnect the wiring connector and washer hose from the headlight pump as the reservoir is removed. Wash off any spilled fluid with cold water.
4   Installation is the reverse of removal ensuring that the washer hose(s) are securely connected.

Washer pump

5   Empty the contents of the reservoir or be prepared for fluid spillage as the pump is removed.
6   Remove the battery as described in Chapter 5.
7   Disconnect the wiring connector and washer hose(s) from the pump.
8   Carefully ease the pump out from the reservoir and recover its sealing grommet. Wash off any spilled fluid with cold water.
9   Installation is the reverse of removal, using a new sealing grommet if the original one shows signs of damage or deterioration. Refill the reservoir and check the pump grommet for leaks.

Windshield washer jets

10   Open up the hood and disconnect the washer hose from the base of the jet. Where necessary, also disconnect the wiring connector from the jet. Carefully ease the jet out from the hood, taking great care not to damage the paintwork.
11   On installation, securely connect the jet to the hose and clip it into position in the hood; where necessary also reconnect the wiring connector. Check the operation of the jet. If necessary, adjust the nozzle using a pin, aiming the spray to a point slightly above the center of the swept area.

Liftgate washer jet

Refer to illustration 16.12
12   Unclip the cover from the wiper arm spindle to gain access to the

18.3a Insert the removal tools into position . . .

washer jet and unclip the jet from the center of the spindle **(see illustration)**.
13   On installation ensure that the jet is clipped securely in position. Check the operation of the jet. If necessary adjust the nozzle using a pin, aiming the spray to a point slightly above the center of the swept area.

17   Headlight washer system components - removal and installation

Washer system reservoir

1   Refer to Section 16.

Washer pump

2   Remove the washer reservoir as described in Section 16.
3   Carefully ease the pump out from the reservoir and recover its sealing grommet. Wash off any spilled fluid with cold water.
4   Installation is the reverse of removal, using a new sealing grommet if the original one shows signs of damage or deterioration. Refill the reservoir and check the pump grommet for leaks.

Washer jets

5   On Sedan models, pry out the trim cover from the front of the washer jet then undo the retaining screws and remove the jet assembly from the bumper.
6   On Hatchback models, unclip the cover from the top of the jet then undo the retaining screws and remove the jet from the bumper.
7   Installation is the reverse of removal making sure that the washer jets are correctly aimed at the headlight.

18   Radio/cassette player - removal and installation

Warning: *On models equipped with airbags, always disable the airbag system before working in the vicinity of the impact sensors, steering column or instrument panel to avoid the possibility of accidental deployment of the airbag, which could cause personal injury (see Section 24).*
Note: *The following removal and installation procedure is for the range of radio/cassette units which VW installs as standard equipment. Removal and installation procedures of non-standard will differ slightly.*

Removal

Refer to illustrations 18.3a and 18.3b
1   Two special tools, obtainable from most car accessory shops, are required for removal. Alternatively suitable tools can be fabricated from 1/8-inch diameter wire, such as welding rod.

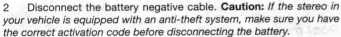

18.3b ... then withdraw the radio/cassette player and disconnect its wiring connectors and antenna lead (arrow)

19.1 Unclip the speaker grille from the top of the dash ...

2   Disconnect the battery negative cable. **Caution:** *If the stereo in your vehicle is equipped with an anti-theft system, make sure you have the correct activation code before disconnecting the battery.*
3   Insert the tools into the slots on each side of the unit and push them until they snap into place. The radio/cassette player can then be slid out of the dash and the wiring connectors and antenna disconnected **(see illustrations)**.

Installation

4   Reconnect the wiring connector and antenna lead then push the unit into the dash until the retaining lugs snap into place.

19   Loudspeakers - removal and installation

Front upper (treble) loudspeaker

Refer to illustrations 19.1 and 19.2
1   Carefully lever the speaker grille out from the top of the dash, taking great care not to mark either component **(see illustration)**.
2   Pry the speaker out of position and disconnect its wiring connector **(see illustration)**.
3   Installation is the reverse of removal making sure the speaker is correctly located.

Front lower (bass) loudspeaker

Refer to illustrations 19.4, 19.5a and 19.5b
4   Using a flat-bladed screwdriver, carefully pry the speaker grille

19.2 ... and carefully pry the front upper loudspeaker out of position

out from the door panel **(see illustration)**.
5   Loosen and remove the retaining screws then remove the speaker from the door, disconnecting its wiring connector as it becomes accessible **(see illustrations)**.
6   Installation is the reverse of removal making sure the speaker is correctly located.

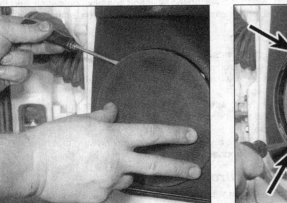

19.4 Unclip the speaker grille ...

19.5a ... then undo the four retaining screws (arrows) and withdraw the speaker from the door trim panel ...

19.5b ... and disconnect its wiring connector

12

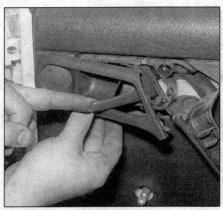

19.8a On four- and five-door models, unclip the trim cover from around the door inner handle . . .

19.8b . . . and disconnect the wiring connector . . .

19.8c . . . then release the retaining clips and remove the speaker

Rear upper (treble) loudspeaker

Refer to illustrations 19.8a, 19.8b and 19.8c

7   On three-door models, carefully pry the speaker out from the trim panel and disconnect it from the wiring connector.

8   On all other models, unclip the upper trim cover from the door armrest handle and remove it from the vehicle, where necessary, disconnecting the wiring connector as it becomes accessible. Unclip the trim cover from around the door inner handle and remove it from the door, disconnecting the wiring connector as it becomes accessible. Unclip the speaker and remove it from the panel **(see illustrations)**.

9   Installation is the reverse of removal.

Rear lower (bass) loudspeaker

10   Refer to Steps 4 to 6.

---

20   Radio antenna - removal and installation

Fender mounted antenna

Removal

1   Open up the hood then loosen the retaining nuts and free the ground strap from the left-hand side of the hood and fender.

2   Undo the retaining screw(s) then unclip the left-hand footwell side trim panel from the vehicle to gain access to the antenna lead connection. Separate the two halves of the antenna lead.

3   Undo the retaining screw from the rear of the left-hand wheelarch liner then release the retaining fasteners and remove the liner; the fasteners are released by pressing out their center pins. If the fastener center pins are not recovered, new fasteners will be required on installation.

4   Pull the antenna lead and ground lead through into the wheelarch, noting their correct routing.

5   Unscrew the nut securing the antenna mounting bracket in position then slide the bracket off from the base of the antenna.

6   Slide the antenna downwards and out of position and recover the mounting grommet from the fender. Inspect the mounting grommet and bracket grommet for signs of damage or deterioration and replace as necessary.

Installation

7   Locate the mounting grommet in the fender and firmly insert the antenna. Slide the mounting bracket into position and securely tighten its mounting nut.

8   Feed the antenna lead and ground lead through their respective apertures in the fender.

9   From inside the vehicle, reconnect the two halves of the antenna lead and install the footwell side trim panel.

10   Reconnect the ground lead to the hood and fender and securely tighten the retaining nuts.

11   Check the antenna operation then install the wheelarch liner making sure it is securely retained by its fasteners and screw.

Roof mounted antenna

Removal

12   Open up the liftgate (where necessary) and carefully pry out the trim strip securing the rear of the headlining to the roof. Carefully peel the headlining back until access is gained to the antenna retaining nut and antenna lead and wiring connectors.

13   Disconnect the antenna lead and wiring connector then undo the retaining nut and remove the antenna from the roof. Recover the antenna sealing grommet.

Refitting

14   On refitting, locate the sealing grommet and antenna in the roof hole.

15   Refit and tighten the retaining nut.

16   Reconnect the antenna lead and wiring connector then clip the headlining trim strip back into position.

---

21   Cruise control system components - removal and installation

Refer to illustration 21.1

1   The cruise control system is a vacuum operated system; the main components being a vacuum pump, an electronic control unit (ECU) and the accelerator pedal position unit. In addition to these there is the operating switch, which is built into the left-hand combination switch, and the vent switch(es) on the clutch and/or brake pedal(s) **(see illustration)**.

Vacuum pump

2   Remove the windshield washer fluid reservoir as described in Section 16 to gain access to the pump.

3   Disconnect the wiring connector then unscrew the pump retaining bolt.

4   Slide the pump assembly to the rear to disengage its retaining pegs. Disconnect the vacuum hose and remove the pump assembly from the vehicle.

5   Installation is the reverse of removal making sure that the vacuum hose is securely reconnected.

Electronic control unit (ECU)

6   The cruise control ECU is located behind the driver's side lower dash panel. Before removing, disconnect the battery negative terminal. **Caution:** *If the stereo in your vehicle is equipped with an anti-theft system, make sure you have the correct activation code before disconnecting the battery.*

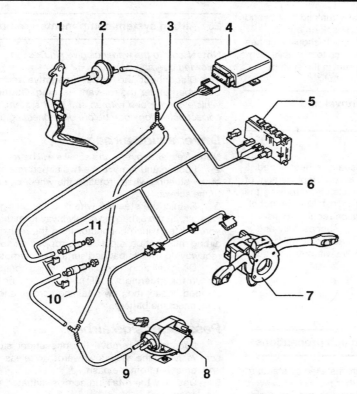

21.1 Cruise control system components

1   Accelerator pedal
2   Accelerator pedal positioning unit
3   Vacuum hose connector
4   Electronic control unit (ECU)
5   Fusebox/relay assembly
6   Wiring harness
7   System operating switch (integral with steering column combination switch assembly)
8   Vacuum pump
9   Vacuum hose
10 Brake pedal vent switch
11 Clutch pedal vent switch

7   Press in the locking buttons and unclip the fusebox cover from the underside of the driver's side lower dash panel.

8   Carefully pry the trim cover from the top of the driver's side lower dash panel then loosen and remove all the panel retaining screws. Carefully move the panel downwards to release it from the dash then remove it from the vehicle.

9   Loosen and remove the retaining nut securing the ECU mounting bracket to the dash and disconnect the ECU wiring connector. **Note:** *On models with an anti-theft alarm system, it will also be necessary to disconnect the wiring connectors from the alarm system ECU as this is mounted on the same plate as the cruise control ECU.*

10 Release the retaining lugs and remove the ECU mounting plate from behind the dash.

11 Undo the retaining screws and separate the ECU and its mounting plate.

12 Installation is a reverse of the removal procedure making sure the wiring connectors are securely reconnected.

Accelerator pedal positioning unit

13 Remove the driver's side lower dash panel as described in Steps 7 and 8.

14 Reach up behind the dash and disconnect the vacuum hose from the unit.

15 Carefully lever the positioning unit rod off its pivot bolt balljoint.

16 Loosen and remove the nut securing the positioning unit to its mounting bracket and maneuver the assembly out from underneath the dash.

17 Installation is the reverse of removal making sure the positioning unit rod is clipped securely onto its balljoint. Prior to installation the dash panel, adjust the unit as follows.

18 Rotate the adjusting sleeve slightly counterclockwise and free it from the front of the positioning unit body. Adjust the accelerator cable as described in Chapter 4 then slide the adjusting sleeve onto the body until there is approximately 1/16-inch of free play in the positioning unit rod. Hold the sleeve in this position and rotate in slightly clockwise to lock it in position. Once the positioning unit rod free play is correctly adjusted, install the lower dash panel.

Pedal vent switch

19 Remove the driver's side lower dash panel as described in Step 8.

20 Disconnect the wiring connector and pull the vacuum hose off from the switch.

21 Remove the vent valve from the pedal bracket.

22 On installation, screw the switch fully into the pedal bracket. With the switch in position, pull the pedal back to the at-rest position; this will automatically adjust the position of the vent switch.

23 Reconnect the vacuum hose and wiring connector and install the lower dash panel.

System operating switch

24 The system operating switch is an integral part of the left-hand combination switch assembly. Refer to Section 4 for removal and installation details.

22 Anti-theft alarm system - general information

Note: *This information is applicable only to the anti-theft alarm system installed by VW as standard equipment.*

Some models in the range are equipped with an anti-theft alarm system as standard equipment. The alarm has switches on all the doors (including the liftgate/trunk lid), the hood and the ignition switch. If the liftgate/trunk lid, hood or either of the doors are opened or the ignition switch is switched on while the alarm is set, the alarm horn will sound and the hazard warning lights will flash. The alarm also has an immobilizer function which makes the ignition (gasoline models) or fuel supply system (diesel models) inoperable while the alarm is triggered.

The alarm is set using the key in the driver's or passenger's front door lock. Simply hold the key in the locking position until the warning light near the driver's door lock button starts to flash. The alarm system will then start to monitor its various switches approximately 30 seconds later.

With the alarm set, if the liftgate/trunk lid is unlocked, the lock switch sensing will automatically be switched off but the door and

12

hood switches will still be active. Once the liftgate/trunk lid is shut and locked again, the switch sensing will be switched back on again.

Should the alarm system become faulty the vehicle should be taken to a VW dealer for examination. They will have access to a special diagnostic tester which will quickly trace any fault present in the system.

23 Heated front seat components - removal and installation

Heater mats

1   On models equipped with heated front seats, a heater pad is equipped on both the seat back and the seat cushion. Replacement of either heater mat involves peeling back the upholstery, removing the old mat, sticking the new mat in position and then installing the upholstery. Note that upholstery removal and installation requires considerable skill and experience if it is to be carried out successfully and is therefore best entrusted to your VW dealer. In practice, it will be very difficult for the home mechanic to carry out the job without ruining the upholstery.

Heated seat switches

2   Refer to Section 4.

24 Airbag system - general information and precautions

Both a driver's and passenger's airbag were installed as standard to some models in the Golf/Jetta range; on other models they were available as an optional extra. Models equipped with a driver's side airbag have the word AIRBAG stamped on the airbag unit, which is fitted to the center of the steering wheel. Models also equipped with a passenger's side airbag also have the word AIRBAG stamped on the passenger's end of the dash. The airbag system comprises of the airbag unit (complete with gas generator) which is fitted to the steering wheel, an impact sensor, the control unit and a warning light in the instrument panel.

The airbag system is triggered in the event of a heavy frontal impact above a predetermined force; depending on the point of impact. The airbag is inflated within milliseconds and forms a safety cushion between the driver and the steering wheel and (where equipped) the passenger and the dash. This prevents contact between the upper body and the wheel/dash and therefore greatly reduces the risk of injury. The airbag then deflates almost immediately.

3   Every time the ignition is switched on, the airbag control unit performs a self-test. The self-test takes approximately 3 seconds and during this time the airbag warning light on the dash is illuminated. After the self-test has been completed the warning light should go out. If the warning light fails to come on, remains illuminated after the initial 3 second period or comes on at any time when the vehicle is being driven, there is a fault in the airbag system. The vehicle should then be taken to a VW dealer for examination at the earliest possible opportunity.

Warning: Before carrying out any operations on the airbag system, disconnect the battery negative terminal. When operations are complete, make sure no one is inside the vehicle when the battery is reconnected.

Caution: If the stereo in your vehicle is equipped with an anti-theft system, make sure you have the correct activation code before disconnecting the battery.

Warning: Note that the airbag(s) must not be subjected to temperatures in excess of 90°C (194°F). When the airbag is removed, ensure that it is stored the correct way up to prevent possible inflation.

Warning: Do not allow any solvents or cleaning agents to contact the airbag assemblies. They must be cleaned using only a damp cloth.

Warning: The airbags and control unit are both sensitive to impact. If either is dropped or damaged they should be renewed.

Warning: Disconnect the airbag control unit wiring plug prior to using arc-welding equipment on the vehicle.

25 Airbag system components - removal and installation

Note: Refer to the warnings given in Section 24 before carrying out the following operations.

1   Disconnect the battery negative terminal then continue as described under the relevant heading. **Caution:** *If the stereo in your vehicle is equipped with an anti-theft system, make sure you have the correct activation code before disconnecting the battery.*

Driver's side airbag

Note: New airbag retaining screws will be required on installation.

2   Loosen and remove the two airbag retaining screws from the rear of the steering wheel, rotating the wheel as necessary to gain access to the screws.

3   Return the steering wheel to the straight-ahead position then carefully lift the airbag assembly away from the steering wheel and disconnect the wiring connector from the rear of the unit. Note that the airbag must not be knocked or dropped and should be stored the correct way up with its padded surface uppermost.

4   On installation reconnect the wiring connector and seat the airbag unit in the steering wheel, making sure the wire does not become trapped. Install the new retaining screws and tighten them securely. Reconnect the battery.

Passenger side airbag

5   Loosen and remove the passenger side dash shelf retaining screws. Move the shelf downwards, to release its upper retaining clips and remove it from the dash.

6   Unscrew the retaining screws, situated along the lower edge of the airbag.

7   Move the airbag assembly downwards to disengage the upper locating pegs from the mounting frame. Remove the airbag unit from the dash, disconnecting the wiring connector as it becomes accessible. Recover the guides from the airbag mounting frame.

8   On installation, ensure that the guides are correctly seated in the mounting frame then maneuver the airbag into position and reconnect the wiring connector.

9   Locate the airbag pegs into the guides then install the retaining screws, tightening them securely.

10   Install the dash shelf and reconnect the battery.

Airbag control unit

11   Remove the center console as described in Chapter 11.

12   Undo the retaining bolts and remove the dash mounting frame center bracket.

13   Loosen the retaining screws and remove the rear footwell duct joining pieces from the base of the air distribution housing.

14   Remove the retaining screw and fastener and remove the front footwell duct assembly from the base of the air distribution housing.

15   Depress the retaining clip and disconnect the wiring connector from the control unit.

16   Unscrew the nuts securing the control unit mounting bracket to the floor and remove the assembly from the vehicle. Note that it may be necessary to cut the carpet to gain access to the mounting nuts.

17   Where necessary, undo the retaining nuts and separate the bracket and control unit.

18   Installation is the reverse of removal making sure the wiring connector is securely reconnected.

Airbag wiring contact unit

19   Remove the steering wheel as described in Chapter 10.

20   Taking care not to rotate the contact unit, undo the three retaining screws and remove it from the steering wheel.

21   On installation, attach the unit to the steering wheel and securely tighten its retaining screws. If a new contact unit is being installed, cut the cable-tie which is installed to prevent the unit accidentally rotating.

22   Install the steering wheel as described in Chapter 10.

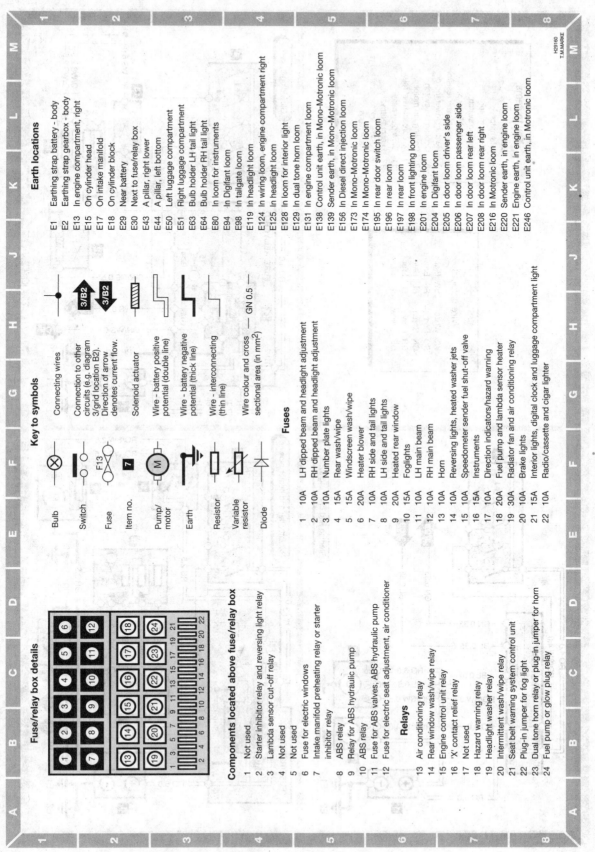

Fuse/relay box details

Components located above fuse/relay box

1 Not used
2 Starter inhibitor relay and reversing light relay
3 Lambda sensor cut-off relay
4 Not used
5 Not used
6 Fuse for electric windows
7 Intake manifold preheating relay or starter inhibitor relay
8 ABS relay
9 Relay for ABS hydraulic pump
10 ABS relay
11 Fuse for ABS valves, ABS hydraulic pump
12 Fuse for electric seat adjustment, air conditioner

Relays

13 Air conditioning relay
14 Rear window wash/wipe relay
15 Engine control unit relay
16 'X' contact relief relay
17 Not used
18 Hazard warning relay
19 Headlight washer relay
20 Intermittent wash/wipe relay
21 Seat belt warning system control unit
22 Plug-in jumper for fog light
23 Dual tone horn relay or plug-in jumper for horn
24 Fuel pump or glow plug relay

Key to symbols

Bulb — Connecting wires

Switch — Connection to other circuits (e.g. diagram 3/grid location B2). Direction of arrow denotes current flow.

Fuse — Solenoid actuator

Item no. — Wire - battery positive potential (double line)

Pump/motor — Wire - battery negative potential (thick line)

Earth — Wire - interconnecting (thin line)

Resistor

Variable resistor — Wire colour and cross sectional area (in mm²) — GN 0.5

Diode

Fuses

1	10A	LH dipped beam and headlight adjustment
2	10A	RH dipped beam and headlight adjustment
3	10A	Number plate lights
4	15A	Rear wash/wipe
5	15A	Windscreen wash/wipe
6	20A	Heater blower
7	10A	RH side and tail lights
8	10A	LH side and tail lights
9	20A	Heated rear window
10	15A	Foglights
11	10A	LH main beam
12	10A	RH main beam
13	10A	Horn
14	10A	Reversing lights, heated washer jets
15	15A	Speedometer sender fuel shut-off valve
16	15A	Instruments
17	10A	Direction indicators/hazard warning
18	20A	Fuel pump and lambda sensor heater
19	30A	Radiator fan and air conditioning relay
20	15A	Brake lights
21	15A	Interior lights, digital clock and luggage compartment light
22	10A	Radio/cassette and cigar lighter

Earth locations

E1 Earthing strap battery - body
E2 Earthing strap gearbox - body
E13 In engine compartment, right
E15 On cylinder head
E17 On intake manifold
E18 On cylinder block
E29 Near battery
E30 Next to fuse/relay box
E43 A pillar, right lower
E44 A pillar, left bottom
E50 Left luggage compartment
E51 Right luggage compartment
E63 Bulb holder LH tail light
E64 Bulb holder RH tail light
E80 In loom for instruments
E94 In Digifant loom
E98 In tailgate loom
E119 In headlight loom
E124 In wiring loom, engine compartment right
E125 In headlight loom
E128 In loom for interior light
E129 In dual tone horn loom
E131 In engine compartment loom
E138 Control unit earth, in Mono-Motronic loom
E139 Sender earth, in Mono-Motronic loom
E156 In Diesel direct injection loom
E173 In Mono-Motronic loom
E174 In Mono-Motronic loom
E195 In rear door switch loom
E196 In rear loom
E197 In rear loom
E198 In front lighting loom
E201 In engine loom
E204 In Digifant loom
E205 In door loom driver's side
E206 In door loom passenger side
E207 In door loom rear left
E208 In door loom rear right
E216 In Motronic loom
E220 Sender earth, in engine loom
E221 Engine earth, in engine loom
E246 Control unit earth, in Motronic loom

H29160
T.M.MARNE

Diagram 1 : Information for wiring diagrams

12

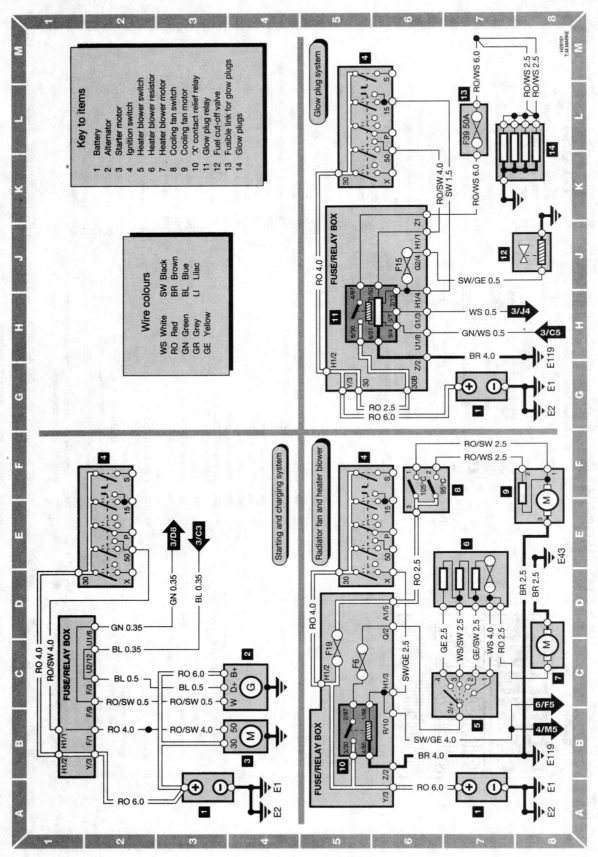

Key to items

1 Battery
2 Alternator
3 Starter motor
4 Ignition switch
5 Heater blower switch
6 Heater blower resistor
7 Heater blower motor
8 Cooling fan switch
9 Cooling fan motor
10 'X' contact relief relay
11 Glow plug relay
12 Fuel cut-off valve
13 Fusible link for glow plugs
14 Glow plugs

Wire colours

WS White	SW Black
RO Red	BR Brown
GN Green	BL Blue
GR Grey	LI Lilac
GE Yellow	

Diagram 2 : Typical starting and charging

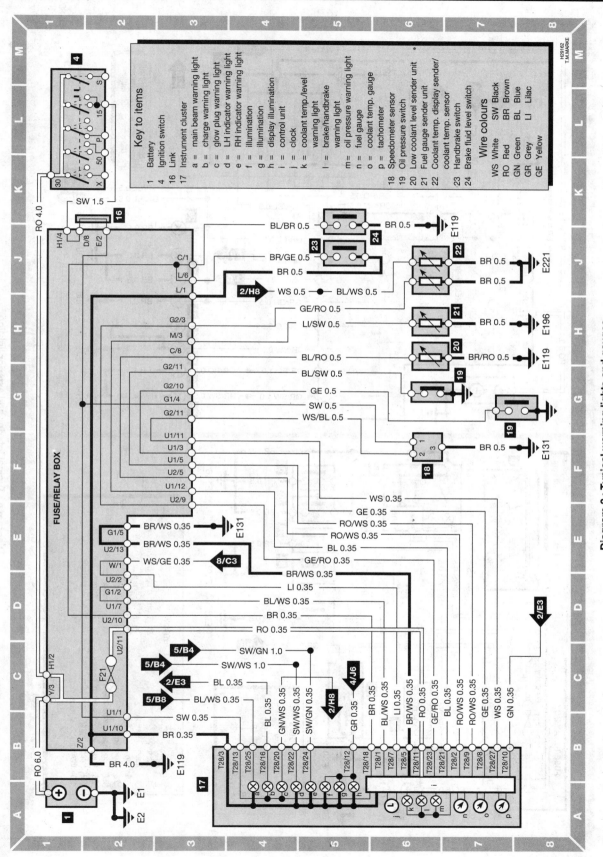

Diagram 3 : Typical warning lights and gauges

Key to items

1 Battery
4 Ignition switch
16 Link
17 Instrument cluster
a = main beam warning light
b = charge warning light
c = glow plug warning light
d = LH indicator warning light
e = RH indicator warning light
f = illumination
g = illumination
h = display illumination
i = control unit
j = clock
k = coolant temp./level
      warning light
l = brake/handbrake
      warning light
m = oil pressure warning light
n = fuel gauge
o = coolant temp. gauge
p = tachometer
18 Speedometer sensor
19 Oil pressure switch
20 Low coolant level sender unit
21 Fuel gauge sender unit
22 Coolant temp. display sender/
      coolant temp. sensor
23 Handbrake switch
24 Brake fluid level switch

Wire colours

WS White   SW Black
RO Red   BR Brown
GN Green   BL Blue
GR Grey   LI Lilac
GE Yellow

12

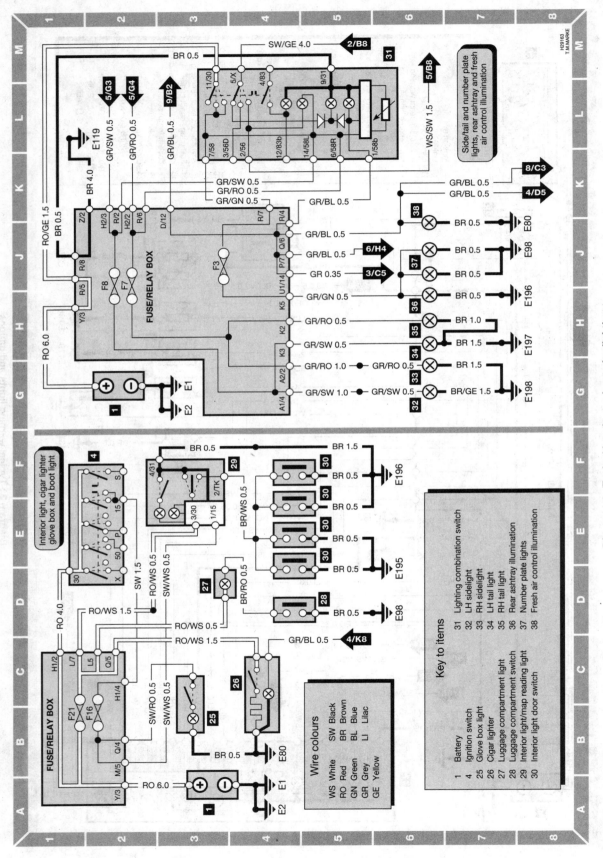

Diagram 4 : Typical interior and exterior lighting

Side/tail and number plate lights, rear ashtray and fresh air control illumination

Interior light, cigar lighter glove box and boot light

Wire colours

WS	White	SW	Black
RO	Red	BR	Brown
GN	Green	BL	Blue
GR	Grey	LI	Lilac
GE	Yellow		

Key to items

1	Battery
4	Ignition switch
25	Glove box light
26	Cigar lighter
27	Luggage compartment light
28	Luggage compartment switch
29	Interior light/map reading light
30	Interior light door switch
31	Lighting combination switch
32	LH sidelight
33	RH sidelight
34	LH tail light
35	RH tail light
36	Rear ashtray illumination
37	Number plate lights
38	Fresh air control illumination

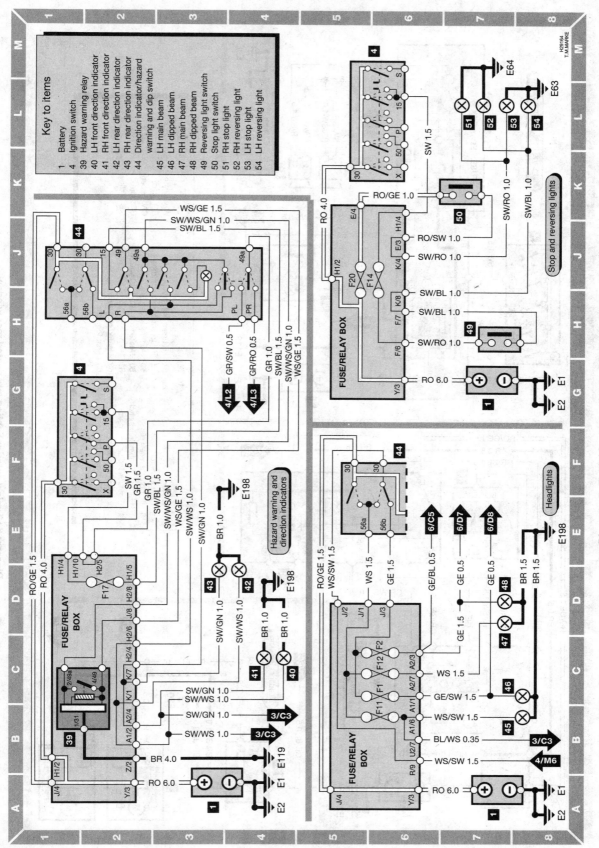

Key to items

1 Battery
4 Ignition switch
39 Hazard warning relay
40 LH front direction indicator
41 RH front direction indicator
42 LH rear direction indicator
43 RH rear direction indicator
44 Direction indicator/hazard
    warning and dip switch
45 LH main beam
46 LH dipped beam
47 RH main beam
48 RH dipped beam
49 Reversing light switch
50 Stop light switch
51 RH stop light
52 RH reversing light
53 LH stop light
54 LH reversing light

Diagram 5 : Typical exterior lighting continued

12

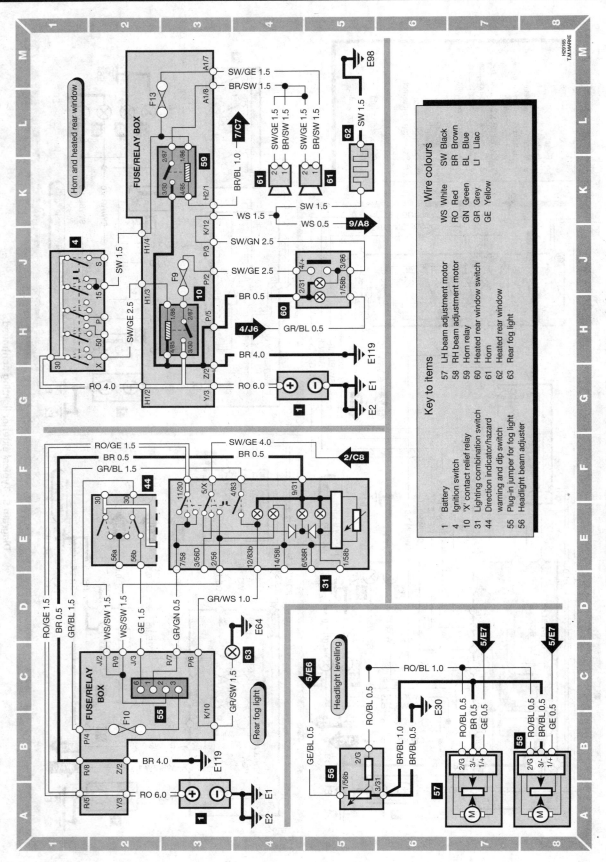

Diagram 6 : Typical exterior lighting and heated rear window

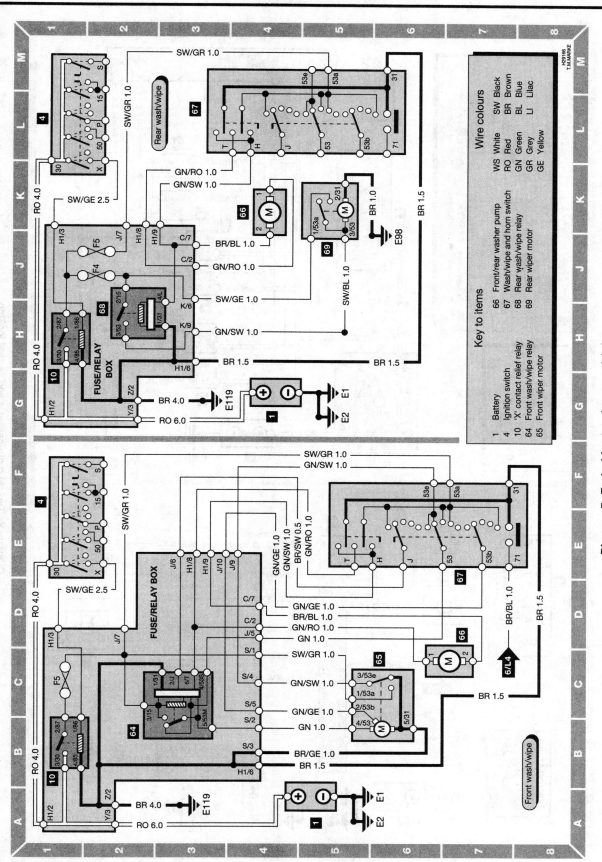

Diagram 7 : Typical front and rear wash/wipe

Wire colours

WS	White	SW	Black
RO	Red	BR	Brown
GN	Green	BL	Blue
GR	Grey	LI	Lilac
GE	Yellow		

Key to items

1 Battery
4 Ignition switch
10 'X' contact relief relay
64 Front wash/wipe relay
65 Front wiper motor
66 Front/rear washer pump
67 Wash/wipe and horn switch
68 Rear wash/wipe relay
69 Rear wiper motor

Rear wash/wipe

Front wash/wipe

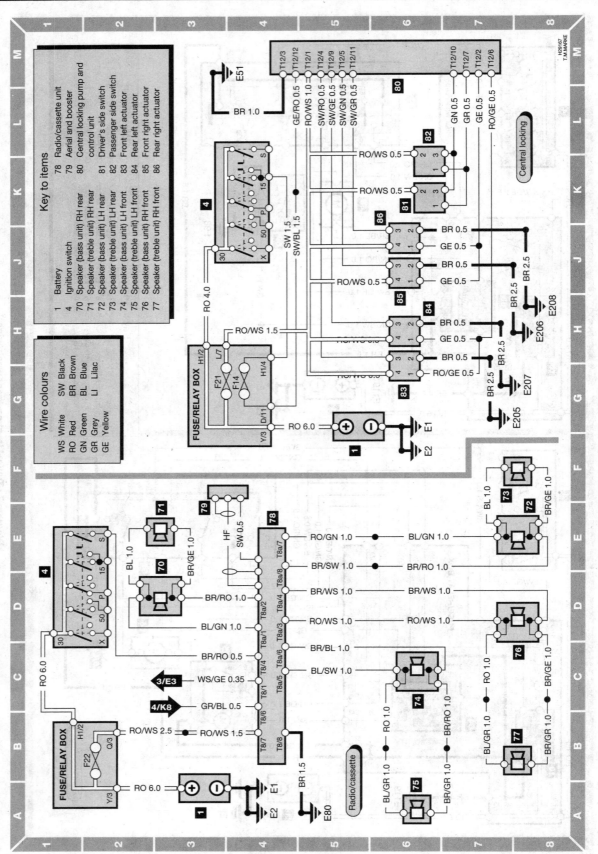

Diagram 8 : Typical radio/cassette and central locking

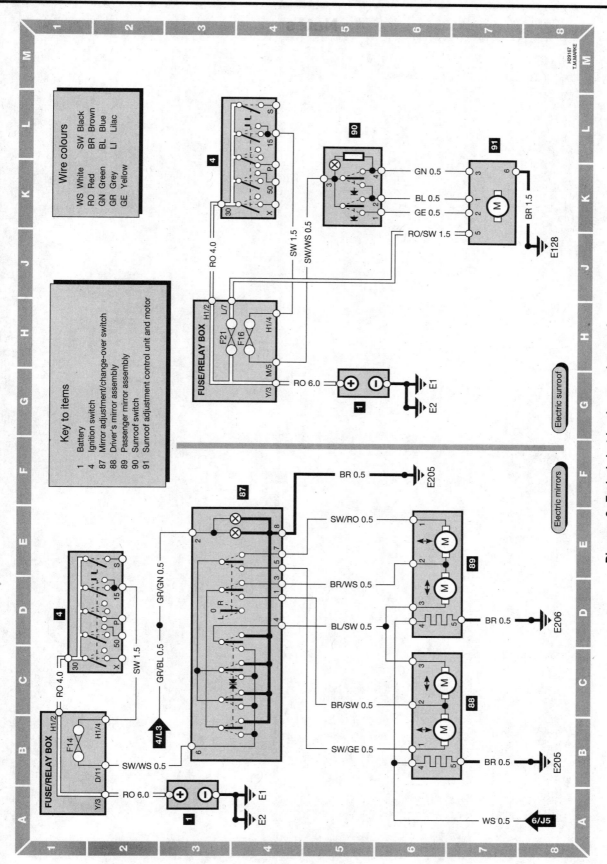

Diagram 9 : Typical electric mirrors and sunroof

Notes

Index

Haynes Automotive Manuals

NOTE: New manuals are added to this list on a periodic basis. If you do not see a listing for your vehicle, consult your local Haynes dealer for the latest product information.

ACURA
*1776 Integra '86 thru '89 & Legend '86 thru '90

AMC
Jeep CJ - see JEEP (412)
694 Concord/Hornet/Gremlin/Spirit '70 thru '83
934 (Renault) Alliance & Encore '83 thru '87

AUDI
615 4000 all models '80 thru '87
428 5000 all models '77 thru '83
1117 5000 all models '84 thru '88

AUSTIN
Healey Sprite - see MG Midget (265)

BMW
*2020 3/5 Series '82 thru '92
276 320i all 4 cyl models '75 thru '83
632 528i & 530i all models '75 thru '80
240 1500 thru 2002 except Turbo '59 thru '77

BUICK
Century (FWD) - see GM (829)
*1627 Buick, Oldsmobile & Pontiac Full-size (Front wheel drive) '85 thru '95
Buick Electra, LeSabre and Park Avenue; Oldsmobile Delta 88 Royale, Ninety Eight and Regency; Pontiac Bonneville
1551 Buick Oldsmobile & Pontiac Full-size (Rear wheel drive)
Buick Estate '70 thru '90, Electra '70 thru '84, LeSabre '70 thru '85, Limited '74 thru '79 Oldsmobile Custom Cruiser '70 thru '90, Delta 88 '70 thru '85,Ninety-eight '70 thru '84 Pontiac Bonneville '70 thru '81, Catalina '70 thru '81, Grandville '70 thru '75, Parisienne '83 thru '86
627 Mid-size Regal & Century '74 thru '87
Regal - see GENERAL MOTORS (1671)
Skyhawk - see GENERAL MOTORS (766)
Skylark '80 thru '85 - see GM (38020)
Skylark '86 on - see GM (1420)
Somerset - see GENERAL MOTORS (1420)

CADILLAC
*751 Cadillac Rear Wheel Drive '70 thru '93
Cimarron - see GENERAL MOTORS (766)

CHEVROLET
*1477 Astro & GMC Safari Mini-vans '85 thru '93
554 Camaro V8 all models '70 thru '81
866 Camaro all models '82 thru '92
Cavalier - see GENERAL MOTORS (766)
Celebrity - see GENERAL MOTORS (829)
24017 Camaro & Firebird '93 thru '96
625 Chevelle, Malibu, El Camino '69 thru '87
449 Chevette & Pontiac T1000 '76 thru '87
Citation - see GENERAL MOTORS (38020)
*1628 Corsica/Beretta all models '87 thru '96
274 Corvette all V8 models '68 thru '82
*1336 Corvette all models '84 thru '91
1762 Chevrolet Engine Overhaul Manual
704 Full-size Sedans Caprice, Impala, Biscayne, Bel Air & Wagons '69 thru '90
Lumina - see GENERAL MOTORS (1671)
Lumina APV - see GM (2035)
319 Luv Pick-up all 2WD & 4WD '72 thru '82
626 Monte Carlo all models '70 thru '88
241 Nova all V8 models '69 thru '79
*1642 Nova/Geo Prizm front wheel drive '85 thru '92
420 Pick-ups '67 thru '87 - Chevrolet & GMC, all V8 & in-line 6 cyl, 2WD & 4WD '67 thru '87; Suburbans, Blazers & Jimmys '67 thru '91
*1664 Pick-ups '88 thru '95 - Chevrolet & GMC, all full-size models '88 thru '95; Blazer & Jimmy '92 thru '94; Suburban '92 thru '95; Tahoe & Yukon '95
*831 S-10 & GMC S-15 Pick-ups '82 thru '93
24071 S-10, Gmc S-15 & Jimmy '94 thru '96
*1727 Sprint & Geo Metro '85 thru '94
*345 Vans - Chevrolet & GMC, V8 & in-line 6 cylinder models '68 thru '96

CHRYSLER
25025 Chrysler Concorde, New Yorker & LHS, Dodge Intrepid, Eagle Vision, '93 thru '96
2114 Chrysler Engine Overhaul Manual
*2058 Full-size Front-Wheel Drive '88 thru '93
K-Cars - see DODGE Aries (723)
Laser - see DODGE Daytona (1140)
*1337 Chrysler/Plym. Mid-size cars '82 thru '95
Rear-wheel Drive - see DODGE (2098)

DATSUN
647 200SX all models '80 thru '83
228 B - 210 all models '73 thru '78
525 210 all models '78 thru '82
206 240Z, 260Z & 280Z Coupe '70 thru '78
563 280ZX Coupe & 2+2 '79 thru '83
300ZX - see NISSAN (1137)
679 310 all models '78 thru '82
123 510 & PL521 Pick-up '68 thru '73
430 510 all models '78 thru '81
372 610 all models '72 thru '76
277 620 Series Pick-up all models '73 thru '79
720 Series Pick-up - see NISSAN (771)
376 810/Maxima all gas models, '77 thru '84
Pulsar - see NISSAN (876)
Sentra - see NISSAN (982)
Stanza - see NISSAN (981)

DODGE
400 & 600 - see CHRYSLER Mid-size (1337)

*723 Aries & Plymouth Reliant '81 thru '89
1231 Caravan & Ply. Voyager '84 thru '95
699 Challenger/Plymouth Saporro '78 thru '83
Challenger '67-'76 - see DODGE Dart (234)
610 Colt/Plymouth Champ '78 thru '87
*1668 Dakota Pick-ups all models '87 thru '96
234 Dart, Challenger/Plymouth Barracuda & Valiant 6 cyl models '67 thru '76
*1140 Daytona & Chrysler Laser '84 thru '89
Intrepid - see Chrysler (25025)
*545 Omni & Plymouth Horizon '78 thru '90
912 Pick-ups all full-size models '74 thru '93
*30041 Pick-ups all full-size models '94 thru '96
*556 Ram D/D50 Pick-ups & Raider and Plymouth Arrow Pick-ups '79 thru '93
2098 Dodge/Ply./Chrysler RWD '71 thru '89
*1726 Shadow/Plymouth Sundance '87 thru '94
*1779 Spirit & Plymouth Acclaim '89 thru '95
*349 Vans - Dodge & Plymouth '71 thru '96

EAGLE
Talon - see Mitsubishi Eclipse (2097)
Vision - see CHRYSLER (25025)

FIAT
094 124 Sport Coupe & Spider '68 thru '78
273 X1/9 all models '74 thru '80

FORD
10355 Ford Automatic Trans. Overhaul
*1476 Aerostar Mini-vans '86 thru '96
268 Courier Pick-up all models '72 thru '82
2105 Crown Victoria & Mercury Grand Marquis '88 thru '96
1763 Ford Engine Overhaul Manual
789 Escort/Mercury Lynx '81 thru '90
*2046 Escort/Mercury Tracer '91 thru '96
*2021 Explorer & Mazda Navajo '91 thru '95
560 Fairmont & Mercury Zephyr '78 thru '83
334 Fiesta all models '77 thru '80
754 Ford & Mercury Full-size, Ford LTD & Mercury Marquis ('75 thru '82); Ford Custom 500,Country Squire, Crown Victoria & Mercury Colony Park ('75 thru '87); Ford LTD Crown Victoria & Mercury Gran Marquis ('83 thru '87)
359 Granada & Mercury Monarch '75 thru '80
773 Ford & Mercury Mid-size, Ford Thunderbird & Mercury Cougar ('75 thru '82); Ford LTD & Mercury Marquis ('83 thru '86); Ford Torino,Gran Torino, Elite, Ranchero pick-up, LTD II, Mercury Montego, Comet, XR-7 & Lincoln Versailles ('75 thru '86)
357 Mustang V8 all models '64-1/2 thru '73
231 Mustang II 4 cyl, V6 & V8 '74 thru '78
*654 Mustang & Mercury Capri incl. Turbo Mustang, '79 thru '93; Capri, '79 thru '86
*36051 Mustang all models '94 thru '97
788 Pick-ups and Bronco '73 thru '79
*880 Pick-ups and Bronco '80 thru '96
649 Pinto & Mercury Bobcat '75 thru '80
1670 Probe all models '89 thru '92
*1026 Ranger/Bronco II all models '83 thru '92
*36071 Ford Ranger '93 thru '96 & Mazda Pick-ups '94 thru '96
*1421 Taurus & Mercury Sable '86 thru '95
*1418 Tempo & Mercury Topaz '84 thru '94
1338 Thunderbird/Mercury Cougar '83 thru '88
*1725 Thunderbird/Mercury Cougar '89 and '96
344 Vans all V8 Econoline models '69 thru '91
*2119 Vans full size '92 thru '95

GENERAL MOTORS
*10360 GM Automatic Trans. Overhaul
*829 Buick Century, Chevrolet Celebrity, Olds Cutlass Ciera & Pontiac 6000 all models '82 thru '96
*1671 Buick Regal, Chevrolet Lumina, Oldsmobile Cutlass Supreme & Pontiac Grand Prix front wheel drive '88 thru '95
*766 Buick Skyhawk, Cadillac Cimarron, Chevrolet Cavalier, Oldsmobile Firenza Pontiac J-2000 & Sunbird '82 thru '94
38020 Buidk Skylark, Chevrolet Citation, Olds Omega, Pontiac Phoenix '80 thru '85
1420 Buick Skylark & Somerset, Olds Achieva, Calais & Pontiac Grand Am '85 thru '95
38030 Cadillac Eldorado & Oldsmobile Toronado '71 thru '85, Seville '80 thru '85, Buick Riviera '79 thru '85
*2035 Chevrolet Lumina APV, Oldsmobile Silhouette & Pontiac Trans Sport '90 thru '95
General Motors Full-size Rear-wheel Drive - see BUICK (1551)

GEO
Metro - see CHEVROLET Sprint (1727)
Prizm - see CHEVROLET (1642) or TOYOTA (1642)
*2039 Storm all models '90 thru '93
Tracker - see SUZUKI Samurai (1626)

GMC
Safari - see CHEVROLET ASTRO (1477)
Vans & Pick-ups - see CHEVROLET

HONDA
351 Accord CVCC all models '76 thru '83
1221 Accord all models '84 thru '89
2067 Accord all models '90 thru '93
*42013 Accord all models '94 thru '95
160 Civic 1200 all models '73 thru '79
633 Civic 1300 & 1500 CVCC '80 thru '83
297 Civic 1500 CVCC all models '75 thru '79
1227 Civic all models '84 thru '91

2118 Civic & del Sol '92 thru '95
*601 Prelude CVCC all models '79 thru '89

HYUNDAI
*1552 Excel all models '86 thru '94

ISUZU
*1641 Trooper '84 thru '91, Pick-up '81 thru '93
Hombre - see CHEVROLET S-10 (24071)

JAGUAR
*242 XJ6 all 6 cyl models '68 thru '86
*49011 XJ6 all models '88 thru '94
*478 XJ12 & XJS all 12 cyl models '72 thru '85

JEEP
*1553 Cherokee, Comanche & Wagoneer Limited all models '84 thru '96
412 CJ all models '49 thru '86
*50025 Grand Cherokee all models '93 thru '95
*50029 Grand Wagoneer & Pick-up '72 thru '91
*1777 Wrangler all models '87 thru '95

LINCOLN
2117 Rear Wheel Drive all models '70 thru '96

MAZDA
648 626 (rear wheel drive) '79 thru '82
*1082 626 & MX-6 (front wheel drive) '83 thru '91
370 GLC (rear wheel drive) '77 thru '83
757 GLC (front wheel drive) '81 thru '85
*2047 MPV all models '89 thru '94
Navajo - see FORD Explorer (2021)
267 Pick-ups '72 thru '93
Pick-ups '94 on - see Ford Ranger
460 RX-7 all models '79 thru '85
*1419 RX-7 all models '86 thru '91

MERCEDES-BENZ
*1643 190 Series 4-cyl gas models, '84 thru '88
346 230, 250 & 280 6 cyl sohc '68 thru '72
983 280 123 Series gas models '77 thru '81
698 350 & 450 all models '71 thru '80
697 Diesel 123 Series '76 thru '85

MERCURY
See FORD Listing

MG
111 MGB Roadster & GT Coupe '62 thru '80
265 MG Midget & Austin Healey Sprite Roadster '58 thru '80

MITSUBISHI
*1669 Cordia, Tredia, Galant, Precis & Mirage '83 thru '93
*2097 Eclipse, Eagle Talon & Plymouth Laser '90 thru '94
*2022 Pick-up '83 thru '96, Montero '83 thru '93

NISSAN
1137 300ZX all models incl. Turbo '84 thru '89
*72015 Altima '93 thru '97
*1341 Maxima all models '85 thru '92
*771 Pick-ups '80 thru '96, Pathfinder '87 thru '95
876 Pulsar all models '83 thru '86
*982 Sentra all models '82 thru '94
*981 Stanza all models '82 thru '90

OLDSMOBILE
Achieva - see GENERAL MOTORS (1420)
Bravada - see CHEVROLET S-10 (831)
Calais - see GENERAL MOTORS (1420)
Custom Cruiser - see BUICK (1551)
*658 Cutlass '74 thru '88
Cutlass Ciera - see GM (829)
Cutlass Supreme - see GM (1671)
Delta 88 - see BUICK Full-size RWD (1551)
Delta 88 Brougham - see BUICK Full-size (1627)
Delta 88 Royale - see BUICK (1551)
Firenza - see GENERAL MOTORS (766)
Ninety-eight Regency - see BUICK Full-size RWD (1551), FWD (1627)
Omega - see GENERAL MOTORS (38020)
Silhouette - see GENERAL MOTORS (2035)
Toronado - see GM (38030)

PEUGEOT
663 504 all diesel models '74 thru '83

PLYMOUTH
Laser - see MITSUBISHI Eclipse (2097)
Other PLYMOUTH titles, see DODGE

PONTIAC
T1000 - see CHEVROLET Chevette (449)
J-2000 - see GENERAL MOTORS (766)
6000 - see GM (829)
Bonneville - see Buick (1627, 1551)
Bonneville Brougham - see Buick (1551)
Catalina - see Buick Full-size (1551)
1232 Fiero all models '84 thru '88
555 Firebird V8 models except Turbo '70 thru '81
867 Firebird all models '82 thru '92
Firebird '93 thru '96 - see CHEVY (24017)
Full-size FWD - see BUICK FWD (1627)
Full-size RWD - see BUICK RWD (1551)
Grand Am - see GM (1420)
Grand Prix - see GM (1671)
Grandville - see BUICK (1551)
Parisienne - see Buick (1551)
Phoenix - see GM (38020)

Sunbird - see GENERAL MOTORS (766)
Trans Sport - see GM (2035)

PORSCHE
*264 911 Coupe & Targa models '65 thru '89
239 914 all 4 cyl models '69 thru '76
397 924 all models incl. Turbo '76 thru '82
*1027 944 all models incl. Turbo '83 thru '89

RENAULT
141 5 Le Car all models '76 thru '83
Alliance & Encore - see AMC (934)

SAAB
247 99 all models including Turbo '69 thru '80
*980 900 including Turbo '79 thru '88

SATURN
*2083 Saturn all models '91 thru '96

SUBARU
237 1100, 1300, 1400 & 1600 '71 thru '79
*681 1600 & 1800 2WD & 4WD '80 thru '89

SUZUKI
*1626 Samurai/Sidekick/Geo Tracker '86 thru '96

TOYOTA
1023 Camry all models '83 thru '91
*92006 Camry all models '92 thru '95
935 Celica Rear Wheel Drive '71 thru '85
*2038 Celica Front Wheel Drive '86 thru '93
1139 Celica Supra all models '79 thru '92
361 Corolla all models '75 thru '79
961 Corolla rear wheel drive models '80 thru '87
*1025 Corolla front wheel drive models '84 thru '92
*92036 Corolla & Geo Prizm '93 thru '96
636 Corolla Tercel all models '80 thru '82
360 Corona all models '74 thru '82
532 Cressida all models '78 thru '82
313 Land Cruiser all models '68 thru '82
*1339 MR2 all models '85 thru '87
304 Pick-up all models '69 thru '78
*656 Pick-up all models '79 thru '95
*2048 Previa all models '91 thru '95
2106 Tercel all models '87 thru '94

TRIUMPH
113 Spitfire all models '62 thru '81
322 TR7 all models '75 thru '81

VW
159 Beetle & Karmann Ghia '54 thru '79
238 Dasher all gasoline models '74 thru '81
*96017 Golf & Jetta '93 thru '97
*884 Rabbit, Jetta, Scirocco, & Pick-up gas models '74 thru '91 & Convertible '80 thru '92
451 Rabbit, Jetta, Pick-up diesel '77 thru '84
082 Transporter 1600 all models '68 thru '79
226 Transporter 1700, 1800, 2000 '72 thru '79
084 Type 3 1500 & 1600 '63 thru '73
1029 Vanagon air-cooled models '80 thru '83

VOLVO
203 120, 130 Series & 1800 Sports '61 thru '73
129 140 Series all models '66 thru '74
*270 240 Series all models '76 thru '93
400 260 Series all models '75 thru '82
*1550 740 & 760 Series all models '82 thru '88

TECHBOOK MANUALS
2108 Automotive Computer Codes
1667 Automotive Emissions Control Manual
482 Fuel Injection Manual, 1978 thru 1985
2111 Fuel Injection Manual, 1986 thru 1996
2069 Holley Carburetor Manual
2068 Rochester Carburetor Manual
10240 Weber/Zenith/Stromberg/SU Carburetor
1762 Chevrolet Engine Overhaul Manual
2114 Chrysler Engine Overhaul Manual
1763 Ford Engine Overhaul Manual
1736 GM and Ford Diesel Engine Repair
1666 Small Engine Repair Manual
10355 Ford Automatic Transmission Overhaul
10360 GM Automatic Transmission Overhaul
1479 Automotive Body Repair & Painting
2112 Automotive Brake Manual
2113 Automotive Detailing Manual
1654 Automotive Eelectrical Manual
1480 Automotive Heating & Air Conditioning
2109 Automotive Reference Dictionary
2107 Automotive Tools Manual
10440 Used Car Buying Guide
2110 Welding Manual
10450 ATV Basics

SPANISH MANUALS
98903 Reparación de Carrocería & Pintura
98905 Códigos Automotrices de la Computadora
98910 Frenos Automotriz
98915 Inyección de Combustible 1986 al 1994
99040 Chevrolet & GMC Camionetas '67 al '87
99041 Chevrolet & GMC Camionetas '88 al '95
99042 Chevrolet Camionetas Cerradas '68 al '95
99055 Dodge Caravan/Ply. Voyager '84 al '95
99075 Ford Camionetas y Bronco '80 al '94
99077 Ford Camionetas Cerradas '69 al '91
99083 Ford Modelos de Tamaño Grande '75 al '87
99091 Ford Modelos de Tamaño Mediano '75 al '86
99095 GM Modelos de Tamaño Grande '70 al '90
99118 Nissan Sentra '82 al '94
99125 Toyota Camionetas y 4-Runner '79 al '95

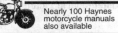

* Listings shown with an asterisk (*) indicate model coverage as of this printing. These titles will be periodically updated to include later model years - consult your Haynes dealer for more information.

Nearly 100 Haynes motorcycle manuals also available

5-97

Haynes North America, Inc., 861 Lawrence Drive, Newbury Park, CA 91320 • (805) 498-6703